ROBOTIC MICROASSEMBLY

ROBOTIC MICROASSEMBLY

Edited by

MICHAËL GAUTHIER
STÉPHANE RÉGNIER

IEEE

IEEE PRESS

WILEY

A JOHN WILEY & SONS, INC., PUBLICATION

For general information on our other products and services or for technical support, please contact our Customer Care Department within the United States at (800) 762-2974, outside the United States at (317) 572-3993 or fax (317) 572-4002.

Wiley also publishes its books in a variety of electronic formats. Some content that appears in print may not be available in electronic formats. For more information about Wiley products, visit our web site at www.wiley.com.

Library of Congress Cataloging-in-Publication Data:

Gauthier, Michaël, 1975-
Robotic micro-assembly / Michaël Gauthier, Stéphane Régnier.
p. cm.
Includes bibliographical references and index.
ISBN 978-0-470-48417-3 (cloth : alk. paper) 1. Robotics. 2. Robots, Industrial.
3. Microfabrication. I. Regnier, Stephane. II. Title.
TJ211.G378 2010
670.42′72–dc22

2009054236

10 9 8 7 6 5 4 3 2 1

CONTENTS

FOREWORD

In 1995, two papers appeared at the 1995 IEEE/RSJ International Conference on Intelligent Robots and Systems (IROS95) that helped precipitate the field of robotic assembly as a topic of research within the robotics community. One of the papers was written by Ron Fearing of the University of California, Berkeley, entitled "Survey of Sticking Effects for Micro-Parts," and the other came out of Toshio Fukuda's group (first author Fumihito Arai) called "Micro Manipulation Based on Micro Physics." Both papers did a wonderful job of illustrating the challenges and the opportunities of manipulating micrometer-size parts automatically, essentially defining the "mechanics of micromanipulation" and encouraging many of us, myself included, to pursue robotic microassembly as a topic of research.

However, one could argue that robotic microassembly got its start about 400 years ago with the invention of the optical microscope in Holland. Imagine being someone like Robert Hooke, one of the first to master the use of these instruments. In the 1650s, while he was at Oxford, Hooke began working with microscopes and discovered a hidden world of small insects and tiny creatures in addition to observing plant structures and various materials. In 1665 he published *Micrographia*, a book illustrating his observations, which is considered one of the greatest of his many considerable achievements. With the publication of *Micrographia*, others rapidly became aware of this secret world, and it was only natural for people to wonder how one could make similar things. Craftsmen learned that by using microscopic techniques, they could see smaller details that enabled them to make finer and finer mechanisms. Manual microassembly became an important industry in much of Europe, where the watchmaking industry grew in places such as France, Switzerland, Germany, and England.

In the late 1940s, the invention of the transistor by Bardeen, Shockley, and Brattain at Bell Labs began another shift in micromanufacturing. Suddenly there was a newfound need to make really, really small things cheaply, driving Kilby and Noyce to the concept of the integrated circuit. Moore's law began, and batch fabrication, not serial assembly, was the obvious way to make small things cheaply, primarily out of silicon, of course. Then in 1982 Kurt Pedersen published his seminal paper "Silicon as a Mechanical Material," a paper that is often cited as representing the beginning of the MEMS era (microelectromechanical systems). While the MEMS community abhorred assembly in the early days, the constraints that microfabrication processes placed on the materials with which microsystems could be made as well as their geometry were extremely limiting. What if we could actually assemble microsystems, instead of relying solely on top-down processes such as photolithography, thermal evaporation, and reactive ion etching? This question was being increasingly asked just as IROS began in Pittsburgh in August of 1995.

These historical trends are what motivate robotic microassembly. Though the field as it is currently defined has been highly active for almost 15 years, this book represents a pioneering achievement by creating, for the first time, a complete view of the field from the physics of micromanipulation, to microassembly, to microhandling in general. A first-class consortium of international authors has been assembled to provide a comprehensive, worldwide view. This effort, which helps further define the field of robotic microassembly, will undoubtedly spur researchers and industry to continue their quest to make small things cheaply.

Bradley Nelson

Zürich, Switzerland
May 2009

PREFACE

This book deals with the current methods developed around the world on robotic microassembly. It is dedicated to Master's and Ph.D. students, and also scientists and engineers involved in microrobotics and also in robotics. As robotic microassembly is a new way to manufacture microelectromechanical systems (MEMS), companies and research institutes involved in this domain will find in this book original methods that can be used to simulate, design, and build new generations of hybrid tridimensional microproducts.

Microproducts are usually divided into two categories by function of the manufactured process used. On the one hand, the standard fabrication using machining or molding is able to produce millimetric and submillimetric pieces (e.g., gears in watches). On the second hand, processes developed initially in microelectronics and based on photolithography have been extended to mechanical structures and are currently used to build MEMS (e.g., air bag sensors). In both cases, the resolution of the details built on the product could be around the micrometer or even less, but the global size of the pieces stays millimetric. The market of miniaturized products, which include always more functionalities in a smaller volume, is increasing very rapidly. In the future, the size of the piece should be reduced below 100 μm, and the microsystem should integrate a large variety of functions including mechanisms, electronic, and control, fluid or optic. It is the reason why a large number of research teams are currently focused on the topic of microassembly.

In MEMS microfabrication, hybridization of technologies is currently obtained using planar assembly (e.g., flip–chip process). However, this method is limited to planar products and does not enable out-of-plane assembly. The advent of a new generation of microsystems based on tridimensional hybrid structures is directly linked with the ability to manufacture microsystems using advanced assembly processes. Two approaches are currently developed in microassembly: self-assembly and robotic microassembly.

Self-assembly consists in creating several minimums of potential energy with a physical field (i.e., electrostatic, capillarity). Microobjects thus need minimum energy and are directly positioned. Self-assembly is a natural process for molecular structures and many examples can be found in nature. These processes are massively parallel but the efficiency and the flexibility still stay low. On the other hand, manipulation robots can be used to assemble micro- or nanopieces. This robotic approach is classically divided into three steps: positioning, handling, and release. This approach is able to reach complex assembly with high flexibility. However, handling and especially release is sensibly disturbed by microscopic peculiarities (i.e., adhesion, deformations of the object, environment). This book focuses on this second approach called *robotic microassembly*.

Robotic assembly is usually carried out on robotic platforms that consist of a gripping device able to grasp the pieces; some sensing systems able to measure position and/or force; a robot able to position the gripper; and a controller to induce automatic movements and tasks. In the microscale, the same functions have to be considered. In *robotic microassembly*, the most critical phase is the gripping task, which depends on interactions between the manipulated object and the end-effector of the gripper. Indeed, at this scale, the behavior of the object is the function of the adhesion and surface forces (i.e., van der Waals, electrostatic forces, etc.), which are predominant compared to the volume effects (i.e., weight, inertia).

The design of robotic assembly platforms must be based on a good understanding and analysis of these adhesion and surface forces, which is the objective of the first part of this book. The physical principles involved in the microscale is developed in order to present the expression of forces in several cases. As the environmental parameters (i.e., humidity, pH in a liquid) highly influence the adhesion and surface forces, a specific chapter is dedicated to the relationship between these forces and the environment. These forces induce specific behavior in microobjects that require specific handling strategies to be handled and assembled.

Based on the knowledge on predominant forces in the microscale, new microhandling methods and prototypes have been developed and are listed in the second part. A lot of handling strategies are presented and compared: hybrid handling strategies based on principles that combine self-assembly and robotic assembly; gripping and release principles in the air; and specific handling strategies dedicated to submerged microobjects. Based on this panorama, the reader will easily understand microscale difficulties and will find methods and information to design microhandling principles.

Even though, handling is a critical phase in microassembly, it is not enough to assemble microobjects. The design of the microobject itself and of the robot structure, which have both to be carefully studied, are presented in the third part. The microassembly requires to design and manufacture micropieces able to be connected together to ensure, at least, mechanical links. Today, electrical and fluid connections in the microscale remain a challenge. Moreover, the required mobility (in terms of number of degree of freedom, DoF) and the required repeatability and precision of the robot require a specific design, calibration, and characterization. This third part focuses on these crucial problems, which are keys to assembly micropieces.

We expect that this book, which proposes a complete overview of the state of the art in robotic assembly, will provide a better understanding of the microscale specificities and methods for robotic microassembly to students, engineers, and scientists. They will be able to apply the models and the methods on microproducts and contribute to the development of the robotic assembly in the microscale.

MICHAËL GAUTHIER
STÉPHANE RÉGNIER

Besançon, France
Paris, France
June 2010

CONTRIBUTORS

Reymond Clavel, Swiss Federal Institute of Technology (EPFL), Lausanne, Switzerland

Mélanie Dafflon, Swiss Federal Institute of Technology (EPFL), Lausanne, Switzerland

Nikolai Dechev, University of Victoria, Greater Victoria, British Columbia, Canada

Michaël Gauthier, FEMTO-ST Institute, Besançon, France

Pierre Lambert, The Free University of Brussels (ULB), Brussels, Belgium

Yusuke Maeda, Yokohama National University, Yokohama, Japan

Akihiro Matsumoto, Toyo University, Kawagoe-shi, Saitama, Japan

Bradley Nelson, Swiss Federal Institute of Technology (ETHZ), Zurich, Switzerland

Dan O. Popa, University of Texas at Arlington, Arlington, Texas

Stéphane Régnier, University of Pierre and Marie Curie, Paris, France

Harry E. Stephanou, University of Texas at Arlington, Arlington, Texas

Veikko Sariola, Helsinki University of Technology, Helsinki, Finland

Kunio Yoshida, AJI Ltd., Yokohama, Japan

Quan Zhou, Helsinki University of Technology, Helsinki, Finland

PART I

MODELING OF THE MICROWORLD

CHAPTER 1

MICROWORLD MODELING IN VACUUM AND GASEOUS ENVIRONMENTS

PIERRE LAMBERT and STÉPHANE RÉGNIER

1.1 INTRODUCTION

1.1.1 Introduction on Microworld Modeling

This first part describes the physical models involved in the description of a micromanipulation task: adhesion, contact mechanics, surface forces, and scaling laws. The impact of surface roughness and liquid is discussed later on in Chapter 2.

The targeted readership of Chapters 1 and 2 is essentially composed of master's degree students and lecturers, Ph.D. students, and researchers to whom this contribution intends:

- To give the theoretical background as far as the physics and scaling laws for micromanipulation are concerned
- To propose design rules for micromanipulation tools and how to estimate the interaction force between a component and the related gripper or between a cantilever tip and a substrate

The goal of developing models may be questioned for many reasons:

- The task is huge and the forces dominating at the micro- and nanoscale can only be modeled very partially: for example, some of them cannot be modeled in a quantitative way (e.g., hydrogen bonds) suitable for robotics

Robotic Microassembly, edited by Michaël Gauthier and Stéphane Régnier

purposes, most of the proposed models are only valid at equilibrium (at least all the models based on the derivation of surface or potential energies).

- The parameters involved in the existing models are sometimes impossible to know, such as, for example, the electrical charge distribution on a dielectric oxide layer.
- Maybe as a consequence of the previous reason—that is, a full characterization is impossible—the micro- and nanoscale specifically suffer from a very large experimental dispersion, which makes the model refinements questionable. According to own experience, experimental results are difficult to keep within a few tens of a percent error interval. Yang and Lin [93] recently write that *the measurements usually show poor reproducibility*, suggesting that *the major causes of irreproducibility can be roughness and heterogeneity of the probe surface and sample*.

Nevertheless the use of—even basic—models helps the microrobotician to get into the nonintuitive physics dominating the microworld—mainly adhesion-related instabilities such as pull-in and pull-out—to give an explicating scheme of the experiments—what is the role of humidity? what is the influence of the coatings—to design at best grippers and tools on a comparative way—no matter the exact value of the force; but a geometries comparison leads to the best design. These advantages will be detailed later on.

Classical adhesion models [20, 41, 67] usually proposed to study adhesion in micromanipulation or atomic force microscopy (AFM) are based on the elastic deformation of two antagonist solids (microcomponent/gripper in micromanipulation, cantilever tip/substrate in AFM). This part will introduce models that are now well known, but they will be introduced in the framework of microassembly. Modern models will refer to recent developments and/or recent papers. The theoretical background proposed in this part aims at detailing:

1. Every phenomenon leading to a force interaction: capillary forces, electrostatic forces (in both liquid and air environment, but restricted to conductive materials), van der Waals forces, and contact forces
2. The influence of surface science concepts such as topography, deformation, and wettability

These elements constitute a basis on which to model adhesion without using empirical global energy parameters such as surface energies.

When dealing with stiction and adhesion problems in micromanipulation, one is often referred to a list of many concepts (van der Waals interaction, capillary force, adhesion, pull-off), which sometimes can recover one another. Lambert and Régnier [53] have proposed to sort out these forces by making the distinction whether there is contact or not. When there is no physical contact between two solids, the forces in action are called distance or surface forces (according to the scientific literature in this domain [12, 22, 76], these latter are electrostatic, van der Waals, and capillary forces). When both solids contact one another, there is

TABLE 1.1. Forces Summary and Their Interaction Distances

Interaction Distance	Predominant Force
Up to infinite range	Gravity
>From a few nm up to 1 mm	Capillary forces
>0.3 nm	Coulomb (electrostatic) forces
0.3 nm < separation distance < 100 nm	Lifshitz–van der Waals
< 0.3 nm	Molecular interactions
0.1–0.2 nm	Chemical interactions

deformation and adhesion forces through the surfaces in contact. In this case, the authors considered contact forces and adhesion or pull-off forces. Electrostatic or capillary effects can be added, but van der Waals forces are not considered anymore because they are thought already involved in the pull-off term. The new idea conveyed in this part is to consider van der Waals, capillary,[1] and electrostatic forces as parts into which the global pull-off force can be split.

Beside these contact or close to contact forces, it is also important to focus on other forces that affect the dynamics of small components. This description can only be done by considering the specificities of the working environment. In liquid environments, for example, we will consider viscous drag (Lenders et al. [58] have recently presented a design of microfeeder using these forces), electrostatic double-layer effects, and (di)electrophoresis. Very recently, a new focus has been found on the effect of gas bubbles in liquid media.

Additionally, we will try to address the question of mechanical contact from two points of view: what are the limits of the Hertz-based models [20, 41, 67] and what is the influence of a liquid environment on this contact?

1.1.2 Microworld Modeling for Van der Waals Forces and Contact Mechanics

The first chapter concerns vacuum or gaseous environments. First in Section 1.2 some well-accepted models are recalled, concerning van der Waals forces, elastic contact mechanics and the related adhesion models, and capillary force models at the submillimetric scale. Second, in Section 1.3 very recent published results are presented together with our own perspective: capillary condensation effects, the influence of surface roughness, and mechanical deformation on electrostatic forces.

Before going through these models, let us mention that many (attractive) effects contribute to adhesion. Based on Lee [57], we propose the schematic forces summary presented in Table 1.1.

Additional effects turn out to be also of importance: Let us cite the Casimir effect, which will not be detailed in this contribution. We refer to Klimchitskaya and Hostepanenko [45].

[1]Capillary forces will be considered at the submillimetric scale [50] and at the nanometric scale [16].

It seems, however, that capillary effect dominates all the microworld from a few nanometers up to tenths of millimeters van der Waals effects turn out to compete with capillary effect but only within the nano range up to a few tens of nanometers (we can consider the limit of the retardation effect as the limit, see later on). We therefore mainly focus on both effects together with the electrostatic adhesion, which comes from either the intense electrostatic fields coming from microrobotic actuation—they can be avoided using thermal actuation—or from the moderate effect of contact potentials.

1.2 CLASSICAL MODELS

1.2.1 Van der Waals Forces

The so-called van der Waals forces are often cited in papers dealing with micromanipulation and microassembly, probably because the founding papers of these bibliography reviews [12, 22] present these forces next to the capillary and the electrostatic forces as being of the utmost importance in the sticking of microparts. Other authors [7] prefer to neglect these forces because they are of a smaller order. The reasons for this opposition do not seem to be clear, all the more so since some authors propose to use it as a suitable gripping principle [3, 23]. The will to clarify this situation is a first reason to study van der Waals forces. A second reason lies in the fact that most force expressions used in the literature on microassembly are only approximations of simplified geometries (spheres and planes). If these approximations are sufficient for experimental case studies, the influence of more complex geometries (nonsymmetrical geometries) including roughness profiles should be studied for many applications. We propose to briefly present the physical underlying phenomena that explain these forces and to explain the way(s) they can be calculated. An overview of the approximations from the literature is proposed in the conclusion of this section.

A good and very didactic introduction to the subject can be found in Israelachvili [38], while a more exhaustive description of the van der Waals (VDW) forces has also been proposed [1, 26, 39]. In order to explain, at least from a qualitative point of view, the power law describing the van der Waals interaction energy, let us start from the potential energy of an electric charge q (Eq. 1.1) and that of a permanent dipole p made of two charges q and $-q$ separated by a distance l (Eq. 1.2 states if $l << r$), in both cases in a point P at a separation distance r and in vacuum (see Fig. 1.1):

$$U(P) = \frac{1}{4\pi\epsilon_0}\frac{q}{r} \tag{1.1}$$

$$U(P) = \frac{1}{4\pi\epsilon_0}\frac{p\cos\theta}{r^2} \tag{1.2}$$

We see that the potential depends on the inverse of the first power of the separation distance in the case of a charge and on the inverse of the second

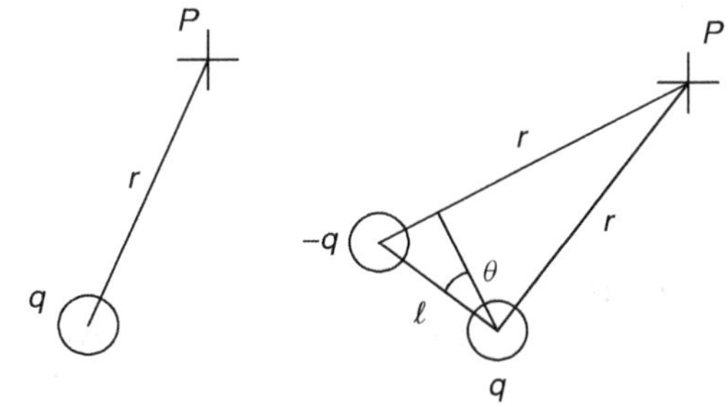

Figure 1.1. Illustration of potentials of a charge and of a permanent dipole.

power in that of a permanent dipole. If we now consider the interaction potential $w(r)$ of two permanent dipoles p_1 and p_2 separated by a distance r, it can be shown [89] that $w(r)$ also depends on the inverse of the third power of the separation distance:

$$w(r) \approx \frac{1}{4\pi\epsilon_0}\frac{p_1 p_2}{r^3} \tag{1.3}$$

We can now introduce the underlying idea to explain the van der Waals forces. Let us consider two molecules, separated by a distance r. If these two molecules are polar (which means that there is a permanent electric dipole inside the molecule due to the fact that the gravity center of the positive charges does not fit with that of the negative forces), their interaction energy can be described by Eq. 1.3. Actually, the van der Waals forces also act between totally neutral atoms and molecules such as helium, methane, and carbon dioxide. This is due to the fact that even in a nonpolar atom, the gravity center of the positive and negative charges are not instantaneously superposed, leading to an instantaneous dipole p_1, with a characteristic charge in the order of the electronic charge e and a separation distance of about one atom radius a_0 (note that this explanation was first applied by D. Tabor to the interaction between two Bohr atoms, a_0 known as the first Bohr radius):

$$p_1 \approx a_0 e \tag{1.4}$$

If the considered molecules are polarizable, this instantaneous dipole will polarize the neighboring atom, and consequently produce a dipole p_2 given by

$$p_2 \approx \alpha \frac{1}{4\pi\epsilon_0 r^3} a_0 e \tag{1.5}$$

where α is the polarizability of the second atom, defined by

$$\alpha \approx 4\pi\epsilon_0 a_0^3 \tag{1.6}$$

The two instantaneous dipoles p_1 and p_2 given by Eqs. 1.5 and 1.6 lead to an interaction potential described by Eq. 1.3:

$$w(r) \approx \frac{1}{4\pi\epsilon_0}\frac{p_1 p_2}{r^3} \approx \frac{\alpha a_0^2 e^2}{(4\pi\epsilon_0)^2}\frac{1}{r^6} \div \frac{1}{r^6} \tag{1.7}$$

This power law holds as far as the orientation (Keesom), the induction (Debye) and the dispersion (London) terms are concerned. Moreover, by assuming these interactions to be additive (= by assuming they do not depend on the surrounding molecules), these three terms can be regrouped:

$$w(r) = \left(-\frac{K_K}{r^6}\right) + \left(-\frac{K_D}{r^6}\right) + \left(-\frac{K_L}{r^6}\right) = -\frac{K}{r^6} \tag{1.8}$$

The so-called retardation effect occurs when the separation distance between the instantaneous dipole and the induced dipole increases over a cut-off length of the order of 5–10 nm: In this case, the traveling time of the electromagnetic wave from the instantaneous dipole and the induced dipole become bigger and, consequently, both dipoles lose their coherence, leading to an energy reduction. The decrease with the separation distance occurs faster and it is assumed that it can be described according to

$$w(r) = -\frac{K_R}{r^7} \tag{1.9}$$

The fast decrease of the van der Waals forces explains that they seem to be limited to the atomic domain. Nevertheless, this decrease occurs more slowly when we consider the interaction between two macroscopic bodies (i.e., a body with a very large number of molecules, including bodies that have a size in the order of a few micrometers and that are consequently considered microcomponents when dealing with microassembly terminology). Therefore, it is not so obvious to choose whether these forces have to be dealt with or not.

Let us now have a look on the ways to compute the van der Waals interaction between two macroscopic bodies: The first one is known as the microscopic or Hamaker approach, and the second one is called the macroscopic or Lifshitz approach. From a strictly theoretical point of view, the van der Waals forces are nonadditive, nonisotropic, and retarded. However, London [60] proposed a straight and powerful way to establish the potential interaction by assuming a pairwise additivity of the interactions. Moreover, this approach does not consider the retardation effect. The results are therefore limited to separation distances between an upper limit of about 5–10 nm (because we neglect the retardation effect) and a lower limit of about one intermolecular distance [because Eq. 1.2 that $l << r$. This lower boundary is reinforced by the value of the equilibrium distance (about 0.1–0.2 nm) arising from the Lennard-Jones potential: for separation distances smaller than 0.1–0.2 nm, very strong repulsive forces occur that can no longer be neglected]. This lower limit is sometimes called the van der

Waals radius [38]. We should keep in mind that even with these restrictions, the results are not exactly correct for the interaction of solids and liquids because of the pairwise summation assumption. However, Israelachvili [39] and Russel et al. [78] consider that these approximations are useful in several applications. We will illustrate this method in what follows.

The Lifshitz method, also called macroscopic approach, consists in considering the two interacting objects as continuous media with a dielectric response to electromagnetic fields. The dispersion forces are then considered the mutual interaction of dipoles oscillating at a given frequency. When the separation distance becomes bigger than a cut-off length depending on this frequency and the light speed, the attraction tends to decrease because the propagation time becomes of the same order as the oscillation period of the dipoles, the field emitted by one dipole interacting with another dipole with a different phase. This effect has first been pointed out by Casimir and Polder [15] and computed by Lifhitz using the quantum field theory [59]. Although this approach is of the greatest complexity, similar results can be obtained by using the Hamaker results, on the condition to replace the Hamaker constant by a pseudoconstant involving more parameters. This method is out of our scope, which is to roughly evaluate the importance of the van der Waals forces in microassembly and to investigate the influence of geometry, roughness, and orientation on the manipulation of microcomponents. We will therefore limit ourselves to the Hamaker method, despite its limitations. The interested reader will find further information about the Lifshitz approach in Adamson and Gast [1], Chapter VI, and in Israelachvili [39].

We present the Hamaker method to calculate the van der Waals forces in the case of the interaction between two spheres, a sphere and a infinite half-space, an infinite half-space limited by a smooth plane, and a rectangular box that has faces that are parallel or perpendicular to that plane. This last example is a good introduction for taking into account the influence of roughness. These results have been published in Lambert and Régnier [53].

In each case the Hamaker method consists in first determining the interaction potential W between two macroscopic objects [while $w(r)$ denotes the potential interaction between microscopic dipoles] and then in deriving it with respect to the separation distance D ($F = -dW/dD$).

1.2.1.1 Interaction Potential Between a Sphere and a Volume Element

The interaction potential $W_{(S,dV)}$ between a sphere S [center O, radius R, number density n_1 (m^3), volume element $d\Omega$] and a volume element dV (number density n_2) located at a distance D from the sphere is given by (see Fig. 1.2)

$$W_{(S,dV)} = -Kn_1n_2\,dV \int_{\Omega} \frac{1}{d^6} d\Omega \tag{1.10}$$

where d is the distance between dV and the volume element $d\Omega$ of S. Let us choose a spherical coordinate frame centered in O and a polar axis linking O and

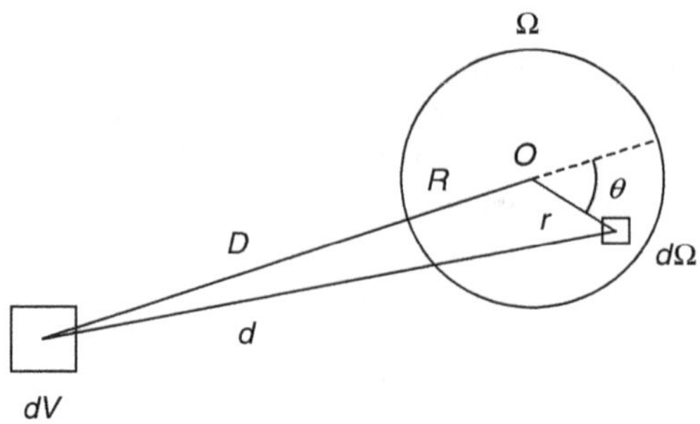

Figure 1.2. Interaction potential between a sphere and a volume element.

dV: Consequently, $d\Omega$ is located in the sphere by its distance r from O and the angle θ (the problem is symmetric as far as the azimutal angle ϕ is concerned). As a consequence, d is given by

$$d^2 = (D + R)^2 + r^2 + 2r(D + R)\cos\theta \tag{1.11}$$

and if we note $x = D + R$ the integral of Eq. 1.10 can be rewritten into

$$\int_\Omega \frac{1}{d^6} d\Omega = \int_0^{2\pi} d\phi \int_0^{\pi} d\theta \int_0^R \frac{r^2 \sin\theta}{(r^2 + x^2 + 2rx\cos\theta)^3} dr \tag{1.12}$$

$$= 2\pi \int_0^R dr \int_0^{\pi} \frac{r^2 \sin\theta}{(r^2 + x^2 + 2rx\cos\theta)^3} d\theta \tag{1.13}$$

The integral with respect to θ can be solved by assuming $\cos\theta = u$ (and thus $-\sin\theta\, d\theta = du$), leading to

$$\int_0^{\pi} \frac{r^2 \sin\theta}{(r^2 + x^2 + 2rx\cos\theta)^3} d\theta = \frac{r}{4x}\left[\frac{1}{(r - x)^4} - \frac{1}{(r + x)^4}\right] \tag{1.14}$$

and Eq. 1.13 is now given by

$$\int_\Omega \frac{1}{d^6} d\Omega = 2\pi \int_0^R \frac{r}{4x}\left[\frac{1}{(r - x)^4} - \frac{1}{(r + x)^4}\right] dr \tag{1.15}$$

$$= -\frac{4\pi}{3}\frac{R^3}{(R^2 - x^2)^3} \tag{1.16}$$

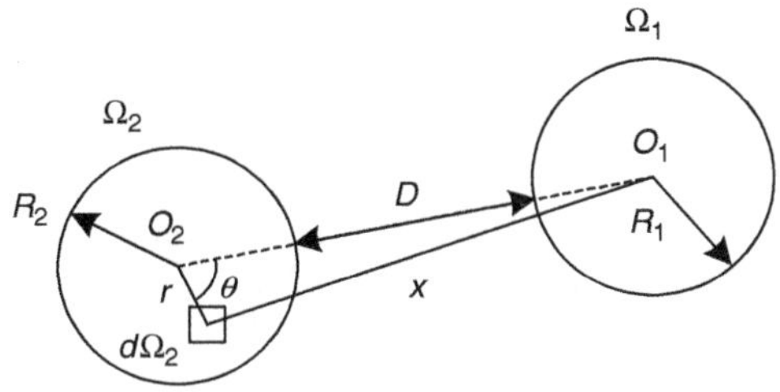

Figure 1.3. Interaction potential between two spheres.

Consequently, with the Hamaker constant $A \equiv Kn_1n_2\pi^2$[J], the interaction potential $W_{(S,dV)}$ of Eq. 1.10 is given by

$$W_{(S,dV)} = \frac{4AR^3}{3\pi[R^2 - (D+R)^2]^3}dV \quad (1.17)$$

1.2.1.2 Interaction Potential Between Two Spheres

In order to determine the interaction potential $W_{(S_1,S_2)}$ between two spheres S_1 (radius R_1, number density n_1, center O_1) and S_2 (radius R_2, number density n_2, center O_2) separated by a distance D (see Fig. 1.3), the interaction potential $W_{(S,dV)}$ of Eq. 1.17 must now be integrated over the second sphere:

$$W_{(S_1,S_2)} = \frac{4}{3\pi}AR_1^3 \int_{\Omega_2} \frac{1}{(R_1^2 - x^2)^3}d\Omega_2 \quad (1.18)$$

where $d\Omega_2$ is the volume element of S_2 and x is the distance between $d\Omega_2$ and O_1. Let us choose a spherical coordinates frame centered in O_2 with polar axis linking O_1 and O_2. The position of the volume element $d\Omega_2$ is defined by r, the distance between $d\Omega_2$ and O_2 and by θ, the angle between O_1O_2 and $O_2d\Omega_2$. The problem is again axially symmetric as far as the azimutal angle ϕ is concerned. As a consequence, by noting $R = R_1 + R_2 + D$, the distance x between O_1 and the volume element $d\Omega_2$ is given by

$$x^2 = R^2 + r^2 - 2rR\cos\theta \quad (1.19)$$

and Eq. 1.18 can be rewritten into[2]

$$W_{(S_1,S_2)} = \frac{4}{3\pi}AR_1^3 \int_0^{2\pi} d\phi \int_0^{R_2} dr \int_0^{\pi} \frac{r^2 \sin\theta}{(R_1^2 - R^2 - r^2 + 2rR\cos\theta)^3}d\theta \quad (1.20)$$

$$= \frac{8}{3}AR_1^3 \int_0^{R_2} dr \int_0^{\pi} \frac{r^2 \sin\theta}{(R_1^2 - R^2 - r^2 + 2rR\cos\theta)^3}d\theta \quad (1.21)$$

[2] $\log = \log_e = \ln \neq \log_{10}$.

$$= \frac{8}{3} A R_1^3 \int_0^{R_2} \frac{r}{4R} \left\{ \frac{1}{[R_1^2 - (R+r)^2]^2} - \frac{1}{[R_1^2 - (r-R)^2]^2} \right\} dr \quad (1.22)$$

$$= -\frac{A}{6} \left[\log \frac{R^2 - (R_1 + R_2)^2}{R^2 - (R_1 - R_2)^2} + \frac{2R_1 R_2}{R^2 - (R_1 + R_2)^2} + \frac{2R_1 R_2}{R^2 - (R_1 - R_2)^2} \right] \quad (1.23)$$

Equation 1.23 can also be written as follows:

$$W_{(S_1,S_2)} = -\frac{A}{6} \left[\log \frac{(R_1 + R_2 + D)^2 - (R_1 + R_2)^2}{(R_1 + R_2 + D)^2 - (R_1 - R_2)^2} + \frac{2R_1 R_2}{(R_1 + R_2 + D)^2 - (R_1 + R_2)^2} + \frac{2R_1 R_2}{(R_1 + R_2 + D)^2 - (R_1 - R_2)^2} \right] \quad (1.24)$$

$$= -\frac{A}{6} \left[\log \frac{D(2R_1 + 2R_2 + D)}{(2R_1 + D)(2R_2 + D)} + \frac{2R_1 R_2}{D(2R_1 + 2R_2 + D)} + \frac{2R_1 R_2}{(2R_1 + D)(2R_2 + D)} \right] \quad (1.25)$$

where D is the separation distance between the two spheres.

1.2.1.3 Potential Interaction Between a Sphere and an Infinite Half-Space

The interaction potential $W_{(S,HS)}$ between a sphere and an infinite half-space can be calculated as the limit of Eq. 1.24 when R_2 tends toward infinity:

$$W_{(S,HS)} = -\frac{A}{6} \left(\log \frac{D}{D + 2R} + \frac{R}{D} + \frac{R}{2R + D} \right) \quad (1.26)$$

where D is the distance between the infinite half-space and the sphere and R is now the radius of the sphere.

1.2.1.4 Force Between Two Spheres

The force is calculated by deriving the interaction potential $W_{(S_1,S_2)}(D)$ given by Eq. 1.24 with respect to the separation distance D:

$$F_{(S_1,S_2)}(D) = \frac{A}{3}(R_1 + R_2 + D) \left[\frac{D(2R_1 + 2R_2 + D) - 2R_1 R_2}{D^2(2R_1 + 2R_2 + D)^2} - \frac{D^2 + 2D(R_1 + R_2) + 6R_1 R_2}{(2R_1 + D)^2(2R_2 + D)^2} \right] \quad (1.27)$$

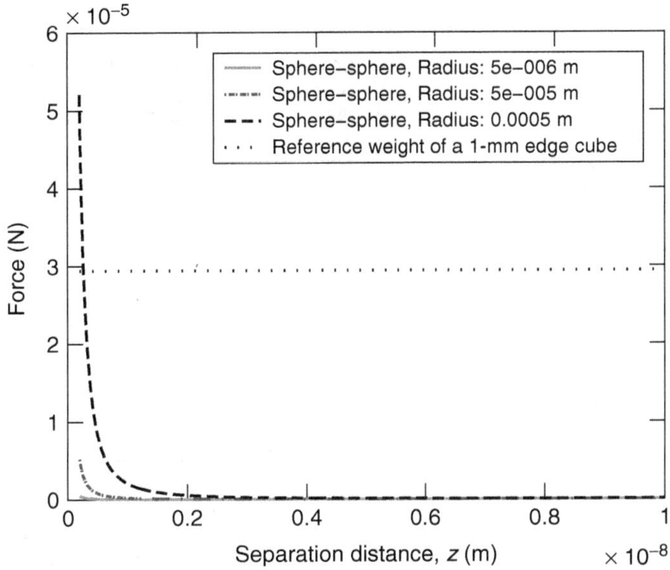

Figure 1.4. van der Waals force between two spheres with equal radii, Hamaker constant $= 5 \times 10^{-20}$ J. (As a comparison, the horizontal strip line represents the weight of a cube with a 1-mm edge and a density equal to 3000 kg m^{-3}, i.e., a bit heavier than aluminum.)

Moreover, if the separation distance D tends toward zero ($D << R_1$ and $z << R_2$), an approximation of $F_{(S_1,S_2)}(D)$ is given by

$$F_{(S_1,S_2)}(D) \approx -\frac{A\rho}{6D^2} \tag{1.28}$$

where $\rho = 1/R_1 + 1/R_2$.

An interesting result is that the force now depends on the inverse of the second power of the separation distance. The decrease consequently occurs more slowly, and the influence of van der Waals forces can be more seriously considered between two macroscopic objects (where macroscopic means "having a considerable number of molecules" but is still related to micrometric components). In order to have an idea of its order of magnitude, we plot the van der Waals forces as a function of the separation distance in Figure 1.4.

The numerical comparison between the analytical expression and its approximation is plotted in Figure 1.5: It can be concluded that the approximations can be widely used since the relative error is small: For objects with a characteristic size larger than a few microns and for separation distances smaller than 10nm (i.e., cut-off length to avoid the retardation effects that are not modeled in the described method), the relative error is smaller than 0.4% in all cases.

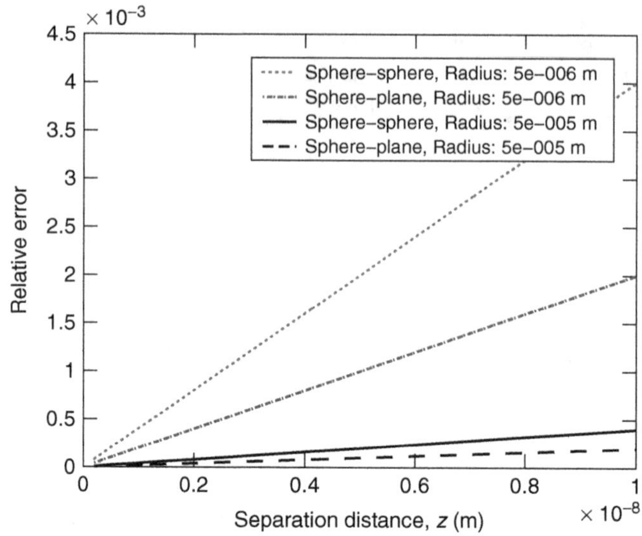

Figure 1.5. Relative errors between the analytical expressions and the approximations of the van der Waals forces.

1.2.1.5 Force Between a Sphere and an Infinite Half-Space

The force is calculated by deriving the interaction potential $W_{(S,HS)}(D)$ given by Eq. 1.26 with respect to the separation distance D:

$$F_{(S,HS)}(D) = -\frac{dW_{(S,HS)}(D)}{dD} = \frac{A}{6}\left[\frac{1}{D} - \frac{1}{2R+D} - \frac{R}{D^2} - \frac{R}{(2R+D)^2}\right] \tag{1.29}$$

Moreover, if D tends toward zero ($D << R$), an approximation of $F_{(S,HS)}$ is given by

$$F_{(S,HS)}(D) \approx -\frac{AR}{6D^2} \tag{1.30}$$

Note the similarity between Eqs. 1.28 and 1.30.

1.2.1.6 Interaction Between an Infinite Half-Space and a Rectangular Box

First, let us consider the interaction between a volume element dV_1 and an infinite half-space separated by a distance D such as the situation represented in Figure 1.6.

From Eq. 1.8, the interaction potential $w(z)$ between the volume element dV_1 containing n_1 molecules in cubic meters and a volume element dV_2 of the infinite

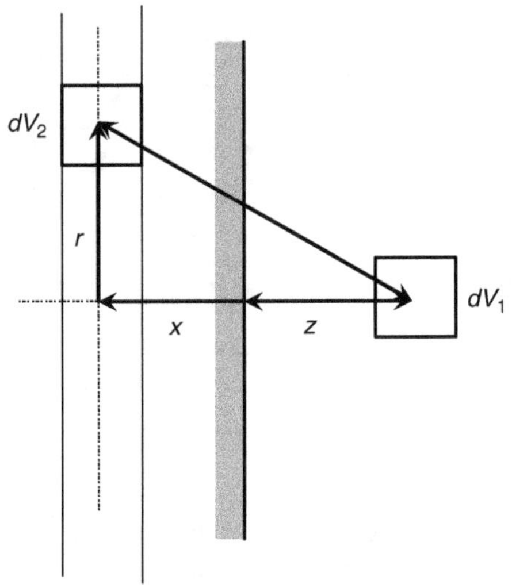

Figure 1.6. Interaction between a volume element and an infinite half-space.

half-space limited by a smooth plane and containing n_2 molecules in cubic meters is given by

$$w(z) = -\frac{n_1 n_2 K\, dV_1\, dV_2}{d^6} \tag{1.31}$$

where d is the separation distance between dV_1 and dV_2. By choosing a coordinates frame centered in dV_1 whose z axis is perpendicular to the plane and by noting that $d^2 = z^2 + r^2$, this leads to the potential interaction between dV_1 and the half-space:

$$\begin{aligned} w(D) &= -2\pi K n_1 n_2\, dV_1 \int_{z=D}^{z=\infty} dz \int_{r=0}^{r=\infty} \frac{r\,dr}{(z^2+r^2)^3} \\ &= -\frac{A}{2\pi}\int_D^{\infty} \frac{dz}{z^4} = -\frac{A}{6\pi D^3} dV_1 \end{aligned} \tag{1.32}$$

where A is the well-known Hamaker constant already defined in a previous subsection. Henceforth, the force $f(z)$ between the half-space and the volume element is given by

$$f(z) = -\frac{\partial w}{\partial D} = -\frac{A}{2\pi D^4} dV_1 \tag{1.33}$$

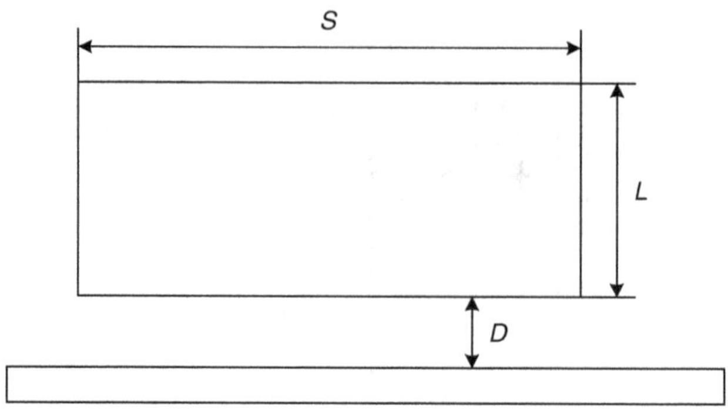

Figure 1.7. Interaction between rectangular box and infinite half-space.

We are now able to calculate the force between a given volume V_1 located near an infinite half-space limited by a smooth plane:

$$F = -\frac{A}{2\pi} \int_{V_1} \frac{1}{D^4} dV_1 \tag{1.34}$$

In the case of a rectangular smooth box with two faces of section S that are parallel to the plane (see Fig. 1.7), the force can be written as a function of materials (A), section (S), thickness (L), and separation distance (D):

$$F(A, S, L, D) = -\frac{AS}{2\pi} \int_{D}^{D+L} \frac{1}{z^4} dz = \frac{AS}{6\pi} \left[\frac{1}{(D+L)^3} - \frac{1}{D^3} \right] \tag{1.35}$$

Note that if $D << L$, Eq. 1.35 can be rewritten as a classical approximation [1]:

$$F(D) \approx -\frac{AS}{6\pi D^3} \tag{1.36}$$

Note that when the geometries become less obvious, the summation can no longer be achieved analytically. A method based on the Green identity is proposed in order to study the influence of the relative orientation of the objects and that of their roughness.

It proceeds as follows: The van der Waals force is computed by replacing the volume integral by a surface integral using the Green identity, as illustrated with the interaction between an infinite half-space and a rectangular box separated by a distance D [see Fig. 1.8(b)]. This problem has an analytical solution given by Eq. 1.35 that can be used to validate the method. This result will now be used in combination with the Green identity $\int\int\int_{\Omega} \operatorname{div} \overline{u}\, d\Omega = \oint_{\partial\Omega} \overline{u}.\overline{n}\, d(\partial\Omega)$. Let

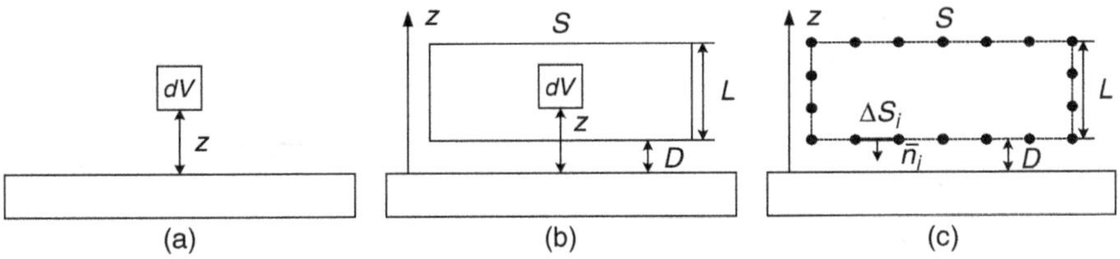

Figure 1.8. Geometry of the rectangular block: (a) infinite half-space and volume element, (b) geometry, and (c) mesh.

us assume a vector field given by $\overline{u} = -(1/3z^3)\overline{1}_z$. Its divergence is given by $\operatorname{div}\overline{u} = 1/z^4$. Consequently, Eq. 1.34 can now be rewritten as

$$F(D) = \frac{A}{2\pi} \oint_{\partial V_1} \frac{n_z}{3z^3}\, dS \tag{1.37}$$

Then, by meshing the surface of the considered object [see Fig. 1.8(c)] into N surface elements, the ith element being characterized by a normal vector with a z-component n_{zi}, the integral in Eq. 1.37 is replaced by a discrete sum:

$$F(D) = \frac{A}{6\pi} \sum_{i=1}^{N} \frac{n_{z_i}}{z^3} \Delta S_i \tag{1.38}$$

Examples of this method can be found in Lambert and Régnier [53] concerning the influence of the relative tilt of two parts and the influence of surface roughness modeled by a bearing curve.

As a summary of this section, let us indicate in Table 1.2 some useful approximations from the literature: additional references exist about the interaction between a sphere and a cylindric pore [73], between a sphere and a spherical cavity [85], and between two rough planes [33, 34].

As a conclusion there exist models: (1) without roughness no orientation [139], (2) with roughness but without orientation [2, 53, 90], and (3) without roughness but with orientation [23, 53]. Note that we have not found any description of a configuration that includes both roughness and orientation. Ideally, these forces should be computed again taking into account the mechanical deformations at contact. The proposed theory should be regarded as a first step.

To close this section, let us recall some useful references: [1, 12, 24, 38, 39, 53].

1.2.2 Capillary Forces

Capillary forces between two solids arise from the presence of a liquid meniscus between both solids. The presence of this liquid is due either to the

TABLE 1.2. Comparison Between the approximations from the Literature (*D*, separation distance, *R* the sphere radius, and *A* is the Hamaker constant)[a]

Object 1	Object 2	Expression	Reference
Plane	Plane //	$W \approx -\dfrac{A}{12\pi D^2}$; $F \approx \dfrac{A}{6\pi D^3}$ (by surface unit)	1,39,88
Cylinder	Cylinder //	$W \approx \dfrac{AL}{12\sqrt{2}D^{3}/2}\left(\dfrac{R_1R_2}{R_1+R_2}\right)^{1/2}$; $F \approx \dfrac{AL}{8\sqrt{2}D^{5/2}}\left(\dfrac{R_1R_2}{R_1+R_2}\right)^{1/2}$ (L, cylinders length; R_i, cylinders radii)	39, own results
Cylinder	Cylinder ⊥	$W \approx -\dfrac{A\sqrt{R_1R_2}}{6D}$; $F \approx \dfrac{A\sqrt{R_1R_2}}{6D^2}$	1,39, own results
Sphere	Plane	$W \approx -\dfrac{AR}{6D}$; $F \approx \dfrac{AR}{6D^2}$	1,88
Sphere	Sphere	$W \approx -\dfrac{AR}{6D}$; $F \approx \dfrac{AR}{6D^2}$ (including conical and spherical asperities)	1,39,88

[a]Note that the minus sign of the forces has been omitted: they must be considered attractive.

user—who puts liquid to provoke an adhesion force, for example, to pick up a component—or due to the condensation of the surrounding humidity—either spontaneously due to environmental conditions or due to the cooling of a gripper, for example [18]. On a more general note, these forces arise from the surface tension of the interface between two media: water–air, water–oil, or oil–air. Therefore, they are also called surface tension forces or surface tension effects. They are of the utmost importance in the microworld because they clearly dominate all the other effects but maybe in some cases at a few nanometer scale the van der Waals forces with which they compete on a balanced manner.

Many aspects are worth mentioning: the underlying concepts, the models, the experimental measure, the applications, and the perspectives. Nevertheless, and it is not the scope of this book to detail all these aspects. The interested reader will find throughout this section many useful references on these topics. More generally, we refer to Lambert [50] for a detailed description of capillary forces in microrobotics (modeling, measurement, application to microassembly).

1.2.2.1 Key Concepts

The key concepts to the understanding and the modeling of capillary forces are the *surface energy*, *surface tension*, the *contact angles*, and *wettability* together with the *Young–Dupré equation*, the *pressure drop* across the interface described by the so-called *Laplace* equation, and the *curvature* of a surface in the three-dimensional (3D) space. Additional concepts are the *contact angle hysteresis*, the

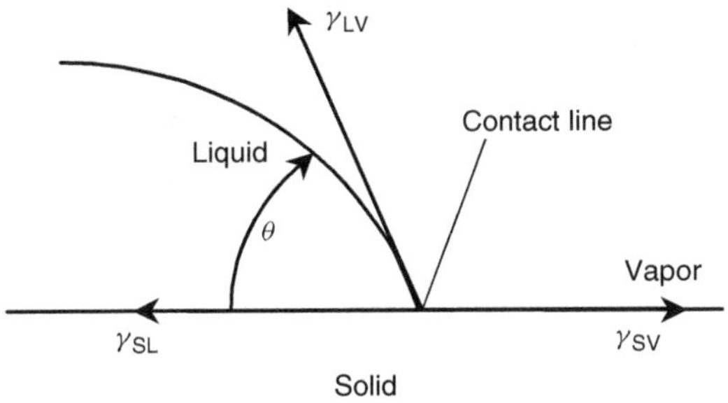

Figure 1.9. Illustration of the Young–Dupré equation.

surface impurities and *heterogeneities*, and the *dynamic spreading* of a liquid on a substrate.

Usually, if a liquid is not contained, it spreads out. However, when we look at soap bubbles or small water droplets, we observe that they behave as if their surface was an elastic membrane, characterized by a surface tension that acts against their deformations.[3] The concept of surface energy (or surface tension), which has the dimensions of an energy surface unit (J m^{-2}). The mechanical point of view considers the surface tension a tensile force by length unit (N m^{-1}). The surface tension is denoted by γ and its numerical value depends on the molecular interactions: in most oils, the molecular interaction is van der Waals interaction, leading to quite low surface tensions ($\gamma \approx 20$ mN m^{-1}). As far as water is concerned, due to the hydrogen bonding, the molecular attraction is larger ($\gamma \approx 72$ mN m^{-1}). Typical values for conventional liquid range from 20 mN m^{-1} (silicone oil) to 72 mN m^{-1} (water at 20°). For example, de gennes et al. [19] gives the following values for ethanol (23 mN m^{-1}), acetone (24 mN m^{-1}), and glycerol (63 mN m^{-1}).

Not only can the interface between a vapor and a liquid be characterized by an interfacial tension, denoted by γ and expressed as an energy surface unit or as a force by length unit, but the interfacial tensions can also be defined at the interfaces between a liquid and a solid (γ_{SL}) and between a solid and a vapor (γ_{SV}). Typical values of γ_{SV} are given in the literature [71]: nylon (polyamid) 6.6 (41.4 mN m^{-1}), high-density Polyethylene (PE) (30.3–35.1 mN m^{-1}), low-density PE (32.1–33.2 mN m^{-1}), Polyethylene terephthalate (PET) (40.9–42.4 mN m^{-1}), poly(methyl) methalcrylate (PMMA) (44.9–45.8 mN m^{-1}), Polypropylene (PP) (29.7), Polytetrafluoroethylene (PTFE) (20.0–21.8 mN m^{-1}).

The surface tension γ will indifferently be denoted by γ_{LV}. When a droplet is posed on a solid substrate (see Fig. 1.9), the liquid spreads out and we can distinguish three phases (vapor, liquid, solid) separated by three interfaces that join one another at the triple line, also called contact line.

At this triple line, the liquid–vapor interface makes an angle θ with the substrate. If the contact line is at equilibrium, θ is called the static contact angle,

[3]This is presented in a didactic way in de gennes et al. [19].

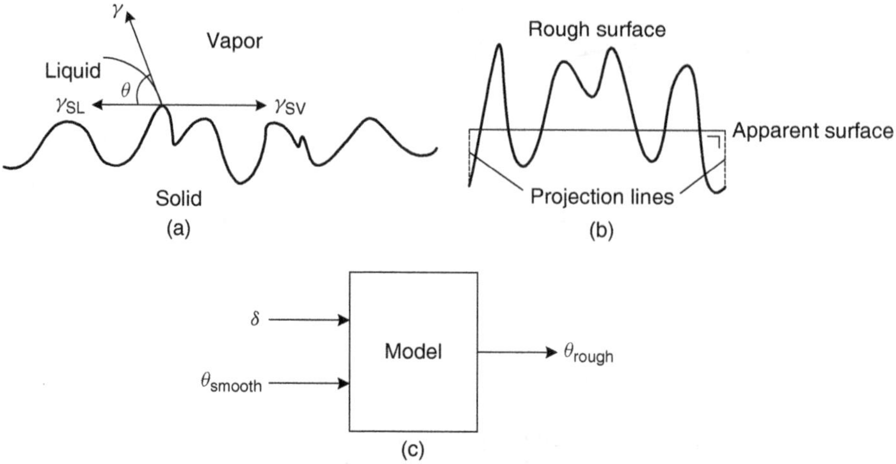

Figure 1.10. Influence of surface roughness: (a) contact line on a rough substrate, (b) actual and apparent surfaces, and (c) model to modify contact angle.

which is linked to the interfacial tensions by the Young–Dupré equation [1, 39]:

$$\gamma_{\mathrm{LV}} \cos\theta + \gamma_{\mathrm{SL}} = \gamma_{\mathrm{SV}} \tag{1.39}$$

This equation can be written immediately by considering the balance of the forces acting on the contact line. A second approach is based on the fact that at equilibrium the energy must be extremal and that any displacement of the contact line leads to an energy variation equal to zero:

$$\begin{cases} \Delta G = \delta\, \Delta\, A(\gamma_{\mathrm{SL}} - \gamma_{\mathrm{SV}}) + \Delta A \gamma_{\mathrm{LV}} \cos\theta \\ \lim_{\Delta A \to 0} \frac{\Delta G}{\Delta A} = 0 \end{cases} \tag{1.40}$$

where ΔA and ΔG are the variation of interface area and energy during the considered displacement. Let us now assume a heterogeneous surface containing two materials 1 and 2. A fraction f_1 of this surface is characterized by a surface energy leading to a contact angle θ_1, and the other part of the surface (fraction $f_2 = 1 - f_1$) leads to the contact angle θ_2. The theoretical contact angle given by the Young equation (1.39) is modified into an effective contact angle θ_C given by the Cassie equation [1, 40]:

$$\cos\theta_C = f_1 \cos\theta_1 + f_2 \cos\theta_2 \tag{1.41}$$

Another expression has been proposed by Israelachvili and Gee [40], but it seems that for the same values of θ_1, θ_2, f_1, and f_2, it will always predict a smaller contact angle than that obtained with Eq. (1.41):

$$(1 + \cos\theta_C)^2 = f_1(1 + \cos\theta_1)^2 + f_2(1 + \cos\theta_2)^2 \tag{1.42}$$

Let us assume a droplet placed on a rough substrate: Due to the roughness asperities, the actual area is bigger than the apparent one. Let us now introduce δ, the ratio of the actual interface area to the apparent one. The area of the actual (i.e., rough) area of the solid–vapor (solid–liquid) interface is denoted by A_{SV} (A_{SL}). The apparent surface is a projection of the rough surface:

$$\delta = \frac{A_{SL}}{A_{Apparent}} = \frac{A_{SV}}{A_{Apparent}} \tag{1.43}$$

Using δ, Eq. 1.40 can now be rewritten into

$$\begin{cases} \Delta G = \delta \Delta A_{Apparent} \gamma_{SL} - \delta \Delta A_{Apparent} \gamma_{SV} + \Delta A_{Apparent} \gamma \cos\theta \\ \lim_{\Delta A \to 0} \dfrac{\Delta G}{\Delta A} = 0 \end{cases} \tag{1.44}$$

Combining Eqs. 1.44 and the expression of the contact angle given by the Young equation, the effective contact angle θ_{rough} can be expressed as a function of the surface ratio δ and the contact angle θ_{smooth} made of the liquid on a plane smooth substrate made of the same material:

$$\cos\theta_{rough} = \delta \cos\theta_{smooth} \tag{1.45}$$

This approach was first proposed by Wenzel [92] and more detailed information can be found in Adamson and Gast [1] and Hao et al. [29]. Henceforth, Eq. 1.45 can feed the previous simulation with contact angles corresponding to actual rough surfaces. That is important if the simulation is used to design gripper tips that usually present roughness profiles.

From Eq. 1.45, we see that angles lower than 90° are decreased by roughness, while the angle increases if θ is larger than 90°.

It must be noted that surface roughness can lead to condensing humid air in small cavities of the surface and hence to an attractive force L_{cp} due to liquid bridging [46]:

$$L_{cp} = \frac{A_l \gamma}{r_k} \tag{1.46}$$

where A_l is the surface area where meniscus formation occurs and r_k is the Kelvin radius given by the Kelvin equation [1][4]:

$$r_k = \frac{\gamma v}{RT \log(p_0/p)} \tag{1.47}$$

where v is the molar volume of the liquid, R is the perfect gas constant, T is the absolute temperature, p_0/p is the relative vapor pressure (= relative humidity for water). Israelachvili [39] gives $\gamma v/RT = 0.54$ nm for water at 20°C.

[4] $\log = \log_e = \ln \neq \log_{10}$.

When the contact line is about to move, one observes the contact angle changing. The receding angle is smaller than the static angle while the observed angle, when moving forward, is larger than the static contact angle. A model has been proposed by Zisman (see Adamson and Gast [1]), who observed that cos θ_A (advancing angle) is usually a monotonic function of γ. Henceforth, he proposed the following equation:

$$\cos\theta_A = a - b\gamma \tag{1.48}$$

Gutowsky [29] cited Johnson and Dettre [43] for a detailed study of the effect of roughness on contact angle hysteresis. This hysteresis implies that even at equilibrium, the contact angle value is not unique. The contact angle also depends on the velocity of the contact line. This phenomenon is described in Hoffman [36].

Due to the surface tension, there exists a pressure difference across the interface between a liquid and a gas. In the case of a soap bubble, for example, the pressure inside the bubble is bigger, to compensate the outside pressure and to overcome the tension effect. In a more general case, the pressure difference is linked to the curvature of the interface according to the Laplace equation [1]:

$$2\gamma H = 2\gamma\left(\frac{1}{R_1} + \frac{1}{R_2}\right) = p_{\text{in}} - p_{\text{out}} \tag{1.49}$$

where H is the mean curvature and R_1 and R_2 are two principal curvature radii.

1.2.2.2 Models of Capillary Forces

Models of capillary forces found in the literature are usually valid only at equilibrium. Before detailing them, let us now consider two solids linked by a liquid bridge,[5] also called meniscus (Fig. 1.11). In order to link this to the general frame of micromanipulation, let us call the upper solid the "tool" or the "gripper" (it will be used as a gripper) and the lower one as the object (it will be used as a micropart or a microcomponent). Since axial symmetry is assumed, it can be seen in Figure 1.11 that the contact line between the meniscus and the object (the gripper) is a circle with a radius r_1 (r_2). The pressure inside the meniscus is denoted by p_{in} and that outside the meniscus by p_{out}. The contact angle between the object and the meniscus is θ_1 and the angle between the gripper and the meniscus is θ_2. The separation distance (also called the gap) between the component and the gripper is denoted by z. The immersion height is called h. At its neck, the principal curvature radii are ρ' (in a plane perpendicular to the z axis, i.e., parallel to the component) and ρ (in the plane rz).

The object is submitted to the "Laplace" force, arising from the pressure difference $p_{\text{in}} - p_{\text{out}}$, and to the "tension" force, directly exerted by the surface

[5]The presented that is, configuration is axially symmetric, to introduce the capillary force from a "mechanical" point of view, that is, using concepts such as pressure or tension. In a more general case, the configuration is not axially symmetric and an energetic approach has to be implemented; see therefore Lambert [50].

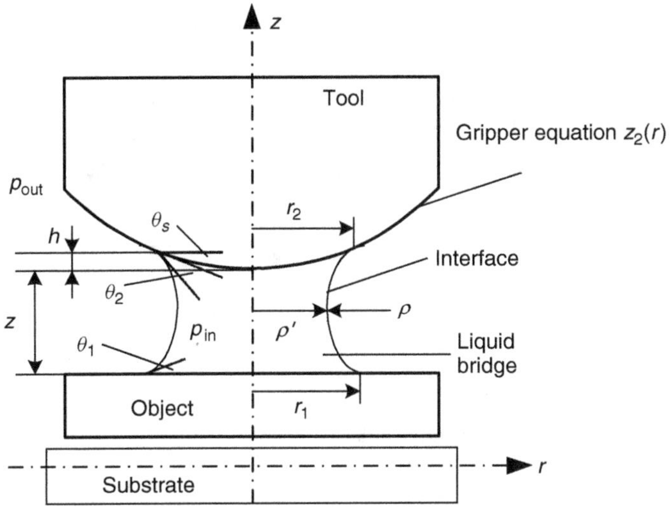

Figure 1.11. Effects of a liquid bridge linking two solid objects (from [52]).

tension. In what follows, we will consider that these two forces constitute what we will call the capillary force.[6] The Laplace force is due to the Laplace pressure difference that acts over an area πr_1^2 (see Fig. 1.12) and can be attractive or repulsive according to the sign of the pressure difference, that is, according to the sign of the mean curvature: A concave meniscus will lead to an attractive force while a convex one will induce a repulsive force.

$$F_{\mathrm{L}} = 2\gamma H \pi r_1^2 \tag{1.50}$$

The tension force implies the force directly exerted by the liquid on the solid surface. As illustrated in Figure 1.13, the surface tension γ acting along the contact circle must be projected on the vertical direction, leading to

$$F_{\mathrm{T}} = 2\pi r_1 \gamma \sin(\theta_1 + \phi_1) \tag{1.51}$$

Therefore, the capillary force is given by

$$F_C = F_T + F_L = 2\pi r_1 \gamma \sin(\theta_1 + \phi_1) + 2\gamma H \pi r_1^2 \tag{1.52}$$

where ϕ_1 denotes the slope of the component at the location of the contact line: It will be considered equal to zero in the following. In a more general way—for example, in the case of nonaxially symmetric geometries—the force is computed from the derivation of the surface energy. Both ways are proven to be equivalent [51]. Let us illustrate both methods in the following examples.

[6]Marmur [62] uses the terms "capillary" force for the term arising from the pressure difference and "interfacial tension force" for that exerted by the surface tension.

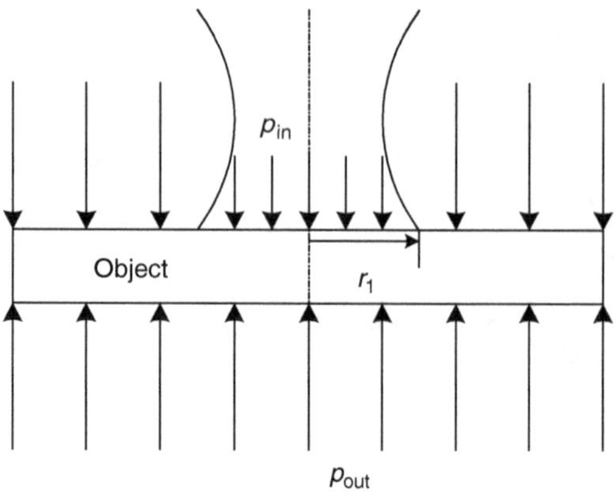

Figure 1.12. Origin of the Laplace force: attractive case (from [52]).

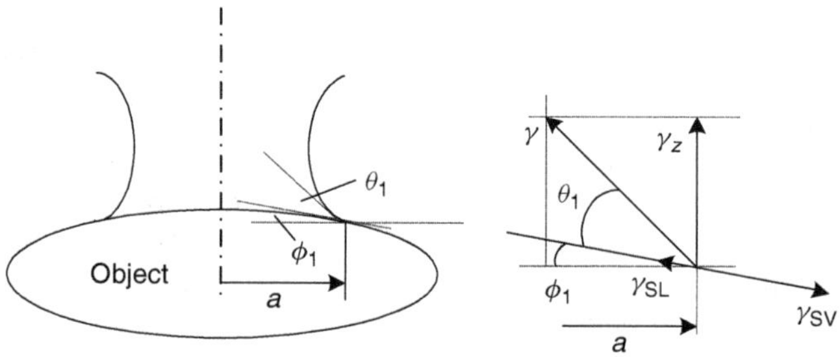

Figure 1.13. Origin of the tension force and detail (from [52]).

Surface Energy Derivation in the Case of a Sphere and a Plate. We detail here the mathematical developments required to calculate the analytical approximations of the capillary forces, based on energetic approach. Let us define preliminary mathematical formulations:

1. Definitions:

$$A(\phi) \equiv \frac{2\pi}{3}\left(1 - \frac{3}{2}\cos\phi + \frac{1}{2}\cos^3\phi\right)$$

$$\frac{dA}{d\phi} = \pi \sin^3\phi$$

2. Properties:

$$\cos\phi = 1 - \frac{\phi^2}{2} + \frac{\phi^4}{24} + O(\phi^6)$$

$$\cos^2\phi = 1 - \phi^2 + \frac{\phi^4}{3} + O(\phi^6)$$

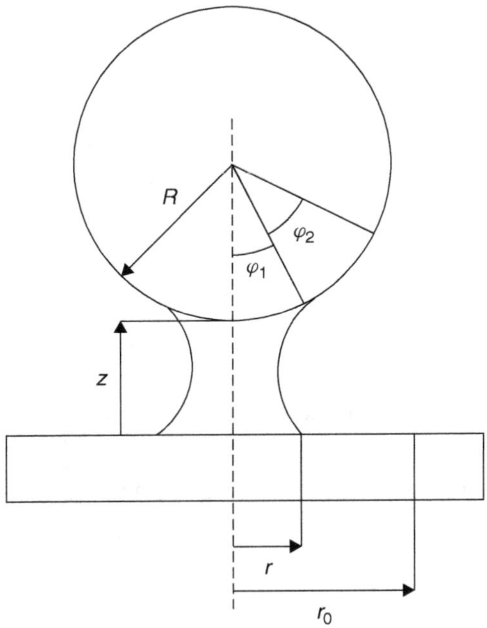

Figure 1.14. Studied configuration.

$$\cos^3\phi = 1 - \frac{3}{2}\phi^2 + \frac{7}{8}\phi^4 + O(\phi^6)$$

$$\sin\phi = \phi - \frac{\phi^3}{6} + O(\phi^5)$$

$$\sin^2\phi = \phi^2 - \frac{\phi^4}{3} + O(\phi^6)$$

$$\sin^3\phi = \phi^3 + O(\phi^5)$$

$$A(\phi) = \frac{\pi}{4}\phi^4 + O(\phi^6)$$

$$\frac{dA}{d\phi} = \pi\phi^3 + O(\phi^5)$$

$$1 - \cos\phi \approx \frac{\sin\phi^2}{2} \approx \frac{\phi^2}{2}$$

Now, let us compute the force between a sphere and a plane: The notations are defined in Figure 1.14. In this figure, ϕ_0 and r_0 are arbitrary constants. Their exact value does not play any role because the force will be calculated by deriving the interfacial energy W with respect to the gap z between the sphere and the plane [39]:

$$F = -\frac{dW}{dz} \tag{1.53}$$

Let us write the interfacial energy of the system:

$$\begin{aligned} W(z) &= A_{\mathrm{SL}}\gamma_{\mathrm{SL}} + A_{\mathrm{SV}}\gamma_{\mathrm{SV}} + \Sigma\gamma \\ &= \gamma_{\mathrm{SL}}\pi r^2 + \gamma_{\mathrm{SV}}\pi(r_0^2 - r^2) + \gamma 2\pi r\big[z + R(1 - \cos\phi)\big] \\ &\quad + \gamma_{\mathrm{SL}} 2\pi R^2(1 - \cos\phi) + \gamma_{\mathrm{SV}} 2\pi R^2\big[(1 - \cos\phi_0) - (1 - \cos\phi)\big] \end{aligned} \tag{1.54}$$

Since ϕ is assumed to be small, W can be rewritten as

$$\begin{aligned} W(z) &= \pi r^2(\gamma_{\mathrm{SL}} - \gamma_{\mathrm{SV}}) + \gamma 2\pi r z + \gamma\pi r R \sin^2\phi + \gamma_{\mathrm{SV}}\pi r_0^2 \\ &\quad + \pi R^2 \sin^2\phi(\gamma_{\mathrm{SL}} - \gamma_{\mathrm{SV}}) + \gamma_{\mathrm{SV}}\pi R^2 \sin^2\phi_0 \end{aligned}$$

and, by considering the Young–Dupré equation ($\gamma \cos\theta = -\gamma_{\mathrm{SL}} + \gamma_{\mathrm{SV}}$):

$$\begin{aligned} W &= -2\pi R^2 \sin^2\phi\gamma \cos\theta + \gamma_{\mathrm{SV}}\pi r_0^2 + \gamma 2\pi r z \\ &\quad + \gamma\pi R^2 \sin^3\phi + \gamma_{\mathrm{SV}}\pi R^2 \sin^2\phi_0 \end{aligned} \tag{1.55}$$

Let us now consider the derivative of W:

$$\begin{aligned} \frac{dW}{dz} &= -4\pi R^2 \sin\phi \cos\phi\gamma \cos\theta\frac{d\phi}{dz} + \gamma 2\pi R \sin\phi \\ &\quad + \gamma 2\pi z R \cos\phi\frac{d\phi}{dz} + 3\gamma\pi R^2 \sin^2\phi \cos\phi\frac{d\phi}{dz} \end{aligned} \tag{1.56}$$

or, by assuming $\sin\phi \approx \phi$ and $\cos\phi \approx 1$:

$$\frac{dW}{dz} = -4\pi R^2\phi\gamma \cos\theta\frac{d\phi}{dz} + \gamma 2\pi R\phi + \gamma 2\pi Rz\frac{d\phi}{dz} + 3\gamma\pi R^2\phi^2\frac{d\phi}{dz} \tag{1.57}$$

The value of $d\phi/dz$ must be evaluated in Eq. 1.57. Therefore, the meniscus volume is assumed to be constant, leading to $dV/dz = 0$. Moreover the meniscus will be assumed to be cylindrically shaped so that the volume is the difference between the external liquid cylinder and the volume of the spherical cap inside the external cylinder:

$$V = \pi r^2\big[z + R(1 - \cos\phi)\big] - \frac{2\pi R^3}{3}\left(1 - \frac{3}{2}\cos\phi + \frac{\cos^3\phi}{2}\right) \tag{1.58}$$

Once again the assumption of small ϕ is made, leading to the following approximation:

$$\frac{2\pi R^3}{3}\left(1-\frac{3}{2}\cos\phi+\frac{\cos^3\phi}{2}\right)=A(\phi)R^3\approx\frac{\pi R^3}{4}\phi^4 \tag{1.59}$$

The final expression for V is now given by

$$V=\pi r^2 z+\frac{\pi r^2 R}{2}\sin^2\phi-\frac{\pi R^3}{4}\phi^4 \tag{1.60}$$

$$=\pi R^2\sin^2\phi z+\frac{\pi R^3}{2}\sin^4\phi-\frac{\pi R^3}{4}\phi^4 \tag{1.61}$$

so that

$$\begin{aligned}\frac{dV}{dz}&=2\pi R^2 z\sin\phi\cos\phi\frac{d\phi}{dz}+\pi R^2\sin\phi\\&\quad+2\pi R^3\sin^3\phi\cos\phi\frac{d\phi}{dz}-\pi R^3\phi^3\frac{d\phi}{dz}\\&=2\pi R^2 z\phi\frac{d\phi}{dz}+\pi R^2\phi^2+\pi R^3\phi^3\frac{d\phi}{dz}\\&=0\\&\Rightarrow\frac{d\phi}{dz}=\frac{-\pi R^2\phi^2}{2\pi R^2\phi z+\pi R^3\phi^3}\\&=\frac{-1}{2z/\phi+R\phi}\end{aligned} \tag{1.62}$$

The total capillary force is then given by substituting this latter result into Eq. 1.57:

$$F=-\frac{4\pi R^2\phi\gamma\cos\theta}{2z/\phi+R\phi}-\gamma 2\pi R\phi+\frac{\gamma 2\pi Rz}{2z/\phi+R\phi}+\frac{3\gamma\pi R^2\phi^2}{2z/\phi+R\phi} \tag{1.63}$$

Since $h=R(1-\cos\phi)\approx(R/2)\sin^2\phi\approx(R/2)\phi^2$:

$$\begin{aligned}F&=-\frac{4\pi R\gamma\cos\theta}{2z/R\phi^2+1}-\gamma 2\pi R\phi+\frac{\gamma 2\pi Rz}{2z/\phi+R\phi}+\frac{3\gamma\pi R\phi}{2z/R\phi^2+1}\\&=-\frac{4\pi R\gamma\cos\theta}{z/h+1}-\gamma 2\pi R\phi+\frac{\gamma 2\pi Rz}{2z/\phi+R\phi}+\frac{3\gamma\pi R\phi}{z/h+1}\end{aligned}$$

The last three terms of this equation represent the contribution of the LV interface to the total interfacial energy. Let us assess their relative importance with respect to the first term. Their sum is given by

$$\frac{\pi R\gamma\phi(R\phi^2-2z)}{R\phi^2+2z} \tag{1.64}$$

The ratio of the first term to the sum of the last three ones is equal to

$$\frac{\frac{4\pi R\gamma \cos\theta}{z/h+1}}{\frac{\pi R\gamma\phi(R\phi^2-2z)}{R\phi^2+2z}} = \frac{4\cos\theta h}{\phi(h-z)} \tag{1.65}$$

If $z = 0$, this ratio tends toward infinity if ϕ tends to zero. Since ϕ cannot be exactly equal to zero, the last three terms can be neglected with the (now) classical assumption $\phi <<$. This leads to the well-known approximation [39]:

$$F_{\max} = -4\pi R\gamma \cos\theta \tag{1.66}$$

If $z \neq 0$ but by neglecting the contribution of lateral area to W, the total capillary force can be rewritten as follows:

$$F = -\frac{4\pi R\gamma \cos\theta}{z/h+1} \tag{1.67}$$

We see that this method requires one to assume a geometric shape for the meniscus. Numerical energy minimization techniques can be used to avoid this assumption, as implemented with finite elements in surface evolver. In this case, the method implementation is exact. Attention should, of course, be paid to the underlying assumptions of the model: constant volume of liquid (i.e., no evaporation), constant contact angles, static modeling, and vanishing Bond number (i.e., the gravity effect on the meniscus shape is neglected). *Note that these assumptions are restrictive for all models presented in this section.*

1.2.2.3 Direct Calculation of the Laplace and Tension Terms in the Case of Two Parallel Plates

Let us consider the configuration shown in Figure 1.15: Two parallel plates separated by a gap D are linked by a meniscus of volume V wetting the lower plate with a contact angle θ_1 and an upper plate with a contact angle θ_2. Since the configuration is axially symmetric, Eq. 1.49 can be rewritten using the expression of the curvature of an axially symmetric surface:

$$-\frac{r''}{(1+r'^2)^{3/2}} + \frac{1}{r(1+r'^2)^{1/2}} = \frac{\Delta p}{\gamma} \tag{1.68}$$

This a second-order nonlinear differential equation with a unknown second member. The initial condition are given by

$$r(z=0) = r_1 \tag{1.69}$$

$$r'(z=0) = -\frac{1}{\tan\theta_1} \tag{1.70}$$

The value of Δp can be adjusted to fit $r'(z=D) = 1/\tan\theta_2$ using a shooting method [50]. The initial radius r_1 can be iteratively guessed to adjust the volume

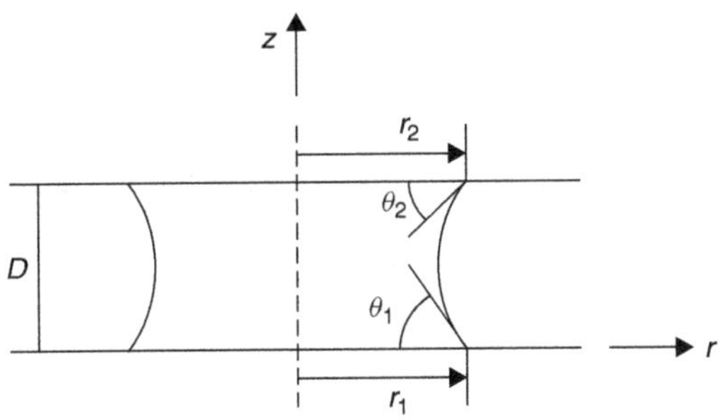

Figure 1.15. Axially symmetric meniscus between two parallel plates.

of liquid of the obtained meniscus equal to V. Thanks to this double iterative scheme, the meniscus shape and the pressure drop Δp can be known. Henceforth the force can be computed according to Equation 1.52.

It is interesting to note that in the case of a 2D configuration (different from the 2D axial symmetry), the curvature along the "extruding" direction perpendicular to the plane of this page is null. Equation 1.68 can be rewritten as

$$-\frac{r''}{(1+r'^2)^{3/2}} = \frac{\Delta p}{\gamma} \tag{1.71}$$

which corresponds to the equation of a circle (i.e., a 2D curve of constant curvature is the definition of a circle). The above-mentioned initial and boundary conditions can then be used to find the circle parameters (center coordinates and radius).

Finally, let us note that in the case of axially symmetric configurations, the meniscus can never be exactly a circle (since the term $1/[r(1+r'^2)^{1/2}]$ is different from zero). Nevertheless, the circle is a quite good approximation when the gap is small because in this case $r''/[(1+r'^2)^{3/2}] > 1/[r(1+r'^2)^{1/2}]$. This is the reason many authors assume the meniscus to be circular.

1.2.2.4 Other Models

Sphere–Sphere. Rabinovich [77] gives an analytical expression for the capillary force between two spheres with radii R_1 and R_2, as a function of the separation distance z:

$$F_{\text{sphere/sphere}} = -\frac{2R\cos\theta}{1+z/(2h)} \tag{1.72}$$

where R is the equivalent radius given by $R = 2R_1R_2/R_1 + R_2$, $2\cos\theta = \cos\theta_1 + \cos\theta_2$, z is the separation distance or gap, and h is the immersion height, approximately given by [77]

$$h = \frac{z}{2}[-1+\sqrt{1+2V/(Rz^2)}] \tag{1.73}$$

where V is the volume of the liquid bridge.

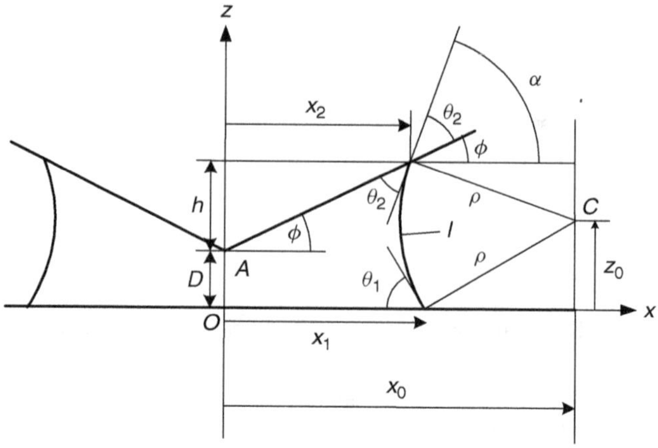

Figure 1.16. Prism–plane configuration.

Prism–Plane. In Lambert et al. [51] a model can be found for the interaction between a prism and a plate (Fig. 1.16). The prism is defined by its length in the y direction, L, and its angular aperture ϕ. Its location is defined by the distance[7] D between its apex A and the plane. Let us assume a volume of liquid V wetting the plane with a contact angle θ_1 and the prism with a contact angle θ_2. Since the curvature of the meniscus in the direction y perpendicular to $0xz$ is equal to zero, the Laplace equation becomes [52]

$$\frac{x''}{(1+x'^2)^{3/2}} = \frac{\Delta p}{\gamma} \tag{1.74}$$

where $x' = dx/dz$.

Assuming a vanishing Bond number, the hydrostatic pressure inside the meniscus is neglected by comparison to the Laplace pressure difference Δp, which is therefore constant in all the meniscus. Therefore, the second term of Eq. 1.74 is constant, and this equation can be integrated twice with respect to z in order to find the relation $x = x(z)$, with two integration constants and the undefined pressure difference Δp. A more straightforward derivation is based on the fact that since one of the curvature radius is infinite and that the total curvature $2H$ is constant, the second curvature radius $(1+x'^2)^{3/2}/x''$ is constant: Let us note it ρ. Therefore, the meniscus profile is a curve with constant curvature, that is, a circle given by the following equation:

$$(x - x_0)^2 + (z - z_0)^2 = \rho^2 \tag{1.75}$$

[7]For the sake of clarity, since z will be used as one of the coordinates, the gap is noted D in the following sections.

where x_0 and z_0 are the coordinates of the circle center. Once again, three parameters are to be determined: x_0, z_0, and ρ. This can be done using three boundary conditions: both contact angles θ_1 and θ_2 and the volume of liquid V.

As preliminary computations, let us express x_0, z_0, and ρ as functions of known data (ϕ,D,θ_1,θ_2) and the immersion height h, which is still unknown at this step, but which will be determined using the condition on the volume of liquid V. Note that x_2 is an intermediary variable and that x_1 will be used later. For the sake of convenience, the notation $\alpha = \theta_2 + \phi$ has been adopted in the following equations:

$$x_2 = \frac{h}{\tan \phi} \tag{1.76}$$

$$\rho = \frac{D + h}{\cos \theta_1 + \cos \alpha} \tag{1.77}$$

$$z_0 = \rho \cos \theta_1 \tag{1.78}$$

$$x_0 = x_2 - (z_0 - D - h) \tan \alpha \tag{1.79}$$

$$x_1 = x_0 - z_0 \tan \theta_1 \tag{1.80}$$

Additional useful relations are the meniscus equation:

$$x = x_0 - \sqrt{\rho^2 - (z - z_0)^2} \tag{1.81}$$

the meniscus slope x':

$$x' = -\frac{z - z_0}{x - x_0} \tag{1.82}$$

and, finally, the rewritten Laplace equation linking Δp and ρ:

$$\Delta p = \frac{\gamma}{\rho} \tag{1.83}$$

and h is still to be determined using the volume of liquid V (see next step).

The volume of liquid can be used to determine the value of the immersion height h, starting from the following expression of V as illustrated in Figure 1.17:

$$V = 2LA \tag{1.84}$$

$$= 2L[x_0(h + D) - A_{\text{I}} - A_{\text{II}} - A_{\text{III}} - A_{\text{IV}}] \tag{1.85}$$

where

$$A_{\text{I}} = \frac{x_2 h}{2} \tag{1.86}$$

$$A_{\text{II}} = \frac{(x_0 - x_2)(D + h - z_0)}{2} \tag{1.87}$$

$$A_{\mathrm{III}} = \frac{z_0(x_0 - x_1)}{2} \tag{1.88}$$

$$A_{\mathrm{IV}} = \frac{\rho^2(\pi - \alpha - \theta_1)}{2} \tag{1.89}$$

Therefore, the equation giving the volume V can be rewritten as follows:

$$V = 2L\Bigg[x_0(D + h) - \frac{x_2 h}{2} \cdots$$

$$\cdots - \frac{\rho^2(\pi - \alpha - \theta_1)}{2} - \frac{(x_0 - x_2)(D + h - z_0)}{2} - \frac{z_0(x_0 - x_1)}{2}\Bigg] \tag{1.90}$$

$$= L\Bigg[2x_2 D + x_2 h \cdots$$

$$\cdots + \rho^2\Bigg[\sin\alpha \cos\alpha + 2\sin\alpha \cos\theta_1 - \pi + \alpha + \theta_1 - \sin\theta_1 \cos(\theta_1)$$

(1.91)

$$= L\left[h^2\left(\frac{1}{\tan\phi} + \mu\right) + 2hD\left(\frac{1}{\tan\phi} + \mu\right) + \mu D^2\right] \tag{1.92}$$

This latter equation can be rewritten as a second-degree equation with respect to the unknown h:

$$h^2 + 2hD + \frac{\mu D^2 - V/L}{\mu + 1/\tan\phi} = 0 \tag{1.93}$$

which leads to

$$h = -D \pm \sqrt{D^2 - \frac{D^2\mu - V/L}{\mu + 1/\tan\phi}} \tag{1.94}$$

The $-$ solution makes no physical sense since the immersion height cannot be negative. Consequently:

$$h = -D + \sqrt{D^2 - \frac{D^2\mu - V/L}{\mu + 1/\tan\phi}} \tag{1.95}$$

and the variation of h with respect to a variation of the separation distance D (it will be used in what follows) is given by

$$\frac{dh}{dD} = -1 + \frac{D}{D + h}\frac{1}{1 + \mu\tan\phi} \tag{1.96}$$

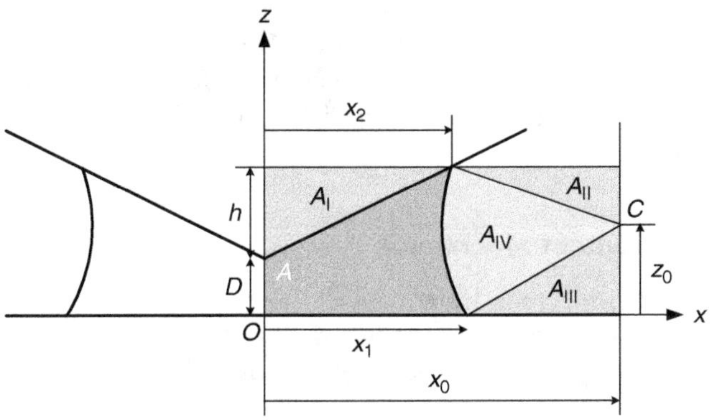

Figure 1.17. Determination of immersion height from volume of liquid.

As it has previously been explained, the capillary force can be written as the sum of a term depending on the Laplace pressure difference Δp and the so-called tension term:

$$F = 2Lx_1\,\Delta p + 2L\gamma\,\sin\theta_1 \tag{1.97}$$

$$= 2L\gamma\left(\frac{x_1}{\rho} + \sin\theta_1\right) \tag{1.98}$$

$$= 2L\gamma\frac{x_0}{\rho} \tag{1.99}$$

$$= 2L\gamma\left(\frac{x_2}{\rho} + \frac{D+h-z_0}{\rho}\tan\alpha\right) \tag{1.100}$$

$$= 2L\gamma\left(\frac{h}{D+h}\,\frac{\cos\theta_1 + \cos\alpha}{\tan\phi} + \sin\alpha\right) \tag{1.101}$$

Using Eq. 1.95, the force can be expressed as a function of the volume of liquid V, the separation distance D, and the angles of the problem: contact angles θ_1 and θ_2 at the one hand and the prism angle ϕ at the other hand. Remember that $\alpha = \theta_2 + \phi$. Lambert [50] explains how to adapt this model to the interaction between a cylinder and a plate. Additional information can be found in the literature [39, 48, 50].

1.2.2.5 Applications and Perspectives

Applications are based on the fact that surface tension is an important parameter in the perspective of a downscaling of the assembly equipment because the force it generates linearly decreases with the size while the weight decreases more quickly. While surface tension has been pointed out as being one of the disturbing effects in microelectromechanical systems (MEMS) (stiction problems [47, 63, 92], other uses have been positively considered [8, 32, 56, 70]. More particularly, surface tension effects have been applied to many fields such as capillary gripping [4, 10, 28, 54, 69, 72, 84], fluidic microvalves [25], actuation [11], and optics [9].

The perspectives in this field are to model the force dynamically (level-set-based simulation packages can do some job), to exploit capillary condensation (see later on in this book).

1.2.3 Elastic Contact Mechanics

This subsection considers the Hertz contact theory and the related adhesion models [20, 35, 41, 67]. In case of a sphere (radius R) on a planar surface, pull-off force is approximately given by JKR (for the lower boundary) or DMT (for the higher boundary) contact models [20, 27]:

$$\frac{3}{2}\pi RW \leq F_{\text{pull-off}} \leq 2\pi RW \tag{1.102}$$

where W is the work of adhesion between the two media. According to Maugis [67], the λ coefficient can be used to choose the most appropriate contact model for a given case. This coefficient is expressed for an interface between two bodies 1 and 2 with

$$\lambda_{12} = 2\sigma_0 \left(\frac{R}{\pi W_{12} K^2}\right)^{1/3} \tag{1.103}$$

where K is the equivalent elastic modulus, calculated using the Poisson ratios μ and Young's modulus E:

$$K = \frac{4}{3}\left(\frac{1-\mu_1^2}{E_1} + \frac{1-\mu_2^2}{E_2}\right) \tag{1.104}$$

and W_{12} is expressed as $W_{12} = \gamma_1 + \gamma_2 - \gamma_{12} = 2\sqrt{\gamma_1\gamma_2}$ with γ_{12} interfacial energy, γ_1 and γ_2 surface energy of the object, substrate, or tip [86]. Using λ, the pull-off force can be estimated with

$$\begin{aligned}
\lambda < 0.1 &\Longrightarrow \text{DMT model} \qquad P = 2\pi RW_{12} \\
\lambda > 5 &\Longrightarrow \text{JKR model} \qquad P = \frac{3}{2}\pi RW_{12} \\
0.1 < \lambda < 5 &\Longrightarrow \text{Dugdale model} \\
P &= \left(\frac{7}{4} - \frac{1}{4}\frac{4.04\lambda^{1/4} - 1}{4.04\lambda^{1/4} + 1}\right)\pi W_{12} R
\end{aligned} \tag{1.105}$$

When two media are in contact, the surface energy W_{12} is equal to

$$W_{12} \simeq 2\sqrt{\gamma_1\gamma_2} \tag{1.106}$$

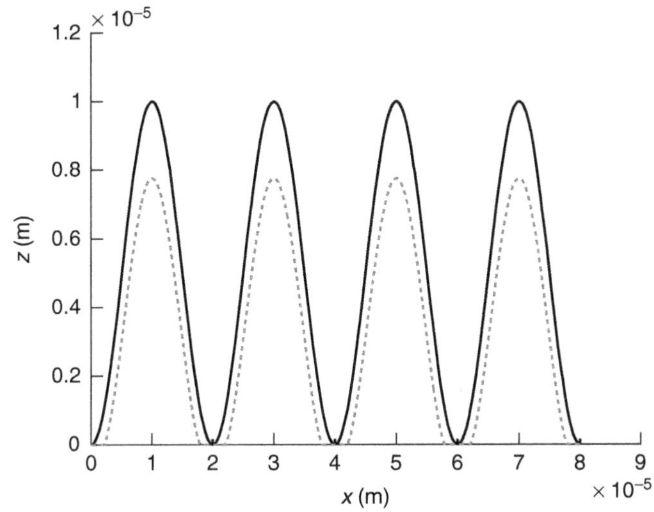

Figure 1.18. Wavy profile (solid line) and its deformation according to the Westegaard model.

with γ_i the surface energy of the body i. From the previous formulas the energy W_{132} required to separate two media 1 and 2 immersed in a medium 3 is given by

$$W_{132} = W_{12} + W_{33} - W_{13} - W_{23} = \gamma_{13} + \gamma_{23} - \gamma_{12}$$

Let us also cite the Westegaard model [42], which allows to compute the elastic deformation of a wavy surface against a infinite half-space. An example of the initial and deformed geometries is shown in Figure 1.18 Nevertheless, these models—which are widely used to interpret AFM measurements or to design grippers and microtools—rely on the elastic deformation assumption, which is usually no longer valid at scales smaller than 1 μm. Indeed, it can be shown from Table 1.3 that for example, in the case of a 0.600-μm diameter ball (the typical size for colloidal probes), the reference pressure p_0 always exceeds the elastic limit of 30 MPa (glass and silicon oxides). Consequently, we conclude the discrepancy of the Hertz model for balls with a diameter smaller that 1 μm. This implies that the other model does not hold either (DMT, JKR). This 1-μm limit cannot be put aside as far as CNT (carbon nanotubes) applications are concerned. Moreover, even for larger components, the roughness details are usually below this limit, which has a considerable impact on adhesion. Current research trends try to study the combined roles of plastic deformation and surface forces. More particularly, the interaction between roughness, plastic deformation, and electrostatic adhesion are fully described in Sausse Lhernould [81].

TABLE 1.3. Results of Some Hertz Model Computations ($E_1 = E_2 = 70$ Gpa, $\nu = 0.3$)

d (μm)	P (μN)	p_0
0.6	50	5 Gpa
0.6	5	2.5 Gpa
0.6	0.5	1.2 Gpa
0.6	0.05	0.5 Gpa
0.6	0.005	0.25 Gpa
60	0.005	11 Gpa
60	0.05	25 Gpa
60	0.5	54 Gpa

1.3 RECENT DEVELOPMENTS

1.3.1 Capillary Condensation

According to Mate [66], *one consequence of the pressure difference across a curved liquid surface is that the vapor pressure over a liquid surface depends on the degree of curvature of the surface. This leads to the phenomenon of capillary condensation, where vapors condense into small cracks and pores at vapor pressures significantly less than the saturation vapor pressure*. Capillary adhesion is an important source of perturbation in miniaturized systems. MEMS breakdown is often caused by adhesion problems [47, 63–65, 91]. Capillary condensation can be modeled thanks to the so-called Kelvin equation:

$$\frac{1}{r_1} + \frac{1}{r_2} = \frac{2}{r_k} = \frac{RT(P/P_s)}{\gamma V_m} \tag{1.107}$$

where r_1 and r_2 are the principal curvature radii, r_k is the so-called Kelvin radius, V_m the molar volume of the condensed liquid, γ the surface tension, R the perfect gas constant at 8.32 $\mathrm{J\,K^{-1}\,mole^{-1}}$, T the absolute temperature, P the vapor pressure over the curved liquid surface, and P_s the saturation vapor pressure over a flat surface. Note that the ratio P/P_s is equal to the relative humidity (RH), which ranges from 0 to 1 (corresponding to a range from 0 to 100%). Thanks to this model, the meniscus curvature can be known from the environmental parameters. It therefore becomes possible to compute a meniscus in the pore or crack or between a sharp AFM tip and the substrate. According to Mate [66], *for water,* $\gamma = 72\,\mathrm{mNm^{-1}}$ *and* $V_m = 18\ \mathrm{cm^3}$, *leading to* $r_k = -10$ nm *for* $P/P_s = 0.9$ *and* $r_k = -1$ nm *for* $P/P_s = 0.34$. *This means that for relative humidity between 34 and 90%, condensation of water vapor only forms a meniscus in nanometer-sized pores or gaps, that is, in those with diameters ranging from 2 to* 20 nm.

Inputing the meniscus curvature in Eq. 1.49, it becomes possible to compute the meniscus shape without knowing the volume of liquid (either the volume

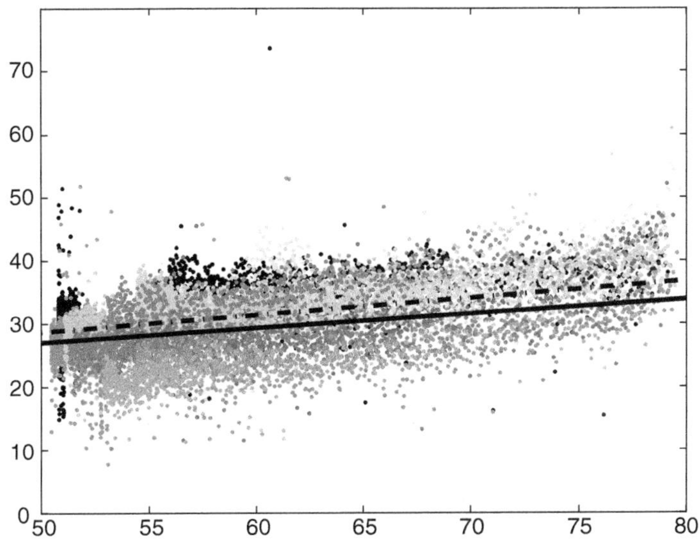

Figure 1.19. Pull-off force (nN) in function of the relative humidity (%): Comparison between model and experiment. The solid line has a slope determined with the model. The vertical position of the line (i.e, b from the $y = ax + b$ equation) has been chosen to best fit the points. The dashed line is the results of data fitting.

or the curvature must be known to solve the equation). As usual, the boundary condition of the differential problem (Eq. 1.49) is known from the contact angles. Using a finite-element resolution, Chau [16] has solved this problem for a full 3D geometry with 6 degrees of freedom of an AFM tip close to a flat surface. The pertinence of this problem has been pointed out by Sang et al. [80], confirming that capillary forces are of the first importance at the nanoscale (but also at all scales below the millimeter scale). The results of Chau [16] are twofold. First the model can predict quite well the experiment *on average*. This means that the large dispersion of the phenomenon can be mastered thanks to a large number of experiments. This allows one to confirm the validity of the Kelvin equation model at the scale of 10–100 nm. Second, the force increases with increasing humidity, and the dependence of the force on humidity is very sensitive to the tilt angle, that is, the relative orientation between the AFM tip and the substrate.

Note that the influence of roughness has not been modeled at this step; however, it is of the utmost importance as shown by very large dispersion of the experimental results.

Recent developments [5] show the influence of relative humidity. (Fig. 1.19) and the influence of the relative orientation between two solids (Fig. 1.20).

The importance of the tilt angle is also pointed out by Sang et al. [80], as indicated in Figure 1.21 (a colloidal probe is an AFM cantilever with a small sphere glued on it).

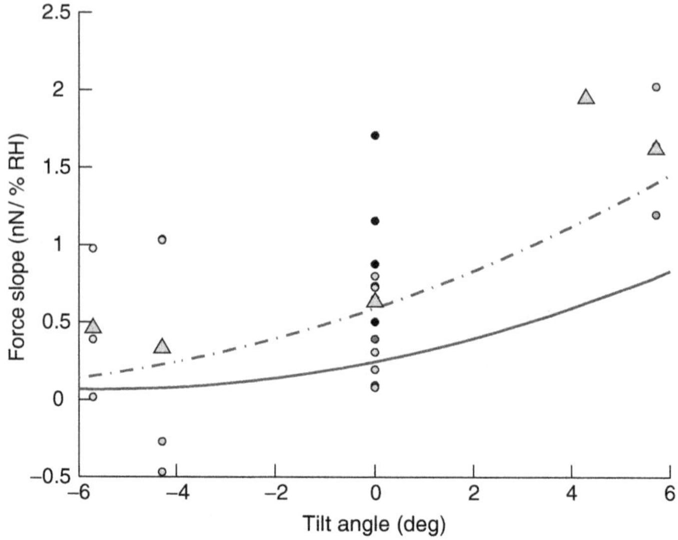

Figure 1.20. Comparison between model and experiment; the points are the slope of the force versus humidity slope; the solid line is the result of model computations for the same geometries. The tilt angle is given relatively to the tip support angle (12°). The dashed line is the result of the model for a tilt of 16°. Each point corresponds approximately to 300 measurements. The triangles are the means for each tilt angle value.

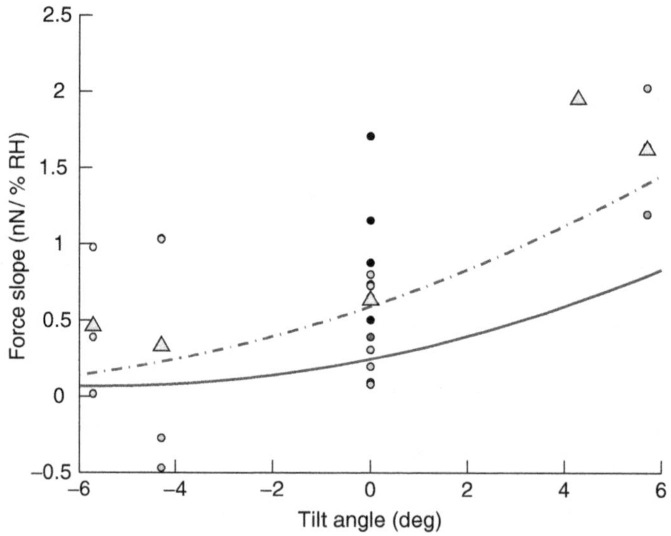

Figure 1.21. Measured pull-off force vs. wedge tilt angle. The result is highly sensitive to small changes of the tilt angle due to the effect of the colloidal probe roughness. A small change in the tilt angle causes a significant change in the pull-off force (from [80]).

TABLE 1.4. Terms Used in Chapter

Term	Definition	Units	Usual Values
ϵ_0	Free space permittivity	$C^{-2}\,N^{-1}\,m^{-2}$	8.85^{-12}
R	Sphere radius	m	10 nm to 100 μm
z	Separation distance	m	1 nm to 100 μm
V	Potential difference	V	0.5–20 V
θ	Cone half-aperture angle	rad	To 10°
L	Length of tip	m	10 μm to 500 μm
A	Area of contact	m^2	
r_{max}	Maximum distance to axis	m	
δ	Truncated cone height	m	
l	Plane width	m	
W	Electrostatic energy	J	
C	Capacitance	F	

1.3.2 Electrostatic Forces

This section could not have been written without the contribution of Marion Sausse-Lhernould [81].

Electrostatic forces between solids are also of importance at micro- and nanoscales. Basically, these forces come from the effect of electric fields on electrical charges. These charges can, for example, be acquired by tribo-electrification. The useful concepts—Coulomb's law, superposition principle, conductivity, permittivity, differences between electrostatics in free space and materials, differences between conductors and insulators, contact charging, polarization, induction, electrical breakdown, method of images also called mirror charges method—have been widely described in the literature [31, 44, 49, 57, 61, 68]. The goal of this section is therefore not to redevelop these theories. We prefer to present a summary of useful analytical models (see Table 1.5). Additionally, we will present some recent results based on Sausse Lhernould's work [81], studying the influence on electrical forces of surface roughness and mechanical deformation at contact.

Before reading through the summary table, let us recall the underlying assumptions. The main one for these analytical models is that surfaces are smooth for models not to take surface topography into account. This is a very strong assumption since, no matter how carefully or expensively a surface is manufactured, it can never be perfectly smooth. The second assumption defines materials as conductive where the potential is uniformly distributed along the surface, the electric field is normal to the surface, and the charges only carried by material surfaces (no volumic charges). The fact that no charge is present between the contacting objects is the third assumption. Table 1.4 summarizes and briefly defines the different terms used. In Figure 1.22 the different geometries involved in this work are presented: plane–plane contact, sphere–plane contact, sphere-ended cone–plane contact, and hyperbole–plane contact.

Let us now detail these analytical models.

TABLE 1.5. Review of Analytical Models

Contact Type	Expression	Reference
Plane–plane	$F_{\text{plane}} = \dfrac{\epsilon_0 V^2 A}{2z^2}$	22
Sphere–plane	$F_{\text{sphere1}} = \dfrac{\pi \epsilon_0 R V^2}{\vec{z}}$ for $R >> z$	6, 12, 14, 22
Sphere–plane	$F_{\text{sphere2}} = \dfrac{\pi \epsilon_0 R^2 V^2}{z^2}$ for $R << z$	14, 37
Sphere–plane	$F_{\text{sphere3}} = \pi \epsilon_0 \dfrac{R^2 U^2}{z(z+R)}$ for $R << z << L$	37, 13
Cylinder–plane	$F_{\text{cyl}}(\text{N/m}) = \dfrac{\pi \epsilon_0 \epsilon_R \sqrt{R} V^2}{2\sqrt{2} z^{3/2}} = \dfrac{\pi \epsilon_0 \epsilon_R \lambda V^2}{4\sqrt{2\pi}\sqrt{A} z^{3/2}}$	87
Conical tip	$F_{\text{ch}} \cong \dfrac{\lambda_0^2}{4\pi\epsilon_0} \ln\left(\dfrac{L}{4z}\right)$ for $R << z$	30
(charged line)	with $\lambda_0 = 4\pi\epsilon_0 V \left[\ln\left(\dfrac{1+\cos\theta}{1-\cos\theta}\right)\right]^{-1}$	
Conical tip	$F_{as} = \pi\epsilon_0 V^2 \left\{ \dfrac{R^2(1-\sin\theta)}{z\,[z+R(1-\sin\theta)]} \cdots \right.$	37
	$\left. +k^2 \left[\ln \dfrac{L}{z+R(1-\sin\theta)} - 1 + \dfrac{R\cos^2\theta\sin\theta}{z+R(1-\sin\theta)}\right]\right\}$	
(asympt. model)	with $k^2 = 1/\{\ln[\tan(\theta/2)]\}^2$	
Hyperb. tip	$F_{\text{hyp1}} = \pi\epsilon_0 V^2 k^2 \left[\ln\left(1+\dfrac{L}{R}\right) - \dfrac{(z - R/\tan\theta^2)L}{z(L+z)}\right]$	55
	with $k^2 = 1/\{\ln[\tan(\theta/2)]\}^2$	
Hyperb. tip	$F_{\text{hyp2}} = 4\pi\epsilon_0 V^2 \dfrac{\ln\left[1 + (r_{\max}/R)^2\left(1+\dfrac{R}{z}\right)\right]}{\ln^2\left(\dfrac{1+\eta_{\text{tip}}}{1-\eta_{\text{tip}}}\right)}$	74, 75
	with $\eta_{\text{tip}} = \sqrt{\dfrac{z}{z+R}}$	

1.3.2.1 Plane–Plane and Sphere–Plane Models

Plane–plane and sphere–plane models are the most encountered in the literature. The expressions have been derived from the electrostatic energy $W(d)$.

$$F_{\text{elec}}(z) = -\frac{\partial W(z)}{\partial z} = -\frac{1}{2}\frac{\partial C}{\partial z} V^2 \tag{1.108}$$

The simple case [22] is the plane–plane contact where two smooth planar surfaces are brought into contact. The surface of contact has an area A and the capacitance

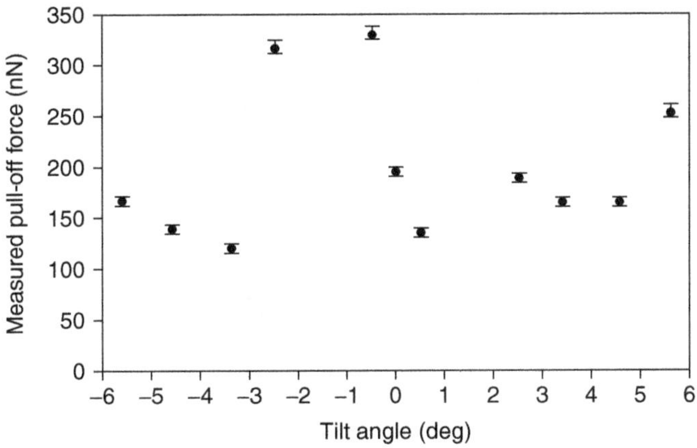

Figure 1.22. Representation of the involved geometries.

is obtained from the well-known plane capacitor case:

$$C(d) = \frac{\epsilon_0 A}{z}$$

$$F_{\text{plane}} = \frac{\epsilon_0 V^2 A}{2z^2} \tag{1.109}$$

This model gives the electrostatic pressure and knowing the area of the surface the electrostatic force can be deduced. Experience shows, however, that it is very difficult to determine the area of contact in real configurations. The planar model is thus very restricted in terms of applications. Moreover, studied objects are rarely totally flat. In application it may thus be used at very close separation distances between objects when the contact can be estimated by flat surfaces.

The sphere models have been developed for more complex shapes and longer separation distances. Many authors such as Krupp [49] have used them when studying the adhesion phenomenon disturbing micromanipulations. These models give an estimation of the electrostatic forces for the contact between a conductive sphere and a conductive plane. As the previous model, they are derived from the electrostatic interaction energy, and the capacitance between a sphere and a plane is given by the following expression:

$$C_{\text{sphere}} = 4\pi\epsilon_0 R \sinh(\alpha) \sum_{n=1}^{\infty} (\sinh n\alpha)^{-1}$$

with $\alpha = \cosh^{-1}[(R + z)/R]$. It is usual to analyze the contact between tip and surface in AFM as a sphere above a conducting plane [12]. The developed expressions depend on the separation distance and more precisely on the ratio between the sphere radius R and the separation distance z. Three models have been developed from the general expression given by Durand [21], depending on the separation distance range.

For small separation distances the electrostatic force is proportional to the inverse of the separation distance [6, 12, 14, 22]:

$$F_{\text{sphere1}} = \frac{\pi \epsilon_0 R V^2}{z} \qquad R >> z \tag{1.110}$$

For large separation distances, the electrostatic force is proportional to the inverse of the squared separation distance [14, 37]:

$$F_{\text{sphere2}} = \frac{\pi \epsilon_0 R^2 V^2}{z^2} \qquad R << z \tag{1.111}$$

For all separation distances [13, 37] a general expression has been developed from Eqs. 1.110 and 1.111:

$$F_{\text{sphere3}} = \pi \epsilon_0 \frac{R^2 V^2}{z(z+R)} \tag{1.112}$$

These models are restricted in their applicable separation distances. They are often used to get a quantitave assesment of the electrostatic forces between the probe and the substrate in scanning probe microscopy.

1.3.2.2 Uniformly Charged Line Models (Conical Tip Models)

The principle consists in replacing the equipotential conducting surfaces by the equivalent image charges. The main hypothesis is that the cone may be approximated by a charged line of constant charge density λ_0 given by Hao et al. [30] for small aperture angle ($\theta \leq \pi/9$) by the expression

$$\lambda_0 = 4\pi \epsilon_0 V \left[\ln \left(\frac{1+\cos\theta}{1-\cos\theta} \right) \right]^{-1} \tag{1.113}$$

In the previous expression, θ is the half-aperture angle of the cone. The hypothesis implies that the charges are uniformly distributed on the conical object. This is only accurate if the objects are sufficiently placed appart from each other but is incorrect at small separation distances. The model is thus restricted to large separation distances:

$$F_{\text{uni}} \cong \frac{\lambda_0^2}{4\pi \epsilon_0} \ln \left(\frac{L}{4z} \right) \qquad R << z << L \tag{1.114}$$

Our validations show that this model fits well the experimentally measured forces at large tip–sample separations.

1.3.2.3 The Asymptotic Model

The principle is to decompose a conical tip into infinitesimal surfaces [37]. The contribution of the apex and the spherical tip are evaluated separately and then added to get the total force. In this method the first step identifies the tip surface as a superposition of infinitesimal surfaces (facets). The first hypothesis is that from an electrostatic point of view, for distances larger than the characteristic facet dimensions, the surface is regular. The second step, which is also the second hypothesis, evaluates the electric field created between the facetted conductor and the plane surface by postulating that the electric field on each infinitesimal surface of the tip is equal to that created by the dihedral capacitance constituted by two infinite planes in the same relative orientation. The tip surface force is obtained by adding the contributions brought by each element. The expression is given by

$$\begin{aligned} F_{\text{asymp}} =& \pi \epsilon_0 V^2 \left[\frac{R^2(1 - \sin\theta)}{z\,[z + R(1 - \sin\theta)]} \right. \\ & \left. + k^2 \left(\ln \frac{L}{z + R(1 - \sin\theta)} - 1 + \frac{R \cos^2\theta / \sin\theta}{z + R(1 - \sin\theta)} \right) \right] \end{aligned} \tag{1.115}$$

with $k^2 = 1/\{\ln[\tan(\theta/2)]\}^2$. The validation has been performed by the authors being able to reestablish the well-known expressions for the sphere–plane contact.

1.3.2.4 The Hyperboloid Model (Hyperboloid Tip Model)

In this model the tip is represented by hyperboles bounded by a maximum distance $r_{\max}$ from the axis. The expression is derived by solving the Laplace equation in a prolate spheroidal coordinate system and by treating the tip–sample geometry as two confocal hyperboloids. Please refer to Patil and co-workers [74, 75] for calculation details. The boundary conditions are: the tip is at a potential V and the sample is grounded. The electric field and charge density are calculated using the boundary conditions. An integration on the surface allows to obtain the force. In this model the electrostatic force between tip and sample is given by

$$F_{\text{hyp}} = 4\pi \epsilon_0 V^2 \frac{\ln\left[1 + \left(\frac{r_{\max}}{R}\right)^2 \left(1 + \frac{R}{z}\right)\right]}{\ln^2\left(\frac{1+\eta_{\text{tip}}}{1-\eta_{\text{tip}}}\right)} \tag{1.116}$$

where $\eta_{\text{tip}} = \sqrt{z/z + R}$ and $r_{\max}$ is the cut-off radius introduced to avoid divergence. The validation is done through our own experimental measures. The theoretical and experimental results are in good agreement over distances ranging from 50 to 350 nm and voltages from 5 to 20 V. The limitation is mainly at very short interaction distances.

1.3.2.5 The Cylindrical Model

This model is different compared to the other because it is two-dimensionnal and not axisymetrical. Using the analytical model for the cylinder–plane contact [87], the electrostatic force is given by

$$F_{\text{elec}}^{\text{nondeformed}}(\text{N/m}) = \frac{\pi\epsilon_0\epsilon_R\sqrt{R}V^2}{2\sqrt{2}z^{3/2}} = \frac{\pi\epsilon_0\epsilon_R\lambda V^2}{4\sqrt{2\pi}\sqrt{A}z^{3/2}} \tag{1.117}$$

1.3.2.6 Tilted Conical Tip Models

The principle of this model is to find an analytical expression for the electrostatic force between a smooth plane and a tilted cantilever with a conical tip in electrostatic force microscopy [17]. The field lines between the objects are assumed to be approximated by segments of circles coming from the tip and ending on a point of the surface (Fig. 1.23). The electric potential decays linearly along these circular segments. If the distance between the two conducting objects is not larger than their physical dimension, the magnitude of the electric field is assumed to be given by $E = V/a$ (a being the arc length of the circular segment). The approximation is valid for small separation distances. The total force is the sum of the contributions brought by the truncated cone and by the spherical apex:

$$F(z) \cong \frac{\epsilon_0 V^2}{2}\int_S \frac{1}{a^2}dS \tag{1.118}$$

$$F_{\text{tilt}}^{\text{apex}}(z) = \frac{\pi\epsilon_0 V^2}{1 + f(2\theta)\times(z/R)^2}\left(\frac{R + z/2}{R - 2z}\right)^2 \times \left\{\frac{R - 2z}{z\left[1 + 2\tan^2(\theta)z/R\right]} + 2\ln\frac{4z}{2z + R + (R - 2z)\cos(2\theta)}\right\} \tag{1.119}$$

$$F_{\text{tilt}}^{\text{cone}}(d) = f_{\text{cone}}\epsilon_0 V^2\left[\ln\left(\frac{z - \delta/2 + L}{z + \delta/2}\right) - \sin\theta\frac{L - \delta}{z - \delta/2 + L}\frac{z - \delta/2}{z + \delta/2}\right] \tag{1.120}$$

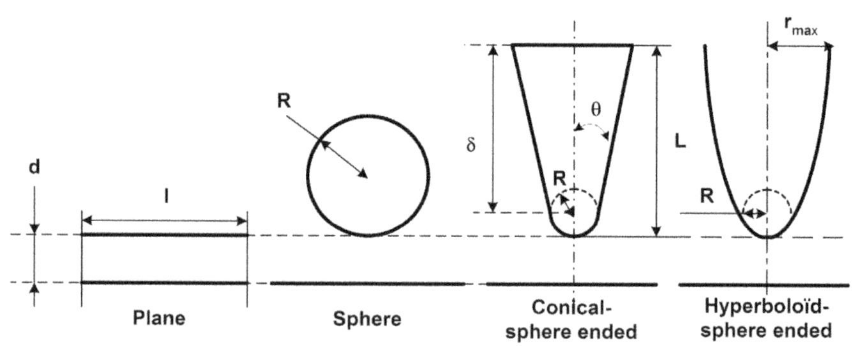

Figure 1.23. Tilted conical tip model, field lines are represented by segments of circle.

where $f_{\text{cone}} = 4\pi/(\pi - 2\theta)^2$, $f(2\theta) = \ln[1/\sin\theta]/([1 - \sin(\theta)][3 + \sin\theta])$, and $\delta = R/\tan^2\theta$.

Work taking into account the contribution of cantilever is found in the literature [17, 55].

1.3.2.7 Application to Scanning Probe Microscopy

The use of two models is recommended by Belaidi et al. [6] for the interaction between a tip and a planar surface in scanning probe microscopy: the spherical model for very short distances and the uniformly charged line model for longer distances. These two approximations correspond, respectively, to the case of an electrostatic force localized on the apex or on the conical side of the tip [55].

As indicated in the underlying assumptions of these analytical models, surface roughness is not taken into account. Sausse Lhernould and co-workers [81–83] have clearly shown the reduction of electrostatic forces in the presence of (even very small) surface roughness.

Indeed, a first comparison between simulation results [81] and literature results [79] is shown in Figure 1.24. The first observation is that even though results are in good correlation, simulated forces are stronger in smooth configuration than what was experimentally obtained. Moreover, the difference between experimental results and simulations increases when the separation distance decreases. This observation was attributed to the fact that even though the spot of contact has been chosen to be smooth, it can never be perfectly smooth. Even a very small roughness may influence the results at such small separation distances. Sausse Lhernould et al. [83] introduced roughness with the generation of a fractal surface using fractal parameters $D = 1.55$ and $G = 1.5 \times 10^{-12}$ for the planar contacting surface in order to have a maximum asperity peaks of 0.8 nm and an average roughness of 0.3 nm (which is often assumed to be negligible). The first observation is that even a roughness as small as this one is influencing the results from simulations, decreasing the electrostatic forces. This is specially true when the tip

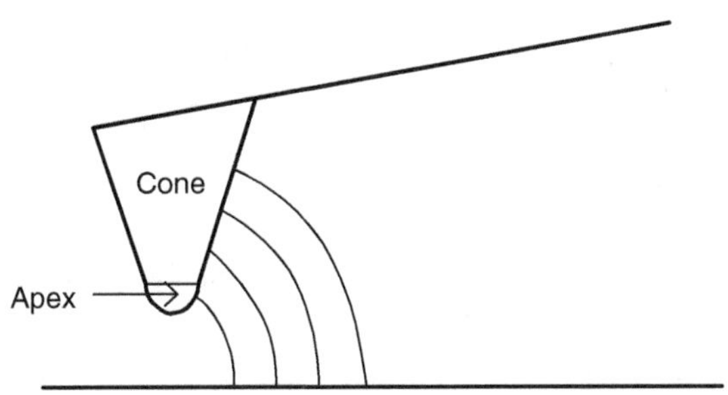

Figure 1.24. Comparison with experimental measures from Sacha et al. [79], who measured electrostatic forces for a sphere-ended conical tip of radius 40 nm and half-aperture angle 10° for different voltages. The characteristics of the tip were found using SEM images. The experimental results are compared with simulations, first without and then with roughness parameters (fractal representation).

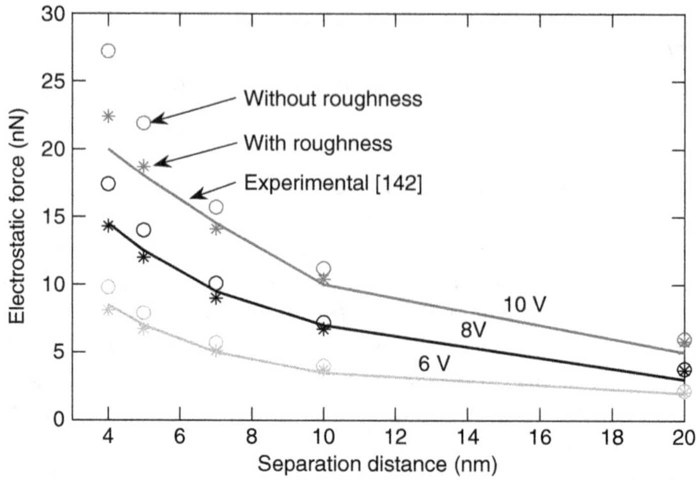

Figure 1.25. Electrostatic normal force (nN) vs. separation distance (nm) for a sphere-ended conical tip of radius 270 nm and half-aperture angle $\theta = 5°$. Plot shows experimental results obtained by Hao et al. [30], simulation results without roughness parameters, and simulation results including different average roughness.

gets closer to the surface. The influence of surface roughness is also more important at higher applied voltages. The results from simulations including roughness are closer to the experimental measures.

Hao et al. [30] also measured electrostatic forces for a sphere-ended conical tip. The tip radius is 270 nm and the half aperture angle is 5°. Figure 1.25 shows experimental results obtained compared with simulations for different roughness parameters. Conclusions are identical to what was observed by Sacha et al. [79].

1.3.2.8 Impact of the Roughness

Sausse Lhernould [81] presents a comparison between numerical simulation and experimental force measurement (see Fig. 1.26). The force is computed and measured between a small conductive 10-μm sphere glued on an AFM cantilever (stiffness $= 0.942\ \text{Nm}^{-1} \pm 17\%$) on the one hand and a conductive plane on the other hand. The sphere is assumed to be perfectly smooth while two different plane samples have been tested:

1. Sample A is: mechanical polishing of a nickel surface: $R_a = 2.49$ nm, $D = 1.218$, $G = 5.989 \times 10^{-20}$.
2. Sample B is: mechanical polishing of a nickel surface: $R_a = 13.55$nm, $D = 1.1355$, $G = 2.2261 \times 10^{-22}$.

We see in Figure 1.26 that both experimental curves indicate lower forces compared to the simulated smooth configuration. Second, we see that the correspondence between experiment and simulation is very good for the more rough sample

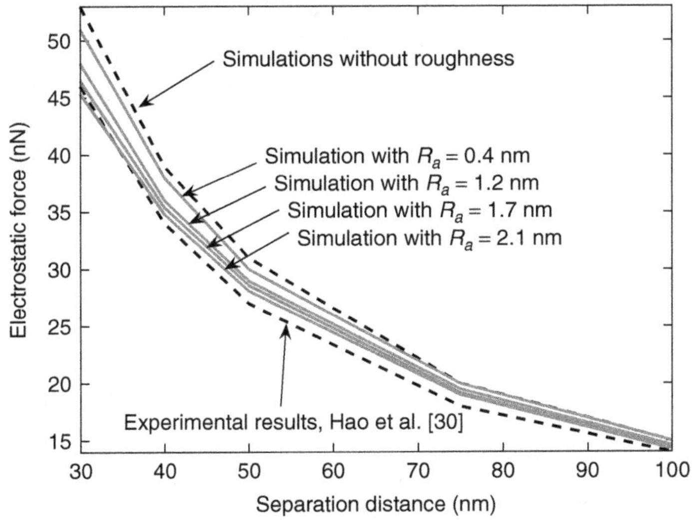

Figure 1.26. Comparison between experimental measures and numerical simulations for a 10-μm sphere approaching sample A and B for a 4-V voltage.

(B). The simulation is based on the Poisson equation solved in the free space around both solids, when a potential difference of 4 V is applied. The rough samples have been scanned and the fractal parameters D and G extracted from these profiles. Then, using the fractal function of Weierstrass–Mandelbrot with D and G, a rough plane geometry could be reconstructed before meshing and solving in COMSOL Multiphysics. In the case of sample A, which is smoother, the correspondence is not as good: the overestimation of the simulated force could come from the fact that the simulated geometry is 2D while the experimental measurement is done on a fully 3D configuration. Consequently, an isolated roughness peak on the actual sample will lead to a peak line in the simulation, overestimating the electrostatic force.

The justification why we claim the 2D simulation to overestimate the force is based on the following argument.

Let us first consider a 2D periodical rough profile, made of periods of length L. A fraction of L, fL, is the roughness peak. Consequently, the complementary $(1 - f)L$ is far away from the contact and its contribution to electrostatic force is neglected, due to the quick decrease of the force with increasing separation distances (cf. Eq. 1.106). Therefore, on a square of edge L, the force is equal to

$$F_{2D} = \frac{\epsilon_0 V^2}{2z^2} f L^2 \tag{1.121}$$

Let us now consider the same periodical surface in 3D: on a square of L^2 area, the peak area is now f^2L^2 while the $(1 - f)^2L^2$ does not contribute to the force (or has a vanishing contribution). Therefore, on this square of edge L, the force

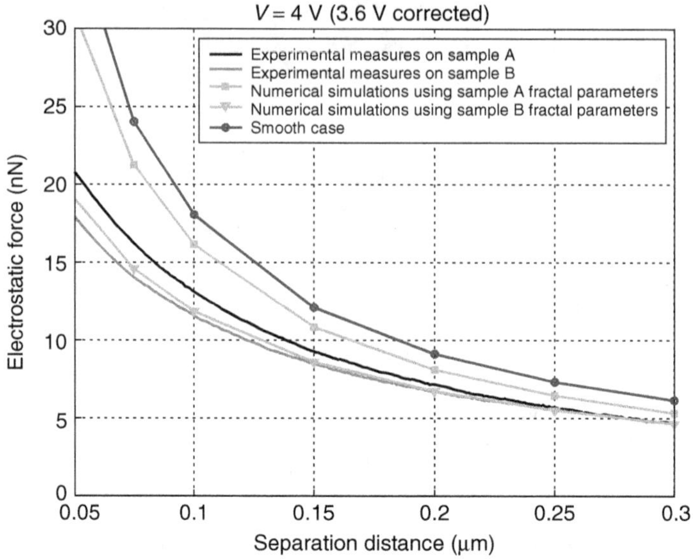

Figure 1.27. Qualitative influence of fractal roughness parameters D and G on the level force in the sphere–plane configuration (separation distance = 0.5 nm) for spherical object between 10 and 100 μm.

is equal to

$$F_{3D} = \frac{\epsilon_0 V^2}{2z^2} f^2 L^2 \tag{1.122}$$

Therefore, the ratio $F_{2D}/F_{3D} = 1/f$ is larger than 1, since f is smaller than 1. The prediction F_{2D} is larger than the actual force F_{3D}.

The influence of roughness on the electrostatic interaction can be estimated from the fractal dimension D and the fractal roughness parameter G (see Section 2.4 for details on roughness parametrization). This influence can be qualitatively estimated using the force reduction factor K given in Figure 1.27. This means that the force given by Eq. 1.110 (only at smaller separation distances) must be divided by K to take into account the effect of roughness. Since the K factor is given for a separation distance between the sphere and the plane equal to 0.5 nm, it cannot be applied to Eqs. 1.111 and 1.112.

Concerning the influence of mechanical deformation, a two-scale numerical study has been led in order to evaluate the effect of the surface roughness deformation on the adhesive electrostatic forces. The initial and deformed shapes of the considered profiles have been used in a numerical simulation to evaluate the adhesive electrostatic forces in a rough contact between two conductors. A large increase in the electrostatic forces between the initial undeformed and the final deformed configuration having a flat portion has been found. The adhesive forces are found to be up to 20 times larger in the deformed configuration (this may lead to difficulties in releasing a manipulated object after micromanipulation). Closed-form expressions are proposed [81].

As a conclusion, we have shown in this section advanced results concerning electrostatic forces at the nano- and microscales. The conclusions to draw are manyfold:

1. The presence of surface roughness reduces the electrostatic force of a factor between 1 and 100. The effect of roughness can be estimated using numerical simulation and the roughness profile can be reconstructed from only two roughness parameters, namely the fractal parameters D and G. Note that the classical roughness parameter R_a has not been used in this study because it is not representative enough (as far as electrostatic effects are concerned).
2. The mechanical contact (which, e.g., occurs when picking a component) crushed the roughness asperities, leading to an electrostatic force increase up to 20×.
3. The correspondence between experiments and models is reasonably good when the potential difference between two *conductive* solids is *imposed*. In case of insulating material, the necessary model parameters such as the charges distribution cannot be known and henceforth no force can be computed.

REFERENCES

1. A. W. Adamson and A. P. Gast, *Physical Chemistry of Surfaces*, 6th ed., Wiley, New York, 1997.
2. F. Arai, D. Ando, T. Fukuda, Y. Nonoda, and T. Oota, Micro Manipulation Based on Micro Physics, in Proc. of IEEE/RSJ Conf. on Intelligent Robots and Systems, Vol. 2, Pittsburgh, 1995, pp. 236–241.
3. F. Arai and T. Fukuda, A New Pick up and Release Method by Heating for Micromanipulation, in IEEE Workshop on Micro Electro Mechanical Systems, Nagoya, Japan, 1997, pp. 383–388.
4. K.-B. Bark, *Adhésives Greifen von kleinen Teilen mittels niedrigviskoser Flissigkeiten*, Springer, New York, 1999.
5. N. Bastin, A. Chau, A. Delchambre, and P. Lambert, Effects of Relative Humidity on Capillary Force and Applicability of These Effects in Micromanipulation, in Poster at the Proc. of the IEEE/RSJ 2008 International Conference on Intelligent Robots and Systems, Nice, September 22–26, 2008.
6. S. Belaidi, P. Girard, and G. Leveque, Electrostatic Forces Acting on the Tip in Atomic Force Microscopy: Modelization and Comparison with Analytic Expressions, *J. Appl. Phys*., 81(3):1023–1029, 1997.
7. L. Benmayor, Dimensional Analysis and Similitude in Microsystem Design and Assembly, Ph.D. Thesis, Ecole Polytechnique Fédérale de Lausanne, 2000.
8. B. Berge, Electrocapillarité et mouillage de films isolants par l'eau, *Acad. Sci*., 317(II):157, 1993.

9. B. Berge and J. Peseux, Variable Focal Lens Controlled by an External Voltage: An Application of Electrowetting, *Eur. Phys. J. E*, 3:159–163, 2000.
10. F. Biganzoli, I. Fassi, and C. Pagano, Development of a Gripping System Based on Capillary Force, in Proceedings of ISATP05, Montreal, Canada, 19–21 July 2005, pp. 36–40.
11. R. T. Borno, J. T. Steinmeyer, and M. M. Maharbiz, Transpiration Actuation: The Design, Fabrication and Characterization of Biomimetic Microactuators Driven by the Surface Tension of Water, *J. Micromech. Microeng*., 16:2375–2383, 2006.
12. R. A. Bowling, A Theoretical Review of Particle Adhesion, in Proc. of Symposium on Particles on Surfaces 1: Detection, Adhesion and Removal, San Francisco, 1986, pp. 129–142.
13. H-J. Butt, B. Cappella, and M. Kappl, Force Measurements with Atomic Force Microscope: Technique, Interpretation and Applications, *Surf. Sci. Rep*., 59:1–152, 2005.
14. B. Cappella and G. Dietler, Force-Distance Curves by Atomic Force Microscopy, *Surf. Sci. Rep*., 34:1–104, 1999.
15. H. B. C. Casimir and D. Polder, The Influence of Retardation of the London–van der Waals Forces, *Phys. Rev*., 73(4):360, 1948.
16. A. Chau, Theoretical and Experimental Study of Capillary Condensation and of Its Possible Use in Micro-assembly—Etude théorique et expérimentale des forces dues Ãă la condensation capillaire, Ph.D. Thesis, Université libre de Bruxelles, 2007.
17. J. Colchero, A. Gile, and A. M. Baró, Resolution Enhancement and Improved Data Interpretation in Electrostatic Force Microscopy, *Phys. Rev. B*, 64:1–11, 2001.
18. M. Dafflon, Préhenseurs, conditions et stratégies pour une micromanipulation de précision, Ph.D. Thesis, EPFL, 2008.
19. P.-G. de Gennes, F. Brochart-Wyard, and D. Quéré, *Gouttes, bulles, perles et ondes*, Belin, 2002.
20. B. Derjaguin, V. M. Muller, and Y. P. Toporov, Effect of Contact Deformations on the Adhesion of Particles, *J. Colloid Interface Sci*., 53(2):314–326, 1975.
21. E. Durand, *Electrostatique: Tome II Problémes Généraux, Conducteurs*, Masson, 1966.
22. R. S. Fearing, Survey of Sticking Effects for Micro Parts Handling, in Proc. of IEEE/RSJ Conf. on Intelligent Robots and Systems, 5–9 August 1995, pp. 212–217.
23. J. T. Feddema, P. Xavier, and R. Brown, Micro-assembly Planning with van der Waals Force, in Proc. of IEEE Int. Symposium on Assembly and Task Planning, Porto, July 1999, pp. 32–38.
24. J. T. Feddema, P. Xavier, and R. Brown, Micro-assembly Planning with van der Waals Force, *J. Micromech*., 1(2):139–153, 2001.
25. Y. Feng, Z. Zhou, X. Ye, and J. Xiong, Passive Valves Based on Hydrophobic Microfluidics, *Sensors Actuators A*, 108:138–143, 2003.
26. R. J. Good, Theory of Adhesive Forces across Interfaces: 1. The Lifshitz–van der Waals Component of Interaction and Adhesion, in Lieng-Huan Lee, Ed., *Fundamentals of Adhesion*, Plenum, New York, 1991.
27. J. A. Greenwood and K. L. Johnson, An Adhesion Map for the Contact of Elastic Spheres, *J. Colloid Interface Sci*., 192(2):326–333, 1997.
28. H. Grutzeck, Fluidisches Greifen in der Mikrosystemtechnik, Ph.D. Thesis, TU Cottbus, 1999.

29. W. Gutowsky, Thermodynamics of Adhesion, in L. H. Lee, Ed., *Fundamentals of Adhesion*, Plenum, New York, 2001.

30. H. Hao, A. M. Baro, and J. J. Saenz, Electrostatic and Contact Forces in Force Microscopy, *J. Vacuum Sci. Technol. B*, 9(2):1323–1328, 1991.

31. W. Harper, *Contact and Frictional Electrification*, Clarenton Press, Oxford, UK, 1967.

32. B. H. W. Hendriks, S. Kuiper, M. A. J. Van As, C. A. Renders, and T. W. Tukker, Electrowetting-Based Variable-Focus Lens for Miniature Systems, *Opt. Rev.*, 12(3):255–259, 2005.

33. M. C. Herman and K. D. Papadopoulos, *J. Colloid Interface Sci.*, 136(2),385–392, 1990.

34. M. C. Herman and K. D. Papadopoulos, *J. Colloid Interface Sci.*, 142(3),331, 1991.

35. H. Hertz, Uber die berihrung fester elastischer kirper, *J. Reine Angew. Math.*, 92:156–171, 1882.

36. R. L. Hoffman, A Study of the Advancing Interface. I. Interface Shape in Liquid–Gas Systems. *J. Colloid Interface Sci.*, 50(2):228–241, 1975.

37. S. Hudlet, M. Saint Jean, and J. Berger, Evaluation of the Capacitive Force between an Atomic Force Microscopy Tip and a Metallic Surface, *Eur. Phys. J. B*, 2:5–10, 1998.

38. J. N. Israelachvili, The Nature of van der Waals Forces, *Contemp. Phys.*, 15(2):159–177, 1974.

39. J. N. Israelachvili, *Intermolecular and Surface Forces*, 2nd ed., Academic, London, 1992.

40. J. N. Israelachvili and M. L. Gee, Contact Angles on Chemically Heterogeneous Surfaces, *Langmuir*, 5(1):288, 1989.

41. K. L. Johnson, K. Kendall, and A. D. Roberts, Surface Energy and the Contact of Elastic Solids, *Proc. R. Soc. London Ser. A*, 324:301, 1971.

42. K. L. Johnson, *Contact Mechanics*, Cambridge University Press, London, 1985.

43. R. E. Johnson and R. H. Dettre, Contact Angle, Wettability and Adhesion, *Adv. Chem. Ser.*, 43(8), 1998.

44. N. Jonassen, *Electrostatics*. Kluwer Academic, Amsterdam, 2002.

45. G. L. Klimchitskaya and V. M. Mostepanenko, Experiment and Theory in the Casimir Effect, *Contemp. Phys.*, 47(3):131–144, 2006.

46. K. Komvopoulos, Surface Engineering and Microtribology for Microelectromechanical Systems, *Wear*, 200(1):305–327, 1996.

47. T. Kondo, S. Juodkazis, and H. Misawa, Reduction of Capillary Force for High-Aspect Ratio Nanofabrication, *Appl. Phys. A*, 81(8):1583–1586, 2005.

48. P. A. Kralchevsky and K. Nagayama, *Particles at Fluid Interfaces and Membranes*, Vol. 10 of *Studies in Interface Science*, Elsevier, New York, 2001.

49. H. Krupp, Particle Adhesion Theory and Experiment, *Adv. Colloid Interface Sci.*, 1:111–239, 1967.

50. P. Lambert, *Capillary Forces in Microassembly: Modeling, Simulation, Experiments, and Case Study*, Microtechnology and MEMS, Springer, Boston, October 2007.

51. P. Lambert, A. Chau, A. Delchambre, and S. Régnier, Comparison between Two Capillary Forces Models, *Langmuir*, 24:3157–3163, 2008.

52. P. Lambert and A. Delchambre, Parameters Ruling Capillary Forces at the Submillimetric Scale, *Langmuir*, 21:9537–9543, 2005.
53. P. Lambert and S. Régnier, Surface and Contact Forces Models within the Framework of Microassembly, *J. Micromech.*, 3(2):123–157, 2006.
54. P. Lambert, F. Seigneur, S. Koelemeijer, and J. Jacot, A Case Study of Surface Tension Gripping: The Watch Bearing, *J. Micromech. Microeng.,* 16(7):1267–1276, 2006.
55. B. M. Law and F. Rieutord, Electrostatic Forces in Atomic Force Microscopy, *Phys. Rev. B*, 66(035402-1), 2002.
56. J. Lee and C.-J. Kim, Surface-Tension-Driven Microactuation Based on Continuous Electrowetting, *J. Microelectromech. Syst.,* 9(2):171–180, 2000.
57. L.-H. Lee, The Chemistry and Physics of Solid Adhesion, in L.-H. Lee, Ed., *Fundamentals of Adhesion*, Plenum, New York, 1991.
58. C. Lenders, J. B. Valsamis, M. Desaedeleer, A. Delchambre, and P. Lambert, Assembly of a Micro Ball-Bearing Using a Capillary Gripper and a Microcomponent Feeder, in *Micro-Assembly Technologies and Applications*, International Precision Assembly Seminar (IPAS), Springer, Boston, 2008.
59. E. Lifshitz, *Sov. Phys.–JETP*, 73(2), 1956.
60. F. London, The London–van der Waals Attraction Between Spherical Particles, *Physica*, 4(10):1058–1072, 1937.
61. J. Lowell and A. Rose-Innes, Contact Electrification, *Adv. Phys.,* 29(6):947–1023, 1980.
62. A. Marmur, Tip-Surface Capillary Interactions, *Langmuir*, 9:1922–1926, 1993.
63. C. H. Mastrangelo, Mechanical Stability and Adhesion of Microstructures under Capillary Forces—Part I: Basic Theory, *J. Microelectromech. Syst.,* 2(1):33–43, 1993.
64. C. H. Mastrangelo, Surface Force Induced Failures in Microelectromechanical Systems, In Bharat Bhushan, Ed., *Tribology Issues and Opportunities in MEM*, pages 367–395. Kluwer Academic, Amsterdam, 1998, pp. 367–395.
65. C. H. Mastrangelo, Suppression of Stiction in MEMS, in *Spring MRS Meeting*, MRS, Boston, 1999.
66. C. M. Mate, *Tribology on the Small Scale*, Oxford University Press, Oxford, 2008.
67. D. Maugis, Adhesion of Spheres: The JKR-DMT Transition Using a Dugdale Model. *J. Colloid Interface Sci.,* 150(1):243–269, 1992.
68. A. D. Moore, *Electrostatics and Its Applications*, Wiley, New York, 1973.
69. K. J. Obata, T. Motokado, S. Saito, and K. Takahashi, A Scheme for Micromanipulation Based on Capillary Force, *J. Fluid Mech.,* 498:113–121, 2004.
70. K. W. Oh and C. Ahn, A Review of Microvalves, *J. Micromech. Microeng.,* 16(5):R13–R39, 2006.
71. U. Ohlerich, Tension superficielle, interfaciale et angle de contact, in Workshop, Anvers, 15 October 2002.
72. C. Pagano, L. Zanoni, I. Fassi, and F. Jovane, Micro-assembly: Design and Analysis of a Gripper Based on Capillary Force, in *Proc. of the 1st CIRP—International Seminar on Assembly Systems*, Stuttgart, Germany, 15–17 November 2006, pp. 165–170.
73. K. D. Papadopoulos and C.-C. Kuo, *Colloids Surf.,* 46(115), 1990.

74. S. Patil and C. V. Dharmadhikari, Investigation of the Electrostatic Forces in Scanning Probe Microscopy at Low Bias Voltage, *Surf. Interf. Anal.,* 33:155–158, 2002.

75. S. Patil, A. V. Kulkarni, and C. V. Dharmadhikari, Study of the Electrostatic Force between a Conducting Tip in Proximity with a Metallic Surface: Theory and Experiment, *J. Appl. Phys.,* 88(11):6940–6942, 2000.

76. J. Peirs, Design of Micromechatronic Systems: Scale Laws, Technologies, and Medical Applications, Ph.D. Thesis, KUL, Belgium, 2001.

77. I. Rabinovich, S. Esayanur, and M. Mougdil, Capillary Forces between Two Spheres with a Fixed Volume Liquid Bridge: Theory and Experiment, *Langmuir*, 21:10992–109927, 2005.

78. W. B. Russel, D. A. Saville, and W. R. Schowalter, *Colloidal Dispersions*, Cambridge University Press, Cambridge, UK, 1989.

79. G. M. Sacha, A. Verdaguer, J. Martinez, J. J. Séenz, D. F. Ogletree, and M. Salmeron, Effective Radius in Electrostatic Force Microscopy, *Appl. Phys. Lett.,* 86:123101, 2005.

80. S. Sang, H. Zhang, M. Nosonovsky, and K. Chung, Effects of Contact Geometry on Pull-off Force Measurements with Colloidal Probe, *Langmuir*, 24:743–748, 200.

81. M. Sausse Lhernould, Theoretical and Experimental Study of Electrostatic Forces Applied to Micromanipulation: Influence of Surface Topography, Ph.D. Thesis, Université libre de Bruxelles, 2008.

82. M. Sausse Lhernould, A. Delchambre, S. Régnier, and P. Lambert, Electrostatic Forces and Micromanipulator Design: On the Importance of Surface Topography Parameters, in Proceeding of the 2007 IEEE/ASME International Conference on Advanced Intelligent Mechatronics, 2007.

83. M. Sausse Lhernould, A. Delchambre, S. Régnier, and P. Lambert, Electrostatic Forces in Micromanipulations: Review of Analytical Models and Simulations Including Roughness, *Appl. Surf. Sci.,* 253:6203–6210, 2007.

84. D. Schmid, S. Koelemeijer, J. Jacot, and P. Lambert, Microchip Assembly with Capillary Gripper, in Proc. of the 5th International Workshop on Microfactories (4 pages), October 2006, pp. 25–27.

85. A. K. Segupta and K. D. Papadopoulos, Vander Waals Interaction between a Colloid and its Host Pore, *J. Colloid Interface Sci.,* 152(2):534–542, 1992.

86. M. Sitti and H. Hashimoto, Teleoperated Touch Feedback from the Surfaces at the Nanoscale: Modelling and Experiments, *IEEE-ASME Trans. Mechatron.,* 8(1):1–12, 2003.

87. W. R. Smythe, *Static and Dynamic Electricity*, McGraw-Hill, New York, 1968.

88. M. van den Tempel, *Adv. Colloid Interface Sci.,* 3(137), 1972.

89. R. Van Hauwermeiren, Cours d'électricité générale, Cours de 2éme candidature, ULB, 1993.

90. B. Vogeli and H. von Kanel, AFM Study of Sticking Effects for Microparts Handling, *Wear*, 238(1):20–24, 2000.

91. K. D. Vora, A. G. Peele, B.-Y. Shew, E. C. Harvey, and J. P. Hayes, Fabrication of Support Structures to Prevent su-8 Stiction in High Aspect Ratio Structures, *Microsyst. Technol.,* 2006.

92. R. N. Wenzel, Resistance of Solid Surfaces to Wetting by Water, *Industrial and Engineering Chemistry*, 28(8):988–994, 1986.

93. D. Wu, N. Fang, C. Sun, and X. Zhang, Stiction Problems in Releasing of 3d Microstructures and Its Solution, *Sensors Actuators A*, 128:109–115, 2006.

94. B. Yang and Q. Lin, A Latchable Microvalve Using Phase Change of Paraffin Wax, *Sensors Actuators A*, 134:194–200, 2007.

CHAPTER 2

MICROWORLD MODELING: IMPACT OF LIQUID AND ROUGHNESS

PIERRE LAMBERT and STÉPHANE RÉGNIER

2.1 INTRODUCTION

This chapter is a complement to the physics introduced in the first chapter. More particularly, it deals with the analysis of the impact of liquid and of roughness on the behavior of the microobjects.

Liquid environments have been tremendously studied these last years because they are supposed to suppress capillary forces since there is no liquid–vapor interface anymore because this environment reduces the electrostatic effect due to higher dielectric constants and so on. These environments nevertheless imply specific phenomena such as the double-layer forces and larger hydrodynamic forces. It is shown how surface forces are actually affected by the liquid environment, including a microscopic analysis based on AFM measurements. Finally, we propose to address the issue of the surface topography and the roughness description (e.g., statistical description versus a fractal one to model the experimental roughness profile).

2.2 LIQUID ENVIRONMENTS

2.2.1 Classical Models

2.2.1.1 Double-Layer Forces

Electrostatic interactions are due to the presence of fixed or induced charges on the surface of particles. In the polar regions (e.g., in liquid environments), the

Robotic Microassembly, edited by Michaël Gauthier and Stéphane Régnier

main electrical charge mechanisms are:

- The dissociation of chemical surface groups
- The adsorption process

The surface ions attract opposite charges contained in the liquid medium. They form an area of counterions around the interface and ensure global electronic neutrality. The length of this characteristic phenomenon, called Debye length, depends on the concentration of ions. In the particular case of liquid colloidal solutions,[1] the concentration usually exceeds 10^{-7} mol/L, so the Debye length is similar to the size of the particles. In this case, an electric double layer (EDL) is formed, consisting of the surface charges and countercharges. EDL affects most dynamic phenomena in colloidal suspensions, as well as their stability. Indeed, the EDL creates a repulsion between two charged colloids, which would aggregate because of van der Waals forces. The van der Waals attraction and EDL repulsion form the basis of the famous DLVO theory on colloidal stability [60]. The DLVO theory is composed of two forces; repulsion due to the electric double layer and attraction due to the van der Waals force.

No comprehensive theory for the quantitative description of EDL has yet been published. Currently, two main approaches exist [1]:

- A statistical thermodynamic approach, the mathematical complexity of which prevents its implementation in practice
- A phenomenological approach based on a local thermodynamic equilibrium, used in the DLVO theory, but which does not take into account the correlations between ions, adielectric saturation, and the finite size of ions

2.2.1.2 Qualitative Models of the Electric Double Layer

EDL models have been designed to describe the change in electric potential at the solid–liquid interface. Several qualitative EDL models exist, differing mainly on how to visualize the spatial distribution of the countercharge. They are, in chronological order, the Helmholtz model (1879), the Gouy–Chapman model (1913), the Stern model (1924), and the Gouy–Chapman–Stern–Grahame model (GCSG) (1947) also called triple layer. First, we will examine how the latter explains EDL. Then by successive approximations, we examine how the older models perform.

The GCSG model breaks down the solid–liquid interface into three layers of charge (see Fig. 2.1):

- The first layer: The surface layer where the ions are adsorbed, determining the potential of solids (e.g., H^+ and OH^- for an oxide metal such as SiO). The charge σ_0 and potential ψ_0 are associated with this layer.

[1]Liquid containing suspended particles the size of which is between 1 nm and 1 μm.

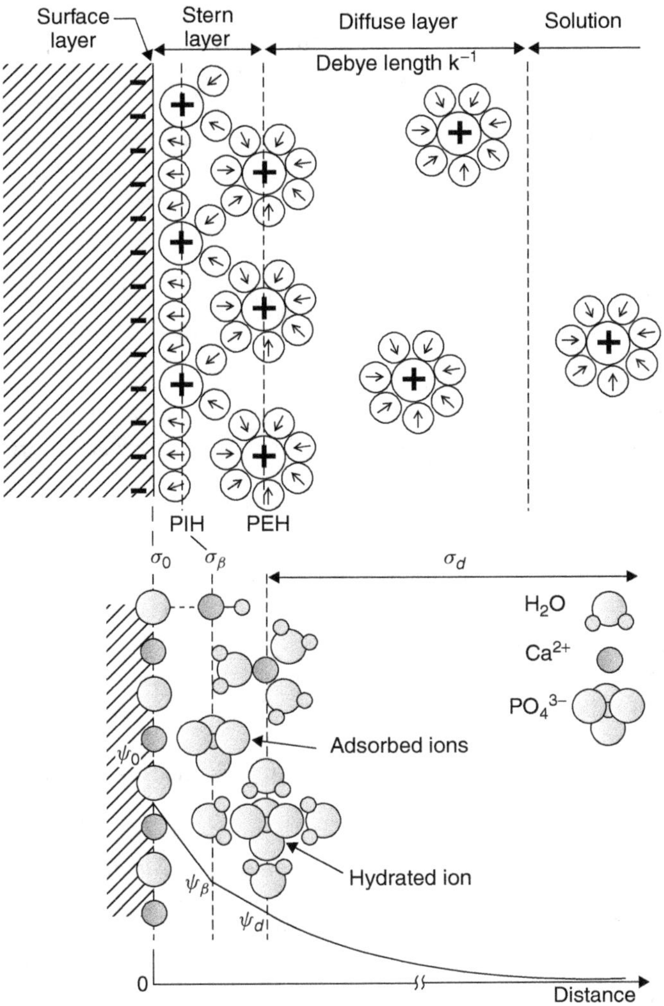

Figure 2.1. Triple layer. Equations are: $\sigma_0 + \sigma_\beta + \sigma_d = 0$, $\psi_0 - \psi_\beta = \sigma_0/C_1$, and $\psi_\beta - \psi_d = (\sigma_0 + \sigma_\beta)/C_2 = -\sigma_d/C_2$. The example shown is the case of a surface covered with calcium phosphate [58].

- The second layer: The compact layer of dehydrated ions interacting strongly with the surface (ions specifically adsorbed). The center of these ions is located at the internal Helmholtz plane (IHP). With this layer, the charge σ_β and the potential ψ_β are defined.
- The third layer: In this layer, the diffuse layer of ions and hydrated counterions are slightly attracted. The plane where the diffuse layer begins is called the external Helmholtz plan (EHP). It combines the charge σ_d and the potential ψ_d.

This model has been designed to take into account the absorption of ions on the surface of inorganic oxides, where the surface charge depends on the pH, dissociation constants, and the adsorption reactions. From the point of view of electrical engineering, separation of charges between the different planes is comparable to three capacities in series, giving a total capacity:

$$\frac{1}{C_t} = \frac{1}{C_1} + \frac{1}{C_2} + \frac{1}{C_d} \tag{2.1}$$

with $\sigma_0 = C_1(\psi_0 - \psi_\beta)$ and $\sigma_\beta = C_2(\psi_\beta - \psi_d)$.

2.2.1.3 *Stern Model*

If C_2 is neglected (i.e., neglecting the finite size of the second layer), then $\psi_\beta = \psi_d$ and $1/C_t = 1/C_1 + 1/C_d$. Adamczyk [1] notes that for interactions of particles, the Stern layer's role is negligible except for very short separations around 5–10 Å.

Two limit cases of this model may arise:

- If the potential and ionic strength are low ($n_0 < 0.01$ mol/L), then $C_1 \gg C_d$ and the model is reduced to the diffuse layer model (Gouy–Chapman model) with $1/C_t = 1/C_d$.
- If the potential and ionic strength are high, then the model is reduced to the constant capacity model (Helmholtz model) with $1/C_t = 1/C_1$.

The model used in this book and these circumstances is the Gouy–Chapman model.

2.2.1.4 *Gouy–Chapman Electric Double-Layer Model*

A flat surface is considered. The charge on that surface influences the ion distribution in nearby layers of the electrolyte. The electrostatic potential, ψ, and the volume charge density, ρ, which is the excess of charges of one type over the other, are related by the Poisson equation:

$$\nabla^2\psi = -\frac{\rho}{\epsilon_0\epsilon_r} \tag{2.2}$$

with ϵ_r relative permitivitty (or dielectric constant) of electrolyte.

The ion distribution in the charged surface region is determined by (i) temperature and (ii) the energy required, w_i, to bring the ion from an infinite distance away (where $\psi = 0$) to the region where the electrostatic potential is ψ. This distribution is given by a Boltzmann equation:

$$n_i = n_i^0 e^{-w_i/(k_B T)} = n_i^0 e^{-z_i e\psi/(k_B T)} \tag{2.3}$$

where n_i^0 is the number of ions of type i per unit volume of bulk solution, $w_i = z_i e\psi$, the energy expended in bringing an ion from an infinite distance from the surface to a point where the potential is ψ and z_i is the valency of ion species i.

The volume charge density at ψ is

$$\rho = \sum_i n_i z_i e e^{-z_i e\psi/(k_B T)} \tag{2.4}$$

Thus the combination of this with the Poisson equation gives the **Poisson–Boltzmann** equation:

$$\nabla^2\psi = -\frac{\sum_i n_i z_i e e^{-z_i e\psi/(k_B T)}}{\epsilon_0\epsilon_r} \tag{2.5}$$

2.2.1.5 Zeta Potential

When an electric field is applied in a suspension containing charged particles, the particles acquire a degree of mobility depending on their charges. Within the double layer, there is a plane delineating the ions from the particle in this movement. The plane, called shear plane, is defined on the basis of hydrodynamic considerations.

However, it is possible to link this plane to the chemical description of the double layer. Indeed, it is usually very close to the plane delineating the compact layer from the diffuse layer—the external Helmoltz plane.

The electric potential in the shear plane is commonly noted as the zeta potential ζ. It is easily accessible by electrokinetic measures (electrophoresis, acoustophoresis, flow potential, etc.). See Hunter [27] for more details. It is often used as a potential surface in the form of an electrostatic interaction.

2.2.1.6 Poisson–Boltzmann Equation

The Poisson–Boltzmann (PB) equation needed to solve the problem is strongly nonlinear due to the exponential terms, which precludes an analytical solution. Some solutions are available in the literature in the form of tables, graph solutions, elliptical integrals, and elliptical functions [39]. Nevertheless, two different approximations can be made to arrive at an analytical resolution.

For a symmetrical electrolyte composed of two types of ions, where $z_1 = z_2 = z$, Eq. 2.5 becomes

$$\nabla^2\psi = -\frac{1}{\epsilon_0\epsilon_r}\left[n^0 z e e^{-ze\psi/(k_B T)} - n^0 z e e^{ze\psi/(k_B T)}\right] \tag{2.6}$$

Using trigonometric relations, we find

$$\nabla^2\overline{\psi} = \kappa^2 \sinh \overline{\psi} \tag{2.7}$$

where $\overline{\psi} = ze\psi/kT$ is the reduced electric potential and

$$\kappa^{-1} = \left(\frac{\varepsilon_0 \varepsilon kT}{2e^2 I}\right)^{1/2} \tag{2.8}$$

where κ^{-1} is the Debye length, $I = z^2 n_0$ is the ionic force of the electrolyte, n_0 its volumic concentration, and z the valency.

Another common form of this equation is the linear PB equation, obtained by linearizing exponential terms, provided that $\max(z_i e\psi/kT) < 1$ or $\max(z_i\psi) <$ 25 mV is at 25°. This is the linear Debye–Heckel approximation:

$$\nabla^2\psi = \kappa^2\psi \tag{2.9}$$

For an order of magnitude, the Debye length κ^{-1} in water and at room temperature varies between about 0,4 nm (for a solution of Na_3PO_4 at 0,1 mol/L) and 30 nm (for a solution of KCl at 10^{-6} mol/L) [1].

2.2.2 Sphere–Sphere and Sphere–Plane Interactions

We consider in this section interaction between a sphere of radius a and a plane or interaction between spheres of radius R_1 and R_2 with h separation distance.

There are two main approaches to calculating an approximation of the interactions between spheres: the method of linear superposition (LSA) restricted to large distances (κh>1 with h the separation distance) and Derjaguin's approximation (1940) appropriate for small distances ($\kappa h < 1$) and small Debye lengths ($\kappa^{-1} < a$). However, Sader et al. [49] obtained a valid formula for any κh, while keeping a simple analytical form.

2.2.2.1 LSA Method

The LSA method [3] postulates that the solution to the PB equation for a system of two particles can be built as the linear superposition of the solutions for isolated particles. This is justified by the fact that the electric potential at distances larger than κ^{-1} drops to low values and its description can be described by the linear PB equation. Therefore, the solution of this equation in this region is calculated assuming the additivity of the potential of isolated particles:

$$\psi_s = \psi_{s1} + \psi_{s2} \tag{2.10}$$

The solution of the linear PB equation for a sphere with a radius a and a small potential $\overline{\psi}_s < 1$ is

$$\psi = \psi_s \frac{a}{r} e^{-\kappa(r-R)} \tag{2.11}$$

where r is the distance from the center. However, in order to be close to the nonlinear solutions, ψ_s can be replaced by the effective potential Y determined by the numerical solution of the exact PB equation for a sphere. For a symmetrical electrolyte, Sader [48] gives a formula valid for any κa and for any ψ_s with $\psi_s < 200$ mV:

$$Y = \frac{kT}{ze}\frac{\overline{\psi}_s + 4\gamma\Omega\kappa R}{1 + \Omega\kappa R} \quad \text{with} \quad \Omega = \frac{\overline{\psi}_s - 4\gamma}{2\gamma^3} \quad \text{and} \quad \gamma = \tanh(\frac{\overline{\psi}_s}{4}) \tag{2.12}$$

For a small potential, we can calculate the force between the particles. We then get for $h > \kappa^{-1}$:

$$F_{ss} = 4\pi\varepsilon_0\varepsilon R_1 R_2 \frac{1 + \kappa r}{r^2} Y_1 Y_2 e^{-\kappa h} \tag{2.13}$$

$$W_{ss} = 4\pi\varepsilon_0\varepsilon \frac{R_1 R_2}{R_1 + R_2 + h} Y_1 Y_2 e^{-\kappa h} \tag{2.14}$$

where $r = R_1 + R_2 + h$ and h is the distance between spheres.

In addition, where a radius tends to infinity, Eqs. 2.13 and 2.14 are reduced to the form describing the sphere–plane interaction:

$$F_{sp} = 4\pi\varepsilon_0\varepsilon\kappa a Y_1 Y_2 e^{-\kappa h} \tag{2.15}$$

$$W_{sp} = 4\pi\varepsilon_0\varepsilon a Y_1 Y_2 e^{-\kappa h} \tag{2.16}$$

Finally, the LSA method can also be applied to the plane–plane interaction. The potential of a single plane for $h > \kappa^{-1}$ is

$$\psi = Y e^{-\kappa h} \tag{2.17}$$

where $Y = 4\tanh(\overline{\psi}_s/4)$ (because $\kappa a \to \infty$ in Eq. 2.12 in the plane case). In the case of an asymmetrical electrolyte, Ohshima [42] gives analytical formulas to calculate Y. The LSA assumption gives the following terms:

$$\Delta\Pi = 2\varepsilon_0\varepsilon\kappa^2 Y_1 Y_2 e^{-\kappa h} \tag{2.18}$$

$$W_{pp} = 2\varepsilon_0\varepsilon\kappa Y_1 Y_2 e^{-\kappa h} \tag{2.19}$$

These two formulas can be found in Israelachvili [28]. It is noted that for a small potential where $Y \approx \psi_s$, Eq. 2.19 corresponds to the solution of the linear DB equation where $\kappa h \gg 1$.

Thanks to their simple mathematical form, Eqs. 2.13–2.19 are widely used in numerical simulations for particle absorption problems. The disadvantage of the LSA method is that it becomes less effective for small separations ($h < \kappa^{-1}$). In this case, Derjaguin's approximation can be used.

2.2.2.2 Derjaguin Method

According to this method [14], the interactions between spheres can be calculated as the sum of the interactions to elementary surfaces (rings) with a plane geometry. The Derjaguin method connects the interaction energy per area unit between two planes W_{pp} and the interaction energy between two curved surfaces W_{ss} with the equation (for Derjaguin, we use classical notation, z interaction distance):

$$W_{ss} = 2\pi G_D \int_{z_{min}}^{\infty} W_{pp}(z)\, dz \tag{2.20}$$

where z_{min} is the minimum distance between the curved surfaces and G_D the Derjaguin factor geometry. The latter is calculated easily for simple geometries, $G_D = R_1 R_2/(R_1 + R_2)$ for two spheres, $G_D = R/2$ for two identical spheres, and $G_D = R$ for the sphere–plane configuration.

However, the approximation assumes that the scope of the interaction energy is much shorter than the radius of curvature of the particles. This means that the interaction energy between the two particles is created in a small region around z_{min} (which admits the limit of infinite integration in Eq. 2.20. The Derjaguin method is valid only if $\kappa R_i \gg 1$. In practice, the approximation remains valid for $\kappa R_i > 5$ [26], which corresponds to micrometer colloidal particles in an electrolyte diluted to approximately 10^{-4} or to globular proteins in an electrolyte with a physiological concentration ($\approx 0,\ 15$ mol/L).

By using Eq. 2.20, we get

$$F_{ss} = \frac{2\pi\varepsilon_0\varepsilon\kappa}{1 - e^{-2\kappa z}} G_D \left[\pm(\psi_{s1}^2 + \psi_{s2}^2)e^{-2\kappa z} + 2\psi_{s1}\psi_{s2}e^{-\kappa z}\right] \tag{2.21}$$

$$W_{ss} = \pi\varepsilon_0\varepsilon G_D \left[\mp(\psi_{s1}^2 + \psi_{s2}^2)\ln(1 - e^{-2\kappa z}) + 2\psi_{s1}\psi_{s2}\ln\frac{1 + e^{-\kappa z}}{1 - e^{-\kappa z}}\right] \tag{2.22}$$

where the $\pm$ sign depends on the conditions at the limit of the resolution: the plus sign for areas with constant charges (c.c.) and the minus sign for surfaces with constant potential (c.p.).

Equation 2.2 is called the HHF formula, named after its authors Hogg, Healy, and Feurstenau [26]. When the surface potential is equal and $\psi_{s1} = \psi_{s2} = \psi_{rms}$, Eq. 2.22 is reduced to the form given by [14]

$$W_{ss} = \mp 4\pi\varepsilon_0\varepsilon G_D \psi_s^2 \ln(1 \pm e^{-\kappa z}) \tag{2.23}$$

Equations 2.22 and 2.23 are often used in the literature to determine the stability criteria of colloidal suspensions. It may also be noted that this method has been generalized to convex bodies of any shape [1]. Figure 2.2 gives a comparison of the LSA model (Eqs. 2.14 and 2.16) and the linear HHF Derjaguin model (Eq. 2.22) with numerical solutions of the nonlinear PB equation.

However, the Derjaguin method becomes less effective for separations exceeding κ^{-1}. This is because the Derjaguin approximation considers the interaction energy between two elementary surfaces that are similar face to face, while interaction energy between two planes is the interaction energy in a point of a plane due to any point of the other plane [4]. This constraint creates an overstatement of interactions and a false asymptotic dependence of $W_{ss} = f(z)$ (in the formula, it lacks dependency on $1/z$ when z increases).

Moreover, the surface potentials must be low. Hogg et al. [26] showed that the approximation was good up to 50 mV (or $\overline{\psi}_{si} \leq 2$). In summary, the conditions to check for using the Hogg, Healy, and Feurstenau (HHF) formula are $\kappa z < 1$, $\overline{\psi}_{si} \leq 2$ and $\kappa R_i > 5$.

2.2.2.3 Improved Formulas

Sader et al. [49] have demonstrated that the HHF formula could be easily amended to apply to any κz while retaining its analytical simplicity. They summarized their analysis from Bell et al. [3] for $\overline{\psi}_{si} < 2$, $\kappa R_i > 5$, and areas with constant potential from which they obtained the amended HHF formula:

$$W_{ss} = \pi \varepsilon_0 \varepsilon \frac{R_1 R_2}{R_1 + R_2 + h} \left[(\psi_{s1}^2 + \psi_{s2}^2) \ln(1 - e^{-2\kappa z}) + 2\psi_{s1}\psi_{s2} \ln \frac{1 + e^{-\kappa z}}{1 - e^{-\kappa z}} \right] \quad (2.24)$$

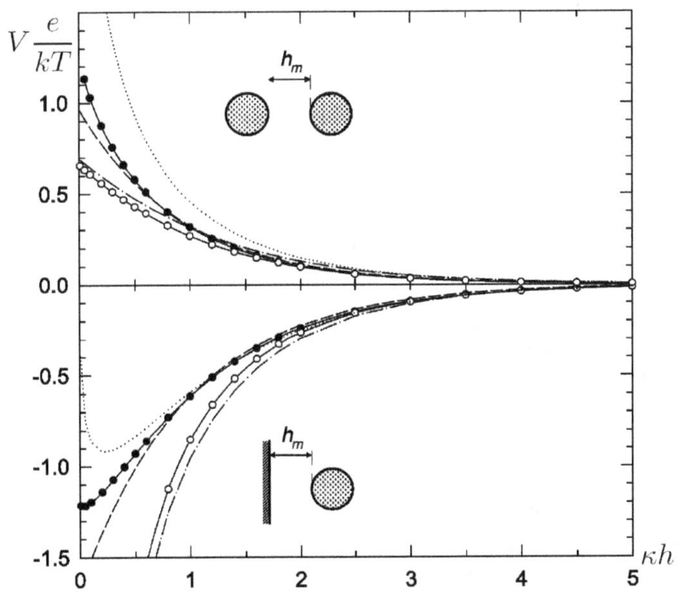

Figure 2.2. Reduced interaction energy $W(e/kT)$ sphere–sphere (low diagram) and sphere–plan (top diagram) computed for $\overline{\psi}_{s1} = 3$, $\overline{\psi}_{s2} = -1.5$, and $\kappa a = 5$: —•—•— exact numerical evaluation for the c.c. case; —◦—◦— iexact numerical evaluation for the c.p. case; ······ linear HHF model for the c.c. case; —·—·—linear HHF model for the c.p. case; ——— LSA model [1].

We note that for a small κz, the amended HHF formula coincides with the HHF formula of Eq. 2.22 and for a large κz, it is reduced to the LSA formula of Eq. (2.14).

Finally, the same authors also offer a simple and effective formula for two identical spheres with moderate to high potential ($\overline{\psi}_{si} \leq 4$ then $\psi_{si} \leq 100$ mV), valid for any κz and $\kappa R_i > 5$. Using the solution of the nonlinear PB equation for a single sphere, they get the expressions:

$$F_{ss} = 4\pi\varepsilon_0\varepsilon\frac{R^2}{r^2}Y^2(z)\left[\ln(1+e^{-\kappa z}) + \kappa r\frac{e^{-\kappa z}}{1+e^{-\kappa z}}\right] \tag{2.25}$$

$$W_{ss} = 4\pi\varepsilon_0\varepsilon\frac{R^2}{r}Y^2(z)\ln(1+e^{-\kappa z}) \tag{2.26}$$

with $Y(z) = 4e^{\kappa z/2}\tanh^{-1}\left[e^{-\kappa z/2}/\tanh(\frac{\overline{\psi}_s}{4})\right]\forall z$ and $Y(z) \approx 4\tanh(\overline{\psi}_s/4)$ for $\kappa z > 2$.

2.2.2.4 DLVO Theory

The DLVO theory is linked to the Stern model and studies, in particular, the diffuse layer. It assumes that the total interaction between two surfaces is the sum of double-layer repulsion and the van der Waals attraction. For example, in the case of an interaction between spheres of radius a with weak potential and $h \ll a$, the potentiel and the force are equal to (using 2.23):

$$\begin{aligned} F_{ss} &= -\frac{A_{132}a}{12h} + 4\pi\varepsilon_0\varepsilon\frac{a}{2}\psi_s^2\left[\ln(1+e^{-\kappa z})\right] \\ W_{ss} &= -\frac{A_{132}a}{12h^2} + 4\pi\varepsilon_0\varepsilon\frac{a}{2}\psi_s^2\frac{\kappa}{1+e^{-\kappa z}} \end{aligned} \tag{2.27}$$

An another example is the interaction between a sphere of radius a and a plane:

$$F_{sp} = -\frac{A_{132}a}{6h} + 4\pi\varepsilon_0\varepsilon a\psi_s^2\left[\ln(1+e^{-\kappa z})\right] \tag{2.28}$$

$$W_{sp} = -\frac{A_{132}a}{6h^2} + 4\pi\varepsilon_0\varepsilon a\psi_s^2\frac{\kappa}{1+e^{-\kappa z}} \tag{2.29}$$

You can use these equations for micromanipulation in a liquid environment, for example. We can also use any expressions previously defined for the EDL. Contrary to the double-layer interaction, the van der Waals interaction is much less sensitive to changes in the electrolyte's concentration and pH. Van der Walls interaction can therefore be considered fixed in the first approximation. Besides, it still exceeds the double-layer repulsion at short distances because $W_{vdW} \propto -1/z^n$. Thus, according to the concentration of the electrolyte and the potential or surfacic density of the charges, different scenarios, schematically illustrated in Figure 2.3 can occur [28]:

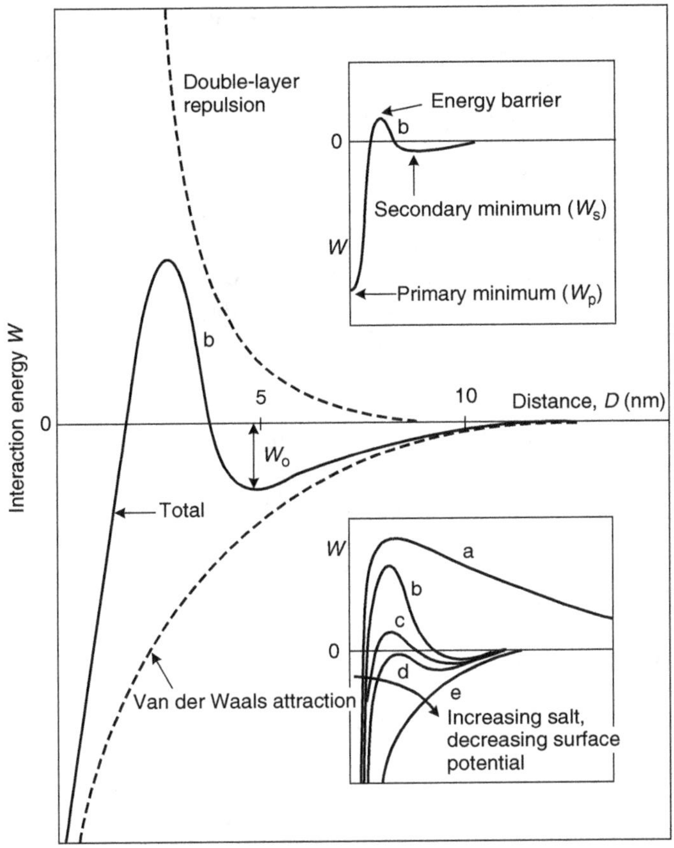

Figure 2.3. Schematic profiles of interaction energy [28].

1. Surfaces repel strongly; colloids are thermodynamically stable.
2. Surfaces can stabilize at a secondary minimum if this minimum is fairly deep and the colloids remain kinetically stable.
3. Surfaces stabilize at the secondary minimum; colloids are slowly bonded. This phenomenon is called flocculation: the formation of agglomerated particles the size of which is sufficient to settle or to float. Simple agitation may cancel this flocculation.
4. At the critical concentration of coagulation, surfaces can remain at the secondary minimum or coagulate; colloids bond quickly.
5. Surfaces and colloids coagulate quickly (formation of a precipitate).

Finally, apart from lyophobe colloids, many differences[2] between DLVO theory and experimental observations have been reported for industrial and natural

[2]DLVO theory predicts stability or instability contrary to experiments.

systems [25]. These differences are due to non-DLVO short-range interactions, such as hydration forces in an aqueous environment [66], steric forces,[3] and hydrodynamic forces. Taking these interactions into account is also part of the *extended DLVO* or *XDLVO* approach [10, 11].

2.2.2.5 XDLVO Model

The XDLVO model is an extended version of the previous model and includes solvation forces (or hydration forces in water). These interactions are also referred to as acid–base Lewis interactions. In water, these forces will be attractive for hydrophobic surfaces and repellent for hydrophilic surfaces. Thus, the components of the surface energy can be classified into two categories [20, 21]:

- The first gathering dispersive interactions, so-called van der Waals Lifshitz interactions, are designated by the exhibitor LW and are mainly due to the van der Waals effect between micro objects.
- The second gathering the interactions linked to the donor–acceptor mechanism of electrons; these interactions form the Lewis acid–base theory, are designated by the exhibitor AB [16], and are mainly due to the liquid effect added by the DLVO theory.

This hypothesis can be written and are follows:

$$\gamma = \gamma^{\mathrm{AB}} + \gamma^{\mathrm{LW}} \tag{2.30}$$

In the Lewis theory, a base is a compound carrying an electronic doublet and is thus an electron donor. An acid is a compound carrying a gap and is thus an electron acceptor. This approach allows us to describe the acid–base property using two parameters, γ^+ and γ^-, where γ^+ is the acceptor electron parameter (or Lewis acid) and γ^- is the electron donor parameter (or Lewis base). The component γ^{AB} surface energy is a function of γ^+ and γ^- according to the relationship

$$\gamma^{\mathrm{AB}} = 2\sqrt{\gamma^+\gamma^-} \tag{2.31}$$

The total surface energy can be written as

$$\gamma = \gamma^{\mathrm{AB}} + \gamma^{\mathrm{LW}} = \gamma^{\mathrm{LW}} + 2\sqrt{\gamma^+\gamma^-} \tag{2.32}$$

with γ^{LW} incorporating the previous DLVO model. We can therefore write the interfacial energy in the case of a contact between a solid and a liquid as

$$\gamma_{\mathrm{SL}} = \gamma_{\mathrm{SL}}^{\mathrm{AB}} + \gamma_{\mathrm{SL}}^{\mathrm{LW}} \tag{2.33}$$

[3]Adsorption of neutral polymers by particles creates a steric repulsion due to the repulsive interactions between the polymer chains.

with expressions of components proposed by Good:

$$\gamma_{SL}^{AB} = 2(\sqrt{\gamma_S^+\gamma_S^-} + \sqrt{\gamma_L^+\gamma_L^-} - \sqrt{\gamma_S^+\gamma_L^-} - \sqrt{\gamma_L^+\gamma_S^-}) \tag{2.34}$$

$$\gamma_{SL}^{LW} = (\sqrt{\gamma_S^{LW}} - \sqrt{\gamma_L^{LW}})^2 \tag{2.35}$$

By combining Eqs. 2.33, 2.34, and 2.35, we get

$$\gamma_{SL} = \gamma_S + \gamma_L - 2\sqrt{\gamma_S^{LW}\gamma_L^{LW}} - 2\sqrt{\gamma_S^+\gamma_L^-} - 2\sqrt{\gamma_L^+\gamma_S^-} \tag{2.36}$$

This equation and the Young–Dupré relationship give the equation for determining the components of the solid's surface energy:

$$\gamma_L[1 + \cos(\theta)] = 2\sqrt{\gamma_S^{LW}\gamma_L^{LW}} + 2\sqrt{\gamma_S^+\gamma_L^-} + 2\sqrt{\gamma_L^+\gamma_S^-} \tag{2.37}$$

This formula gives an equation with three unknowns (each α component of the solid's surface energy) that are solved by measuring the angles of contact with three different liquids [59]. Water is the reference for determining the acid–base components of other liquids:

$$\gamma_{water}^+ = \gamma_{water}^- = 25.5 \text{ mJ/m}^2 \tag{2.38}$$

The term added to the DLVO formulation is due to acid–base interactions. The XDLVO theory for a force between a sphere and a plane can be written:

$$\begin{aligned} F &= F_{vdW} + F_{edl} + F_{AB} = F_{LW} + F_{AB} \\ &= -\frac{A_{132}a}{6h} + 4\pi\varepsilon_0\varepsilon a\psi_s^2\left[\ln(1 + e^{-\kappa z})\right] - \frac{P\exp(z_0 - z)}{\lambda} \end{aligned} \tag{2.39}$$

with P the *pull-off* force previously described with $\frac{3}{2}\pi RW_{132}^{AB} \le P \le 2\pi RW_{132}^{AB}$, $W_{132} = \gamma_{13}^{AB} + \gamma_{23}^{AB} - \gamma_{12}$ (different energies can be estimated using the previous method), z_0 the minimum equilibrium distance $z_0 = 0.157$ nm, and λ the decay length. This length is estimated to 0.6 nm.

The form of the double-layer equation is once again chosen according to the assumptions described above, but for microscopic object you can use this approximation.

2.2.2.6 *Hydrodynamic Forces*

In the microworld, the Reynolds number, which characterizes liquid flow, is usually very low (<1). The flow is thus highly laminar. This part will present the hydrodynamical law valid at the microscopic scale.

For example, in case of a microobject placed in a uniform liquid flow, Stokes's law directly gives the hydrodynamic force applied on the object. This law is valid when the flow's Reynolds number is lower than 1 and can be extrapolated to a Reynolds number lower than 10 with a good approximation.

Stokes's law defines the force applied on an object in a uniform flow of fluid defined by a dynamic viscosity μ_3 and a velocity V:

$$\vec{F}_{\text{hydro}} = -k\mu_3\vec{V} \tag{2.40}$$

where k is a function of the geometry. In case of a sphere with a radius R, k is defined by

$$k = 6\pi R$$

Table 2.1 gives the values of dynamic viscosity μ of both water and air. So, the hydrodynamic force proportional to the dynamic viscosity highly increases in a submerged medium.

2.2.3 Theoretical Comparison Between Air and Liquid

2.2.3.1 *Surface Forces*

When two media are in contact, the surface energy W_{12} is equal to

$$W_{12} \simeq 2\sqrt{\gamma_1\gamma_2} \tag{2.41}$$

with γ_i the surface energy of the body i.

From the previous formulas the energy W_{132} required to separate media 1 and 2 immersed in medium 3 is given by

$$W_{132} = W_{12} + W_{33} - W_{13} - W_{23} = \gamma_{13} + \gamma_{23} - \gamma_{12}$$

For example, in case of a silicon–silicon contact ($\gamma_{\text{silicon}} = 1400\ mJ \cdot m^{-1}$), surface energy in water and in air are:

$$W_{12} = 2800\ \text{mJ}\,\text{m}^{-1} \quad W_{132} = 1670\ \text{mJ}\,\text{m}^{-1} \tag{2.42}$$

TABLE 2.1. Density and Dynamic Viscosity of Water and Air, $T = 20°\text{C}$

Properties	Water	Air
ρ (kg m^{-3})	10^3	1, 2
μ (kg m^{-1} s^{-1})	10^{-3}	18.5×10^{-6}

TABLE 2.2. Values of Hamaker Constant for Some Materials $A/10^{-20}$J

Materials	Vaccum	Water
Gold	40	30
Silver	50	40
Al_2O_3	16.75	4.44
Copper	40	30

In this example, the pull-off force is less in water compared to air. Usually, solid-state surface energy is around 1000 mJ m^{-1}, and this example gives a good overview of the reduction of pull-off forces in a liquid medium.

2.2.3.2 Van der Waals Forces

For contact of two dissimilar materials in the presence of a third media, A_{132} may be estimated by

$$A_{132} = A_{12} + A_{33} - A_{13} - A_{23}$$

Using combination formulas, it follows that

$$A_{132} \approx (\sqrt{A_{11}} - \sqrt{A_{33}})(\sqrt{A_{22}} - \sqrt{A_{33}}) \tag{2.43}$$

Table 2.2 gives the values of the Hamaker constant for some materials in vacuum and in water.

2.2.3.3 Electrostatic Forces

The force applied by an electrostatic surface (σ surface charge density) on an electrically charged particle (q) is given by

$$F = \frac{q\sigma}{2\epsilon_0\epsilon} \tag{2.44}$$

where

$\epsilon =$ relative dielectric constant of the medium
$\epsilon_0 =$ dielectric constant of the vaccum

The relative water dielectric constant ($\epsilon = 80.4$) is greater than the relative air dielectric constant ($\epsilon = 1$). So, in the same configuration of electrical charges (q, σ) the electrostatic force is less in water.

Moreover electrostatic perturbations observed in micromanipulation are caused by triboelectrification. During a microassembly task, friction between manipulated objects induces electric charges on the object's surface. The charge density is a function of the triboelectrification and conductivity of the medium. Effectively, a higher electric conductivity medium is able to discharge the object surface. Water, especially ionic water, has better electric conductivity than air. Consequently, the charge density in water is less. The electrostatic force directly proportional to charge density is therefore less.

In conclusion, electrostatic perturbation forces are much lower in water compared to air.

2.2.4 Impact of Hydrodynamic Forces on Microobject Behavior

We present in this section the impact of the hydrodynamic forces on the behavior of microobjects and especially on the reduction of the objects' jumps.

Because inertial effects are negligible in the microworld, microobject accelerations are usually very high. In this way, microobject velocity is able to increase in a very short time. Consequently, microobjects can reach high velocity, and object trajectory could be difficult to control especially because of the visual feedback. So, in most of cases, velocity limitation in the microworld is not induced by inertial physical limitation but by hydrodynamic physical limitation. From this point of view, a liquid medium is able to reduce maximal microobject velocity.

As a synthetic example we prove in this part that the hydrodynamic forces allow to reduce significantly the jump of microobjects.

To study the diminution of object jump, we have considered the trajectory of a microball (diameter d and density ρ_0), which has an initial velocity $\vec{V}(0)$:

$$\vec{V}(0) = V_x(0)\vec{x} + V_z(0)\vec{z} \tag{2.45}$$

The hydrodynamic force and weight are applied to the object (see in Fig. 2.4). We compare the object trajectories in air and in water. Table 2.1 recalls the density ρ and dynamic viscosity μ values of both water and air

In the case of an object placed in a uniform liquid flow, Stokes's law directly gives the hydrodynamic force applied on the object. Stokes's law is valid when the Reynolds flow number is lower than 1 and can be extrapolated to a Reynolds number lower than 10 with a good approximation.

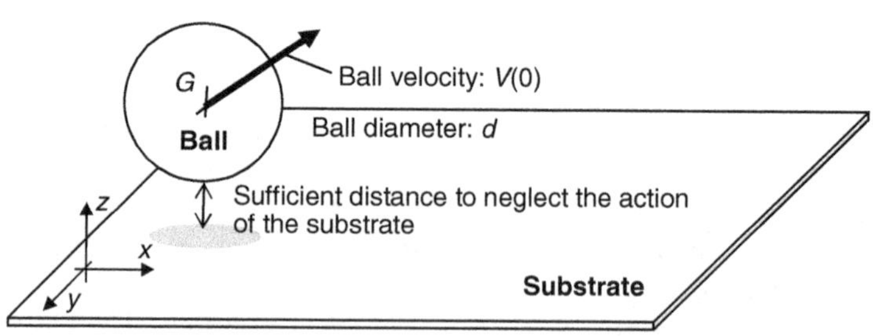

Figure 2.4. Example of hydrodynamic effects on microobject behavior: initial configuration.

We consider that Stokes's law (2.40) is valid. The object trajectory obtained with the ball dynamic equilibrium equation verifies

$$\begin{cases} x(t) & = V_x(0)\tau(1 - e^{-t/\tau}) \\ z(t) & = V_z(0)\tau\left[-\dfrac{t}{\tau_s} + \left(1 + \dfrac{\tau}{\tau_s}\right)(1 - e^{-t/\tau})\right] \end{cases} \tag{2.46}$$

with

$$\begin{cases} \tau_s & = \dfrac{\rho_0}{(\rho_0 - \rho)}\dfrac{V_z(0)}{g} \\ \tau & = \dfrac{\rho_0 d^2}{18\mu} \end{cases} \tag{2.47}$$

Parameter τ represents the time constant associated with the hydrodynamic force. Parameter τ_s represents the time constant associated with gravity (sedimentation).

In water, τ is much smaller than τ_s:

$$\tau \ll \tau_s \tag{2.48}$$

In other words, the impact of the hydrodynamic force on the trajectory is faster than the impact of gravity (sedimentation). So the trajectory is divided into two steps:

$$t \sim \tau \begin{cases} x(t) & = V_x(0)\tau(1 - e^{-t/\tau}) \\ z(t) & = V_z(0)\tau(1 - e^{-t/\tau}) \quad = \dfrac{V_z(0)}{V_x(0)}x(t) \end{cases} \tag{2.49}$$

$$t \sim \tau_s \begin{cases} x(t) & = V_x(0)\tau \\ z(t) & = V_z(0)\tau\left(1 - \dfrac{t}{\tau_s}\right) \end{cases} \tag{2.50}$$

We obtain a "triangle" trajectory:

- First stage: The trajectory is linear with its direction being the initial velocity $\vec{V}(0)$.
- Second stage or sedimentation stage: The trajectory is linear with its direction being gravity ($-z$).

Also the final object position L is

$$\begin{aligned} L &= V_x(0)\tau \qquad \text{from (2.50)} \\ &= \frac{\rho_0 V_x(0) d^2}{18\mu} \qquad \text{from (2.47)} \end{aligned} \tag{2.51}$$

We consider that this position is reached at $t = T = 5\tau$:

$$T = \frac{5\rho_0 d^2}{18\mu} \quad \text{from (2.47)} \tag{2.52}$$

We have chosen to present the following example:

$$\begin{cases} \rho_0 & = 2000\ \text{kg m}^{-3} \\ d & = 50\ \mu\text{m} \\ V_x(0) = V_z(0) & = 35\ \text{mm s}^{-1} \end{cases} \tag{2.53}$$

The final position calculated by Eqs. (2.51) and (2.52) is

$$\text{In water} \qquad L = 10\ \mu\text{m} \quad \text{at} \quad T = 1.4\ \text{ms} \tag{2.54}$$

$$\text{In air} \qquad L = 530\ \mu\text{m} \quad \text{at} \quad T = 75\ \text{ms} \tag{2.55}$$

Object trajectories are described in Figure 2.5. In the case of water [Fig. 2.5(a)], simulated trajectory (2.46) and "triangle trajectory" 2.49 and 2.50 are similar. In the case of air [Fig. 2.5(b)], hypothesis 2.48 is not true and the trajectory is different from the "triangle trajectory." However, Eqs. 2.51 and 2.52 remain valid.

This example illustrates microobject behavioral differences between the liquid medium and the dry medium. The distance L (defined in Eq. mk) is inversely proportional to the dynamic viscosity μ, and is 50 times longer in air than in water. The displacement of the microobject in air is around 10 times its diameter contrary to water, where the displacement is about one fifth of the object's diameter. This example illustrates the impact of the hydrodynamic forces on the reduction of microobject loss.

2.2.4.1 Conclusion

To explain the experimental differences between dry micromanipulation and micromanipulation in liquid, we analyze in this section the theoretical impact of the liquid on surface forces. We will focus this study on water. The current surface forces considered in the microworld are capillary, van der Waals, pull-off, and electrostatic forces. Although the main forces that induce adhesion effects are pull-off and electrostatic forces, we have chosen to present in this section the influence of water only on these two forces.[4] We also present the impact of the hydrodynamic forces on the microobject's behavior. We propose to separate these forces by making the distinction between whether there is contact or not. When there is no physical contact between two solids, the forces in action are called distance or surface forces (according to the scientific literature in this domain [9, 15, 45], the latter are electrostatic, van der Waals, and capillary forces). When

[4]We could also prove that van der Waals forces and capillary forces are, respectively, reduced [62] and decreased in a liquid medium.

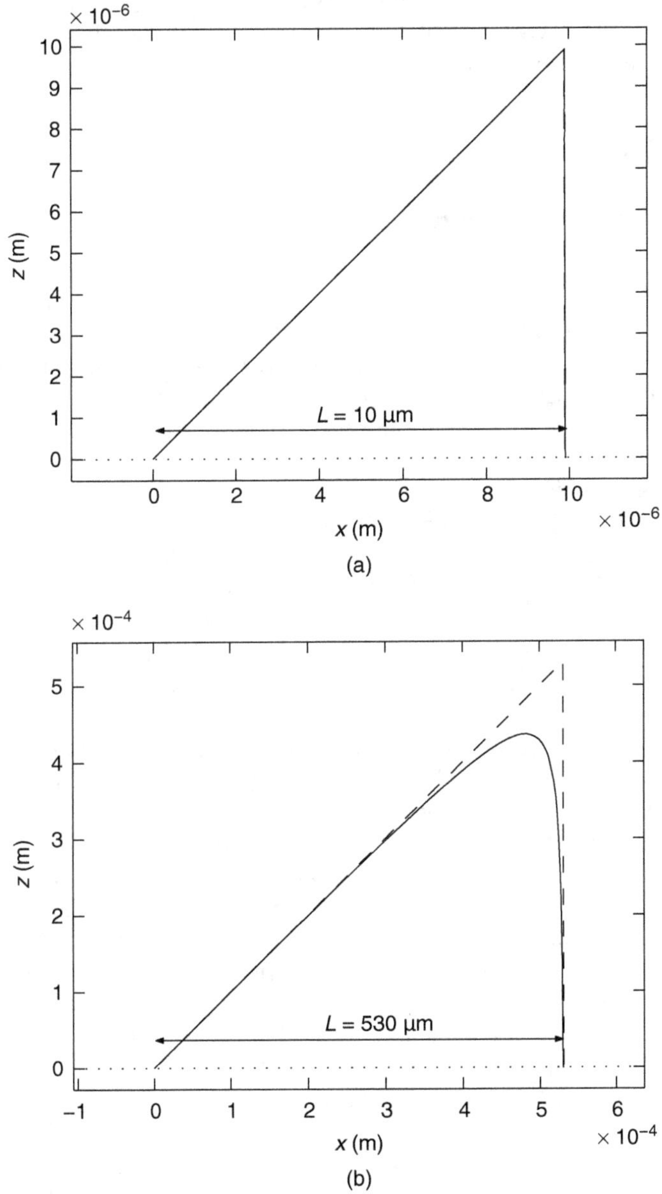

Figure 2.5. Comparison between ball trajectories in water and in air: (a) ball trajectory in water similar to the triangle trajectory and (b) ball trajectory in air compared to a triangle trajectory.

both solids are in contact with one another, there are deformation and adhesion forces through the surfaces in contact. In this case, we consider contact forces and adhesion or pull-off forces. Electrostatic or capillary effects can be added, but van der Waals forces are not considered in this case because they are already included in the pull-off term.

2.3 MICROSCOPIC ANALYSIS

2.3.1 AFM-Based Measurements

The goal of this section is to compare theoretical noncontact and contact surface forces at the microscale with corresponding values measured with an atomic force microscope (AFM). In order to develop new and adequate systems for efficient and reliable manipulation systems for objects at the microscale, it is indeed really necessary to be able to correctly estimate the forces at play through a set of realistic equations describing these forces.

2.3.1.1 Description

A view of the device is given in Figure 2.6(a). This system is based on an atomic force microscope. The microlever of the AFM can be moved in three perpendicular directions XYZ[5] by a piezotube (with X, Y, and Z strokes of, respectively, 45, 45, and 4 μm). Three linear micropositioning stages are also used for the studied sample motions on longer strokes in XYZ directions ($15 \times 15 \times 15$ mm^3, with a repeatability of 0.1 μm). All these motions can be controlled in automatic mode or in manual mode, notably by using a force-feedback joystick. This joystick applies in real time to the operator, with the bending and torsion effects measured on the microlever. These strains are measured by a photodiode (which delivers a corresponding voltage V_m).

Finally, two microscopes with charge-caupled device (CCD) cameras give visual information in real time on, respectively, vertical and lateral views [see Fig. 2.6)(a)]: One is fixed under the glass sample support and the other is placed perpendicularly. Then the first one gives the position of the microlever tip in the sample plan. The second one is used to estimate the vertical distance between the tip and the sample plan.

In this book, the microlever used is a silicon one that is 350 μm in length, 35 μm in width, and 2 μm in thickness. Its tip [see Fig. 2.6(b)] has a curvature radius of less than 10 nm and a height of 15 μm.

2.3.1.2 Measurement Method

Our setup is used here for the production of experimental force curves based on the real-time measurement of the AFM microlever bending (Fig. 2.7). A force curve is a quasi-static trajectory that corresponds to an "approach-and-retract" cycle between the microlever and the sample (in the Z direction). Depending on

[5]The Z axis designating the vertical direction.

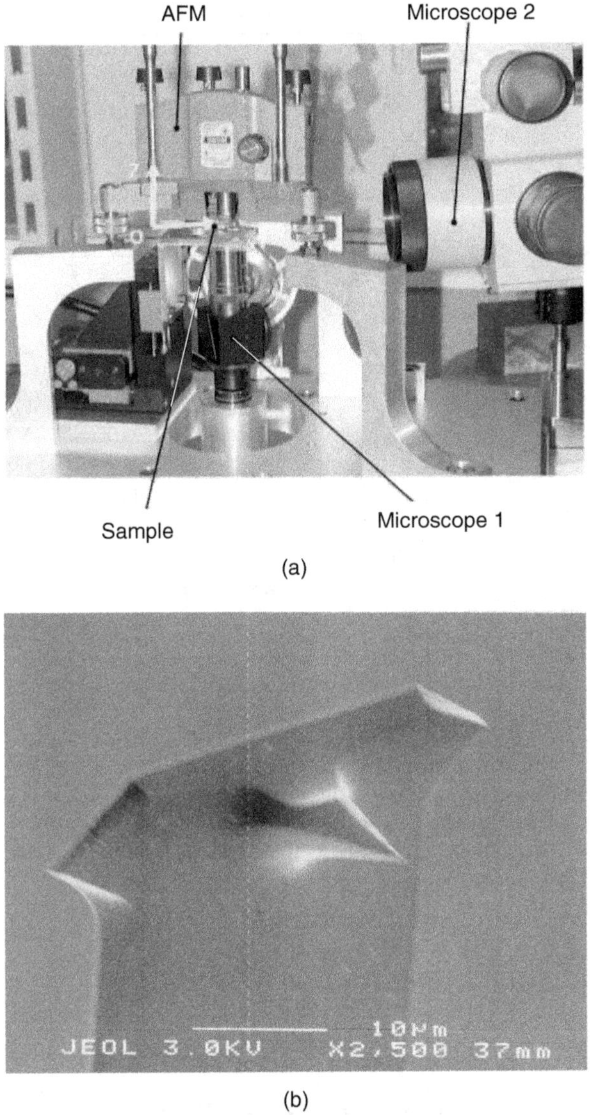

Figure 2.6. Description of atomic force microscope measurement system: View of (a) the AFM and (b) the tip.

the required stroke, these motions are actuated by the piezotube as well as the sample vertical axis.

First, the sample is placed in the AFM system and the operator can start the application: The approach-and-retract cycle is then executed automatically. The acquired data are the microlever bending, that is, the voltage measured from the photodiode, according to the relative vertical motion between the microlever and

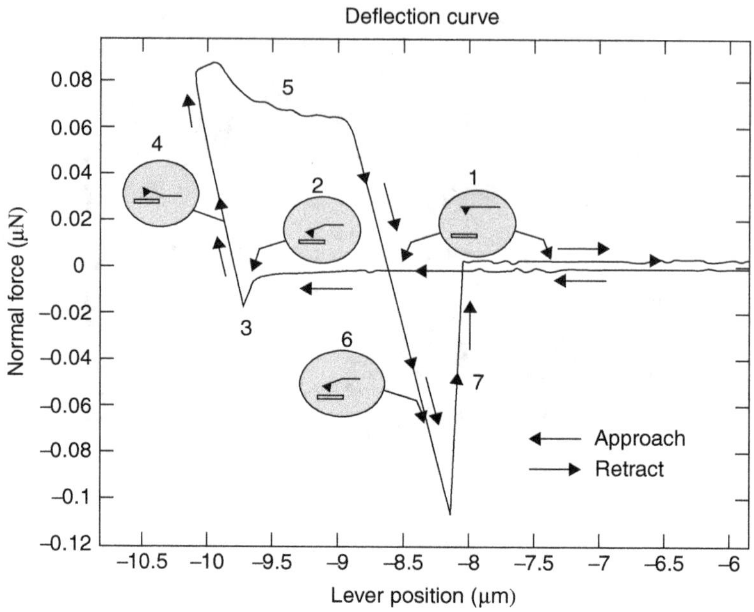

Figure 2.7. Example of force curve.

the sample. Then an adequate processing is done to extract the bending forces F from the measured voltage V_m. A specific identification procedure is used to estimate the coefficient C, which linearly relates V_m to the microlever bending δ and the microlever stiffness k. The stiffness of the microlever used for the experiments is $k = 0.03$ N/m. Then F is obtained by

$$F = k\delta = kCV_m \tag{2.56}$$

An example of such a curve is shown on Figure 2.7

2.3.2 Experiments on Adhesion Forces

2.3.2.1 *Van der Waals Forces and Pull-Off Forces*

First, the AFM measurement device is used to study van der Waals forces (attraction in approach phase of the AFM tip) and pull-off force (breaking load during the withdrawal of the AFM tip). These experiments are carried out with polystyrene and glass substrates.

From these curves [see figs. 2.8(a) and 2.8(b)], an approached value of the pull-off forces for these two interactions is deducted. They are estimated at

$$P_{\text{silicon-polystyrene}} = 26.23 \text{ nN} \tag{2.57}$$

$$P_{\text{silicon-glass}} = 34.70 \text{ nN} \tag{2.58}$$

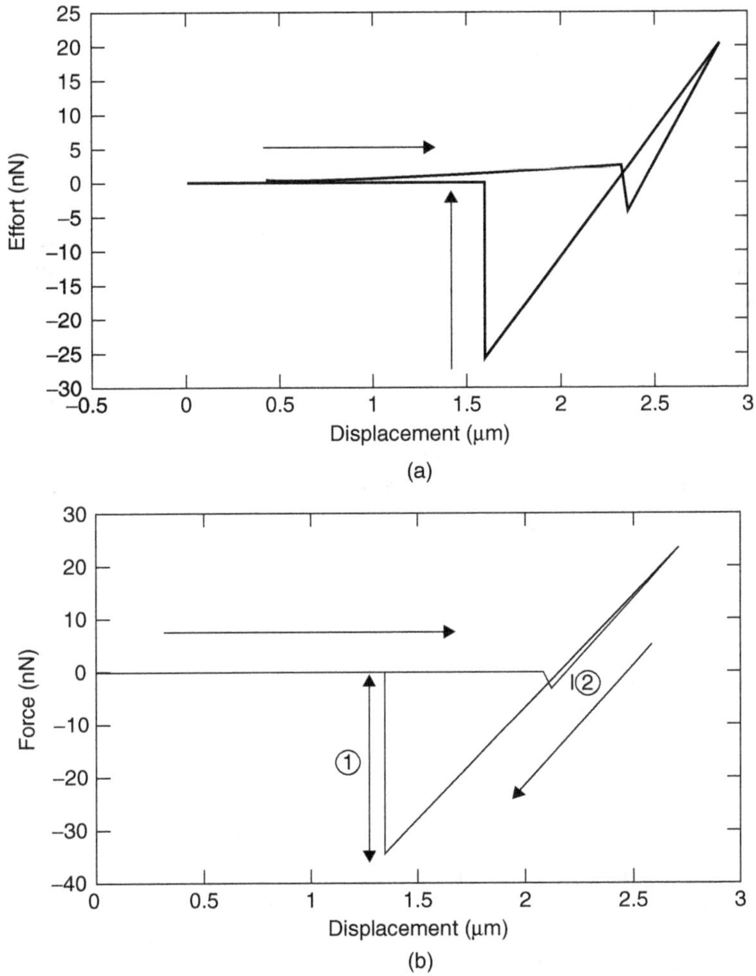

Figure 2.8. Force curves: interactions between AFM tip and (a) a polystyrene substrate and (b) a glass substrate.

The following data are used to determine the theoretical values of pull-off forces:

- Silicon $\gamma = 1400$ mJ/m^2, $A = 25.6 \times 10^{-20}$ J, $\nu = 0.17$, and $E = 140$ MPa.
- Polystyrene $\gamma = 36$ mJ/m^2, $A = 7.9 \times 10^{-20}$ J, $\nu = 0.35$, and $E = 3200$ MPa.
- Glass $\gamma = 170$ mJ/m^2, $A = 6.5 \times 10^{-20}$ J, $\nu = 0.25$, and $E = 69{,}000$ MPa.

From Eq. 1.105, pull-off forces can be estimated:

- Silicon–polystyrene interface $\lambda = 0.33$, and $P = 28.21$ nN.
- Silicon–glass interface $\lambda = 0.54$, and $P = 39.43$ nN.

These values fit very closely to the measurements. Hence, theoretical estimation of pull-off forces can generally be trusted when no direct measurement is possible.

On these force curves, one can observe the attraction phenomena in the approach phase. Though some works suggested different origins to the behavior, a traditional approximation is to consider only van der Waals interaction [43, 61]. These forces are thus estimated at:

- Silicon–polystyrene interface $F = 5.14$ nN.
- Silicon–glass interface $F = 4.03$ nN.

With the assumption of the contact distance $D_0 = 0.165$ nm, the values of Hamaker constants of the interfaces can be deduced from measurements:

- Silicon–polystyrene interface $A = 8.43 \times 10^{-20}$J.
- Silicon–glass interface $A = 6.58 \times 10^{-20}$J.

These measured values can be compared with the values obtained from the equation $A_{12} \approx \sqrt{A_{11}A_{22}}$:

$$A_{\text{silicon-polystyrene}} = 13.87 \times 10^{-20} \text{ J} \tag{2.59}$$

$$A_{\text{silicon-glass}} = 13.01 \times 10^{-20} \text{ J} \tag{2.60}$$

The errors are more significant because it is difficult to determine the various phenomena [53]. Nevertheless, this estimation seems to give a realistic value in order to estimate these forces via the Hamaker constants. A significant remark is that the range of van der Waals forces are of the order of 100 nm for all experiments carried out. Thus, this force seems to be relatively negligible compared to forces for objects of microscopic size.

A last experiment studies an interaction with a glass substrate in an aqueous environment in order to see the influence of the environment (Fig. 2.9).

The pull-off force is thus estimated at

$$P = 5.52 \text{ nN} \tag{2.61}$$

The theoretically calculated pull-off is then

$$P = 8.06 \text{ nN} \tag{2.62}$$

The estimation is still rather precise. Note also that the van der Waals force is almost not perceptible any more by our system. Its influence thus seems negligible in an aqueous environment.

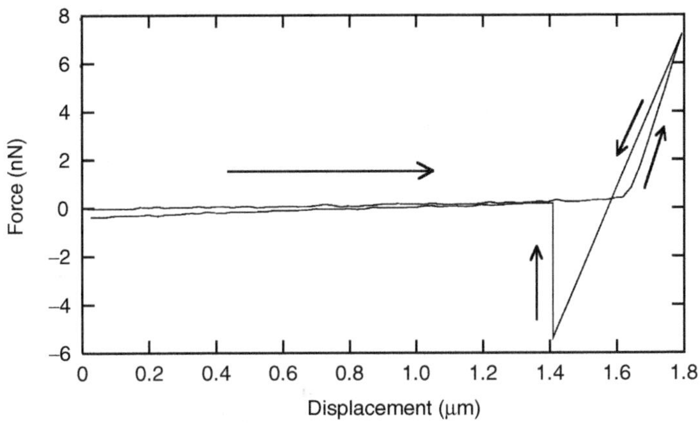

Figure 2.9. Force curve for an interaction between the cantilever and a glass substrate in a aqueous environment.

2.3.2.2 Capillary Forces

For micromanipulation, the capillary forces can represent a key parameter. Two interactions are studied with our system by placing a water droplet on polystyrene and glass substrates (Fig. 2.10). As the height of the droplet exceeds the maximum stroke of the AFM probe actuator, only the table motion is possible. Note that in this case force curves are reversed in terms of displacement and also that this motion reduces precision.

The capillary force is dependent on the reverse of the separation distance, contrary to the van der Waals and electrostatic forces. This force is present only when the AFM probe actually touches the water droplet. This force can thus be estimated using the following contact angles [13]

$$\theta_{\text{glass}} = 37^\circ \qquad \theta_{\text{polystyrene}} = 67^\circ \tag{2.63}$$

The measured forces are about

$$F_{\text{silicon-glass}} = 71.98 \text{ nN} \tag{2.64}$$

$$F_{\text{silicon-polystyrene}} = 37.82 \text{ nN} \tag{2.65}$$

The theoretical forces can thus be calculated from the equation $F_{\text{cap}} = 4\pi R\gamma_l \cos\theta/(1 + D/d)$ [28]:

$$F_{\text{silicon-glass}} = 73.26 \text{ nN} \tag{2.66}$$

$$F_{\text{silicon-polystyrene}} = 35.86 \text{ nN} \tag{2.67}$$

Theoretical models fit the capillary interaction quite well. The following remarks can be proposed for the sensivity of the capillary forces:

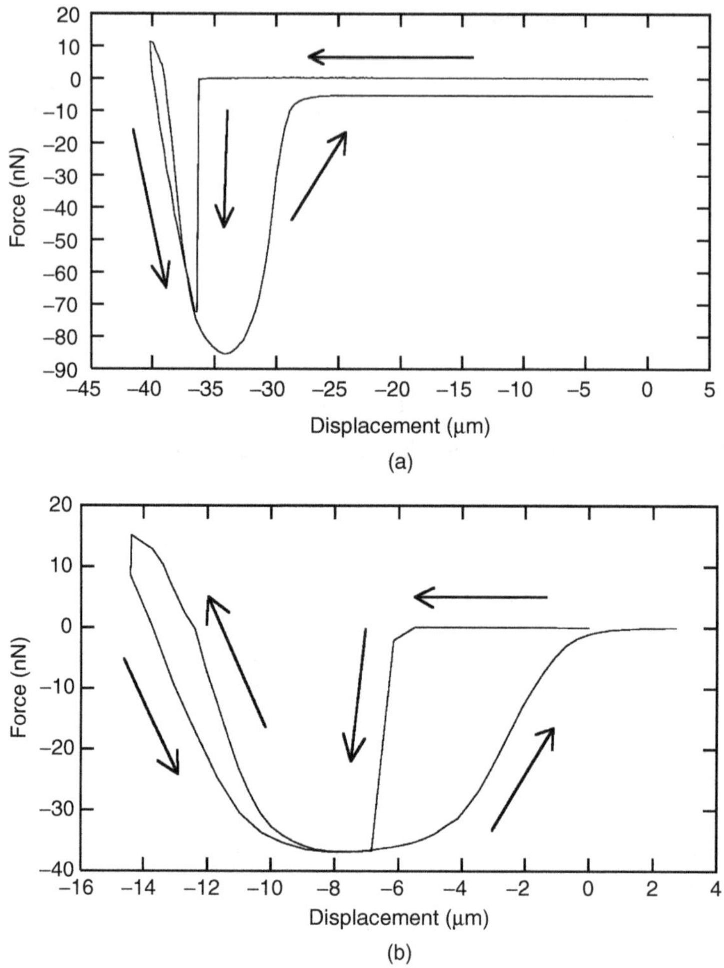

Figure 2.10. Force curves for interactions between the AFM tip and a water droplet: (a) on a polystyrene substrate and (b) on a glass substrate.

- When they exist, the capillary forces are most significant (their module is higher than van der Waals module).
- Their range is almost null.
- They always exist in a laboratory environment via, for example, a layer of oxidation of around of 10 nm on metals [60]. This layer seems negligible compared to the size of the considered object. They are commonly in the expression of pull-off forces.

Moreover, the pull-off force is overbalanced by a viscoelastic force due to the presence of water.

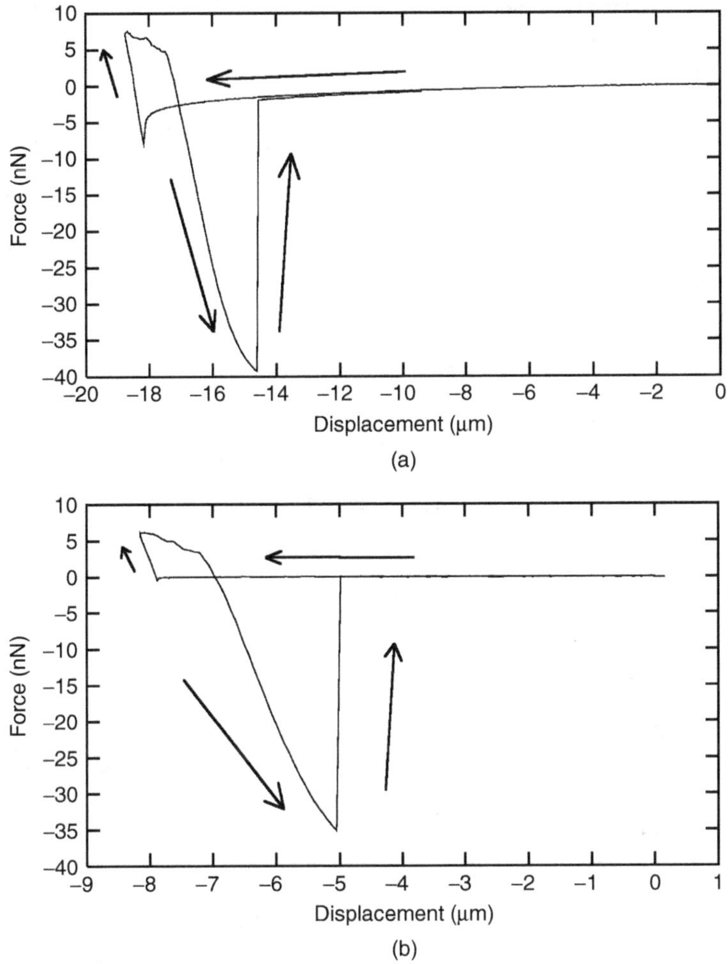

Figure 2.11. Force curves for interactions with (a) gold substrate and (b) grounded substrate.

2.3.2.3 Electrostatic Forces

This last part studies the electrostatic forces in case of contact with conductors and insulators. The AFM tip is made of silicon and is grounded. The first experiment describes a contact with a gold substrate. The electrostatic forces appear at a very significant separation distance compared to the other forces (10 μm).

To avoid this force, the substrate can be grounded as in Figure 2.11(b). In the same way, the van der Waals forces are not measurable (its range is about the resolution of the microtranslator).

The second study is on an insulator of polystyrene substrate. The results are shown in Figure 2.12(a).

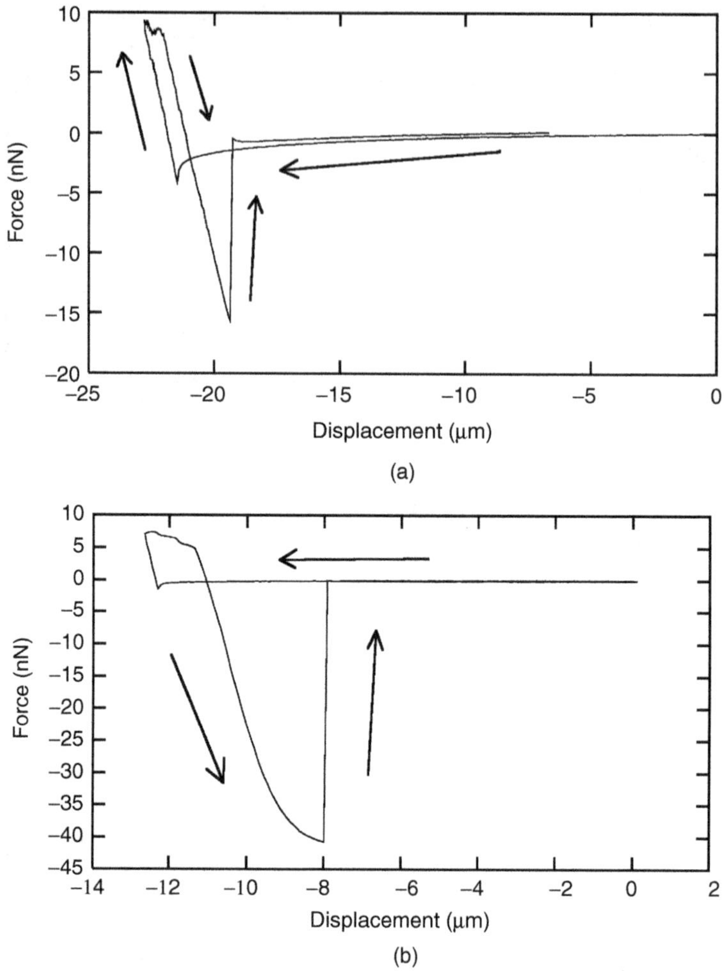

Figure 2.12. Force curves for interactions with polystyrene substrate: (a) interactions with a polystyrene substrate and (b) interactions with a polystyrene substrate after cleaning the substrate with distilled water.

In the same way, to avoid this force, the substrate is cleaned with distilled water. The curve obtained is then represented in Figure 2.12(b).

Electrostatic forces are efficient in the long range, starting at 10 μm. As the electrostatic load of the micropart is very badly known, and gives the highest modules of the distance forces, the most suitable approach for a micromanipulation application is to avoid this phenomenon by grounding for conductor or by the use of distilled water for the insulator. Note that in case of a manipulation involving only conductors, including the gripper, the use of electrostatic force can be an interesting solution.

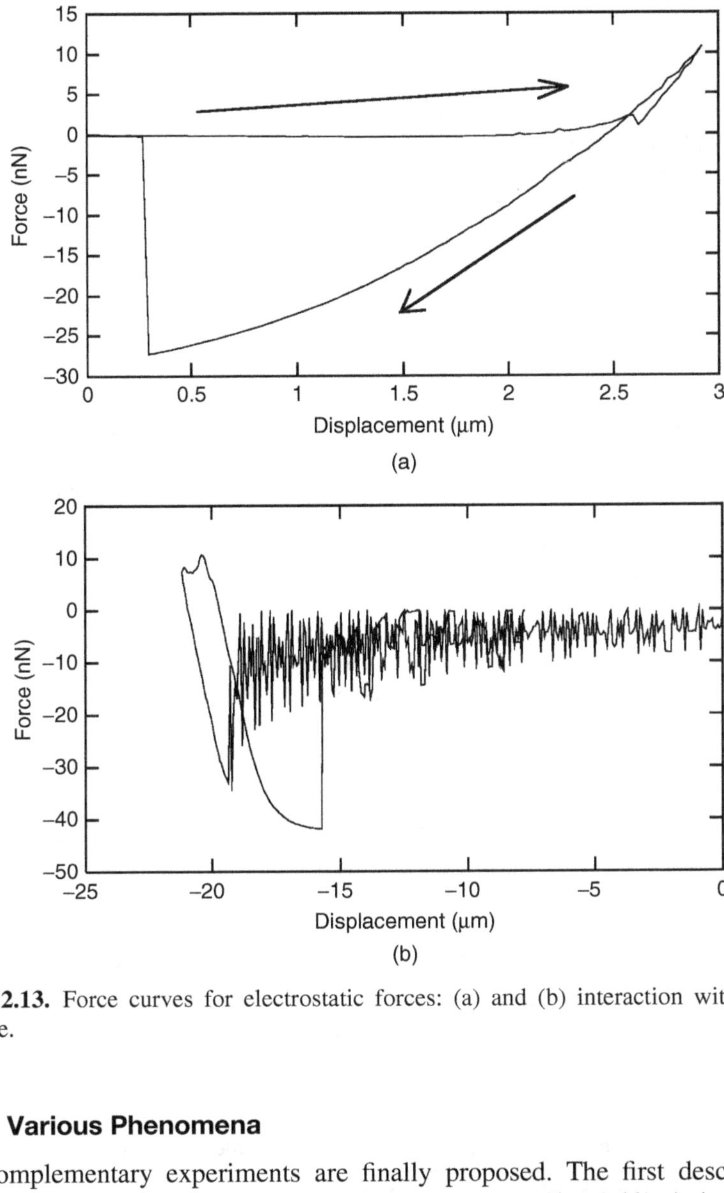

Figure 2.13. Force curves for electrostatic forces: (a) and (b) interaction with a glass substrate.

2.3.3 Various Phenomena

Two complementary experiments are finally proposed. The first describes an interaction with a glass substrate. On this curve (see Fig. 2.13), it is possible to observe at the same time the appearance of the electrostatic force (repulsive force with glass) and close to the contact, the appearance of the attractive van der Waals force. The amplitudes of these forces are comparable. Thus, the capillary forces, if they exist, are the most significant compared to electrostatic and van der Waals forces.

The second experiment studies the approach of the AFM cantilever with a copper substrate initially charged with a 2-V voltage. The approach curve of the

AFM cantilever is then drastically modified (see Fig. 2.13). Phases of attraction/release appear due to phases of discharge of the AFM tip. Moreover, tip effects can be observed, making difficult any identification. The only means of carrying out a discriminating analysis is to use a tipless cantilever. In the same way, this phenomenon disappears as soon as the substrate is grounded.

A lot of measurements are taken in this section. They highlight significant and often ignored phenomena as the influence of the force of pull-off, the weak range of the van der Waals forces, the influence of the capillary forces, and the long range of the electrostatic forces.

2.4 SURFACE ROUGHNESS

This section is a brief introduction to some of the concepts linked to surface roughness (measure, modeling, manufacturing). Additional information on how to include surface roughness in surface forces studies can be found in Sausse Lhernauld [50]. Indeed, it turns out that roughness cannot be avoided (effects on electrostatic forces have been proofed for as small a roughness as a few nanometers [51]). Therefore, it is of the utmost importance for engineers to be able to measure and model it. The following introduction will be restricted to 2D geometries. Details for 3D geometries can be found in tribology books such as Bhushan [6].

2.4.1 Surface Topography Measurements

Experimental techniques used for surface topography measurements can be divided into two broad categories: contact types and noncontact types (see Table 2.3) [6]. Different techniques are briefly reviewed in this section:

- Mechanical stylus method uses an instrument that amplifies and records the vertical motions of a stylus displaced at a constant speed above the surface to be measured. As the stylus rides over the sample, detecting surface deviations by a transducer, it produces an analog signal corresponding to the vertical stylus movement that is amplified, conditioned, and digitized.

TABLE 2.3. Comparison of Roughness Measurement Methods [6]

Method	Spatial Resolution	Vertical Resolution	Limitations
Stylus	15–100 nm	0.1–1 nm	Contact may damage surface.
Optical	0.5 μm–1 mm	0.02–25 nm	
STM	0.2 nm	0.02 nm	Conductive surfaces
AFM	0.2–1 nm	0.02 nm	
SEM	5 nm	10–50 nm	Conductive surfaces

- Optical microscopy uses light waves reflected on the surface. The angle of reflection is equal to the angle of incidence in the case of perfectly smooth surfaces. As roughness increases, the intensity of the specular beam decreases while the diffracted radiation increases in intensity and becomes more diffuse. Optical methods may be divided into geometrical (taper sectioning and light sectioning) and physical methods (specular and diffuse reflexion, speckle pattern, and optical interference).
- Scanning tunneling microscopy (STM) is based on the principle that if a potential difference is applied to two metals separated by a thin insulating film, a current will flow because of the ability of electrons to penetrate a potential barrier. In STM a sharp tip (one electrode of the tunnel junction) is brought close enough to the surface to be investigated (second electrode) so that, at a convenient operating voltage, the tunneling current will vary and will be measured. STM requires the surface to be measured to be conductive.
- Atomic force microscopy (AFM) is capable of investigating surfaces of both conductors and insulators on an atomic scale. AFM measures ultrasmall forces (less than 1 nN) present between the AFM tip and a surface to be investigated. These forces are measured by measuring the motion of a very flexible cantilever beam.
- Scanning electron microscopy (SEM) uses a beam of highly energetic electrons to examine objects on a very fine scale. It functions exactly as its optical counterparts except that SEM uses a focused beam of electrons instead of light to image the specimen and gain information about its structure and composition.

Other methods to measure surface roughness include fluid methods and electrical methods that are mainly used for continuous inspection procedures (quality control) because they are very fast and function without contact with the surface.

A typical encountered problem in roughness measuring is the influence of the curvature radius of the scanning probe (see Figs. 2.14 and 2.15. The probe filters curvatures: Only curvatures smaller than that of the probe can be measured (a large tip cannot scan into small roughness valleys).

2.4.2 Statistical Parameters

Two main approaches can be followed to describe roughness, relying on *statistical* (described in this section) or *fractal* (described in Section 2.4.4) descriptions.

The statistical approach relies on the computing of several statistical amplitude and spatial parameters in order to fit a measured profile $z(x)$. For a profile scanned along a sampling length L, the parameter R_a (center-line average) is given by

$$R_a = \frac{1}{L}\int_0^L |z(x) - \overline{z}|\, dx \tag{2.68}$$

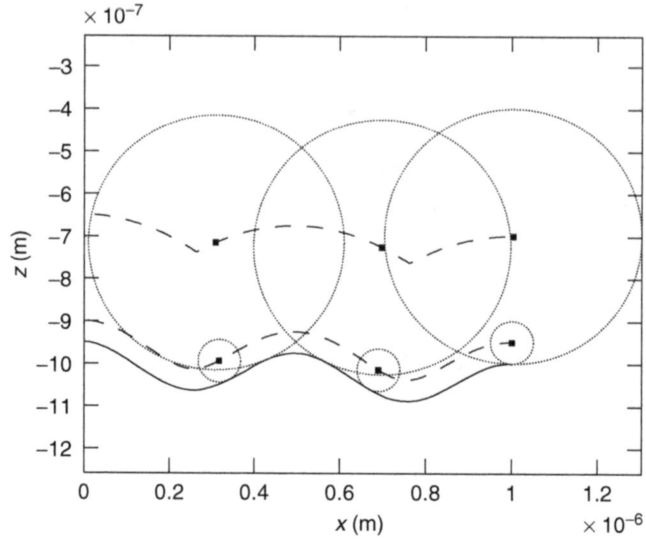

Figure 2.14. Influence of the curvature radius of the scanning probe: actual profile (solid line), measured profile (dashed line), circles indicate three different locations of the probe (curvature radii equal to 300 and 50 nm).

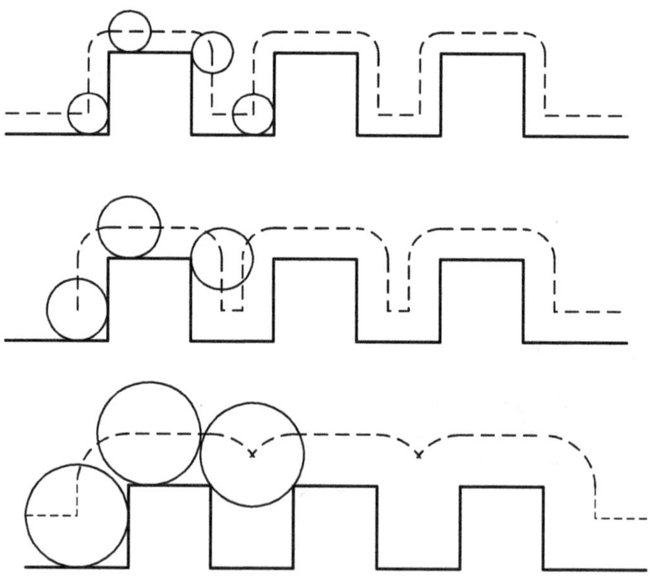

Figure 2.15. Influence of the curvature radius of the scanning probe: actual profile (solid line), measured profile (dashed line).

where $\overline{z}$ is the average height of the profile. The standard deviation of the distribution $z(x)$ is the base of R_q:

$$R_q = \sqrt{\frac{1}{L}\int_0^L (z(x) - \overline{z})^2\, dx} \tag{2.69}$$

Both parameters can be used to define a Gaussian model of the surface, but unfortnuately, not all manufacturing processes lead to such a profile [40]. In this case, additional indicators can be used such as those built on third- (skewness) and fourth- (kurtosis) order moments [7].

Norms also mention the maximum peak-to-peak distance R_t, which is unfortunately not robust (it can be very sensitive to the presence of dust):

$$R_t = \max(z) - \min(z) \tag{2.70}$$

We see from these definitions the sensitivity of these amplitude parameters to the sampling length L [30]: These parameters are not intrinsically related to the surface [63].

One way to supplement amplitude information is to provide some index of crest spacing or wavelength on the surface [6]. Such parameters include the average wavelength and the root-mean-square (rms) wavelength defined by

$$\lambda_a = 2\pi \frac{R_a}{\delta_a} \tag{2.71}$$

$$\lambda_q = 2\pi \frac{R_q}{\delta_q} \tag{2.72}$$

where δ_a and δ_q are the amplitudes of the individual wavelength. Other parameters may include the peak density or the mean spacing between peaks. The rms values are, however, still scale dependent.

Beside these parameters, additional functions can be used: They are also known as surface texture descriptors and referred to as spatial function [5]:

1. The **autocorrelation function** is a way of representing spatial variation. It is the product of two measurements taken on a profile at a distance τ apart.
2. The **structure function** is defined by

$$S(\tau) = \lim_{L\to\infty} \frac{1}{L}\int_0^L [z(x) - z(x+\tau)]^2\, dx \tag{2.73}$$

3. The **power spectrum** (or power spectral density) describes how the power of a signal or time series is distributed with frequency. It is deduced from the square of the modulus of the Fourier transform of the signal:

$$P(\omega) = \lim_{L\to\infty} \frac{1}{L}\left[\int_0^L z(x)\exp(-i\omega x)\, dx\right]^2 \tag{2.74}$$

Inversely, the autocorrelation function is the inverse Fourier transform of the power spectrum:

$$S(\tau) = \int_{-\infty}^{\infty} \{[1 - \exp(i\omega\tau)]P(\omega)\}\, d\omega \tag{2.75}$$

An alternative is to consider a representation of the surface, which is independent on the considered scale, which is precisely one of the features of fractal modeling known as self-similarity. This will be the topic of Section 2.4.4.

2.4.3 Models of Surface Roughness

When studying the adhesion phenomenon, surface roughness is a very important factor. Indeed it is well known that the existence of nanoscale roughness dramatically reduces adhesion between surfaces and that this is due to a decrease in the real area of contact and an increase in the distance between bulk surfaces [33, 57]. For these reasons many attempts have been made to model surface roughness that will be summarized in the following. Please note that much more work has been done in the field of influence of surface roughness on van der Waals forces since roughness has a great influence on these forces, whereas the electrostatic forces are often believed not to change much with asperities [9].

The simplest case [2] only considers the roughness peak and assumes the roughness profile to be equivalent to a smooth profile located at a separation distance $d+R/2$ where R is the height of the highest peak and d is the distance between the plane and the highest peak. This model is, however, not good since it does not take the density of protrusions into account.

Sphere (hemispherical model) and cone models are the most encountered in the literature:

- Vogeli et al. [61] model the roughness profile using several half-spheres with a diameter R.
- Herman et al. [24] determine the effects of spherical and conical asperities on van der Waals and double-layer interactions between two parallel flat plates.
- Suresh et al. [25] model roughness as hemispherical asperities characterized by the average asperity height and the density of asperities on the surface for calculating both electrostatic and van der Waals interaction energies between a rough particle and a flat plate.

Discretized profiles have also been used as a roughness representation:

- In Lambert [34] the Abbott diagram is considered and related to the surface, the roughness profile is discretized into M cites, and his equation is applied to the M-discretized elements (see Fig. 2.16).

- In Shulepov and Frens [52] the roughness of the particle is characterized as a uniform pattern of hills, each hill consisting of steps and plateaus with a characteristic width and height to evaluate the total interaction energy between two rough spheres.

Sinusoidal functions are also a way of representing roughness, knowing the amplitude and the period (or as a sum of sine function):

- In Danuser et al. [12] a sinusoidal surface is modeled, where the density n of protrusions is related to the wavelength λ by $n = 1/\lambda^2$ and the radius of the spheres is $r = \lambda^2/(4\pi^2 h)$ (see Fig. 2.16).
- In Kostoglou and Karabelas [32] a cosine function is used that is an idealized periodic surface characterized by shape, height, and wavelength to calculate the electrostatic repulsive energy between two rough colloidal particles.

The limitation of all the previous models is that the roughness is represented as a very idealized shape. Such approaches provide nonetheless qualitative information about the effects of surface roughness. Unfortunately, little has been said about how well those geometries correlate with known surface roughness profiles, especially at the nanoscale. Moreover, Rabinovich et al. [47] state that the application of a Gaussian distribution to model asperities on a surface may produce errors because large asperities significantly affect the interaction even though their number is low. Another modeling strategy is therefore presented in next section.

2.4.4 Fractal Parameters

2.4.4.1 Introduction to Fractals

Because surface roughness can have such a large effect on the adhesion force, errors introduced by modeling surface roughness with elementary protuberances will cause adhesion force predictions to be inaccurate in many cases [19]. More particularly, a roughness description based only on the R_a parameter puts aside spatial consideration. An alternative is to proceed to a Fourier analysis of the sampled profile in order to extract the main components of the roughness signal, leading to the following mathematical model of the surface:

$$z(x) = \sum_{i=1}^{n_f} A(i) \cos\left(2\pi \frac{x}{\lambda_i}\right) \tag{2.76}$$

where the λ_i is the spatial wavelength of the ith component with amplitude A_i. This approach can be very powerful but still requires $2n_f$ parameters to reconstruct the geometry. Nevertheless, some surfaces exhibit a *self-repeating nature* of surface roughness at different scales: They are known as fractal surfaces. The term *fractal* comes from the Latin *fractus*, meaning irregular or fragmented. Fractals are irregular objects possessing similar geometrical characteristics at all scales. This characteristic is called *self-similarity*. Observing the coastline of Britain,

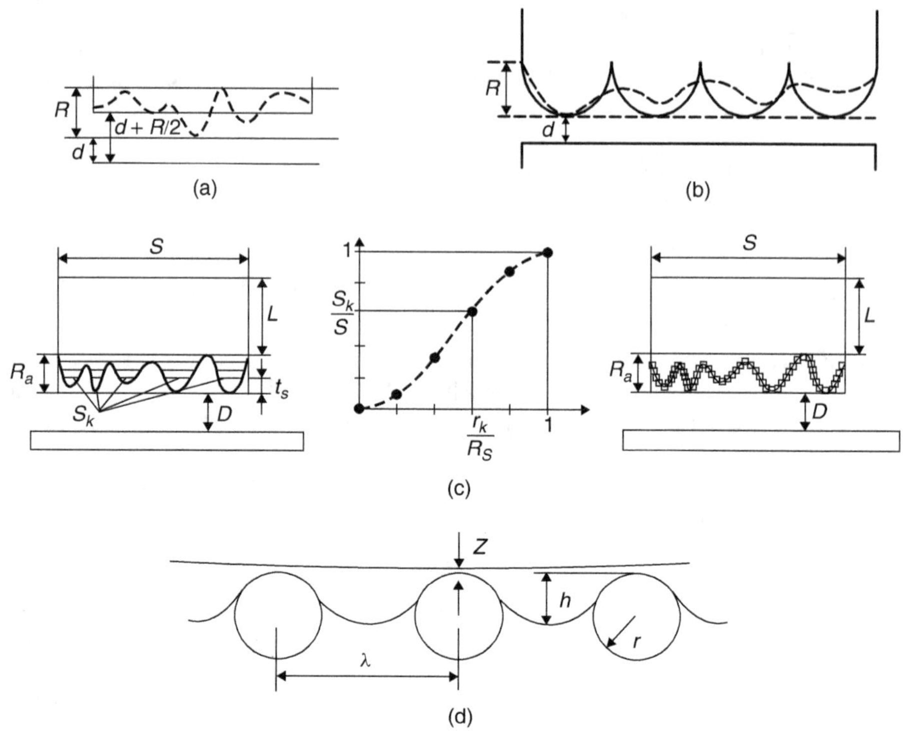

Figure 2.16. Discretized profiles from (a) planar [2], (b) hemisphere [61], (c) discretized [34], and (d) sinus [12].

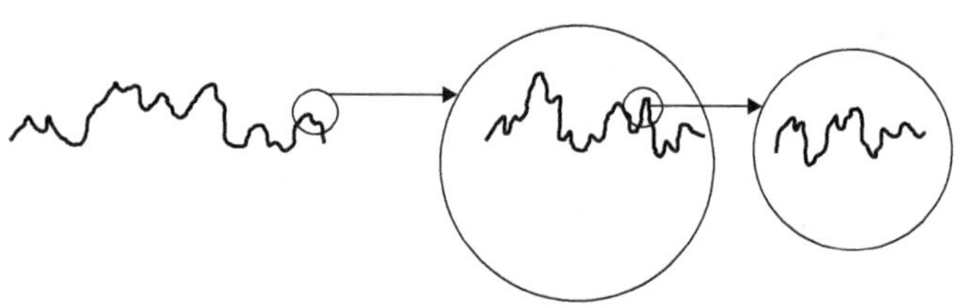

Figure 2.17. Illustration of the self-similarity property [63].

Mandelbrot [38] showed, for example, that the more the coastline is magnified, the more features and details are observed. This illustrates the self-similarity notion (Fig. 2.17).

Geometries of fractal surfaces are also continuous and nondifferentiable. Since the profile of rough surfaces $z(x)$ (typically obtained from stylus measurements) is assumed to be continuous even at the smallest scales and ever-finer levels of detail appear under repeated magnification, the tangent at any point cannot be defined. The profile has thus the mathematical property of being continuous everywhere but nondifferentiable at all points. Surface profiles are also known to have self-affinty in roughness structure. The Weierstrass–Mandelbrot function satisfies the properties of continuity, nondifferentiability, and self-similarity and can therefore be used to simulate such profiles, with only a small number of parameters.

2.4.4.2 Fractal Representation of Roughness

In fractal characterization, the continuity, nondifferentiability, and self-affinity of a two-dimensional surface profile height may be represented by the function [8, 18, 23, 31, 37, 44, 63, 64, 68]

$$z(x) = G^{D-1} \sum_{n=n_1}^{\infty} \frac{\cos(2\pi \gamma^n x + \phi)}{\gamma^{(2-D)n}} \tag{2.77}$$

where D is the fractal dimension ($1 < D < 2$), G is the fractal roughness parameter or scaling parameter, ϕ is random phase, and L is the fractal sample length.

Constant γ is chosen to be 1.5 [23, 31, 37] for phase randomization and high spectral density (in order for the phases of the different modes not to coincide at any given x position, the value of γ must be chosen to be a noninteger). Constant γ determines the density of the frequency spectrum with smaller γ values yielding larger numbers of frequency components.

Equation 2.77 represents the surface profile by a series of cosine functions (Fig. 2.18) with geometrically increasing frequencies starting from the lowest frequency $w_l = 1/L$. Factor n_1 is thus deduced from the length of the sample L: $\gamma^{n_1} = 1/L$.

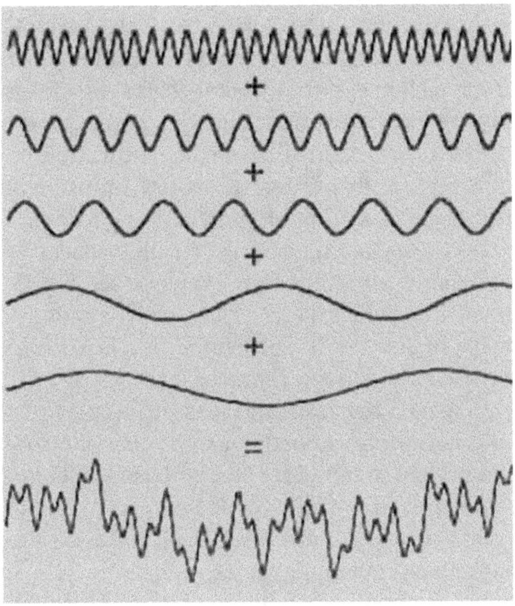

Figure 2.18. Construction of a fractal surface [56].

2.4.4.3 Description of Fractal Parameters

Fractal parameters are D and G. Parameter D can vary from 1 to 2 in two dimensions, whereas G is not limited. As D becomes larger, the number of asperities increases and their height decreases; therefore, it governs the contribution of low- and high-frequency components to the surface (see Figs. 2.19 and 2.24). The amplitude or scale parameter is G (also called fractal roughness). As it increases, the peaks and valleys are amplified. Thus, as the magnitudes of D and G increase, a rougher and more disordered surface topography is produced. An increase of D stretches the profile in the lateral dimension and therefore changes the spatial frequency. On the other hand an increase of G stretches the profile along the vertical dimension. So D controls the relative amplitude of roughness at different length scale, whereas G controls the amplitude of roughness over all length scales. The Hausdorff or fractal dimension, $D + 1$, of rough surfaces in three dimensional simulations is a fraction between 2 and 3 [36]. The specific integer values of 0, 1, 2, and 3 correspond to smooth objects, respectively, point, line, surface, and sphere (or any three-dimensional object), whereas the noninteger values correspond to wiggly and complex objects with self-similar behavior.

2.4.4.4 Advantages/Limitations

Advantages. The advantage of fractals is to offer a set of parameters that are invariant with respect to scale. More specifically, once scale-invariant parameters have been determined, the roughness can be predicted at all length scales with only two parameters (D and G). Characterizing a multiscale surface using traditional statistical parameters would render different results depending on the scale at which the measurements were taken. Of course, all surfaces are not fractal, and obviously the fractal description can only be applied to surfaces that show fractal behavior. According to Morraw [41], most engineering surfaces have been proved to exhibit roughness on different scales. For Komvopoulos [30], the topography of many engineering surfaces may be represented by fractals because similar features can be observed at different magnification of the same surface. In Ling [35], conclusions are that fractal geometry forms an attractive adjunct to Euclidean geometry in the modeling of engineering surfaces. For Majumdar and Bhushan [36], it is necessary to characterize rough surfaces by intrinsic parameters that are independent of all scales of roughness (see Fig. 2.20). This suggests that the fractal dimension, which is invariant with length scales and is closely linked to the concept of geometric self-similarity, is an intrinsic property and should therefore be used for surface characterization.

For Zhou et al. [67] manufactured surfaces produced by electrical discharge, water-jet cutting, and ion-nitriding coating can be characterized by fractal geometry. It is moreover important to introduce fractal-based techniques to study surface engineering and tribology and to apply the techniques to engineering applications such as contacts, wear processes, and friction. It has also been shown that surfaces formed by natural and random courses such as fracture surface of a solid, a deposition surface of a material, or a solidification surface of a liquid have fractal structures [23]. All the previously cited works tend to validate the theory that fractal geometry gives an accurate description of the profile geometry for many engineered surfaces. This point of view is not, however, shared by all authors and is discussed in the next section.

Limitations. The scale independence of the fractal parameters D and G was doubted by Ganti and Bhushan [18]. In a series of measurements of fractal parameters, measures have shown to depend on the method used for surface scanning (i.e., AFM or nontcontact optical profiler) and on the scan size. Majumdar and Bhushan [36] even discovered that some rough surfaces are bifractal, that is, the surface exhibits different D and G values within different scale ranges. These studies bring the conclusion that fractal parameters are relatively scale independent but not absolutely scale independent in all scale ranges [23].

Gelb et al. [19] used fractals to model rough copper surface and polytetrafluoroethylene (PTFE) surfaces. In both cases, the fractal dimension decreased with decreasing the scan size, indicating that fractals should not be used to describe these surfaces.

Because fractals are scale invariant, they are hardly suitable for characterizing features that are not. Only if the surface feature is scale invariant can it be expected to reflect the dimension D. Whitehouse [63] does not agree with the fact that engineered surfaces exhibit fractal properties. The author states that fractal analysis can be used to characterize natural phenomena such as growth mechanisms (e.g., bifurcation or fracture mechanics), but most manufacturing effects are not growth phenomena. The conclusion is that fractals tend to ignore manufacture rather than to clarify it. Whitehouse thinks random process parameters are more suitable to describe manufacturing mechanisms. In fact, whitehouse [63] even states that trying to allot fractal parameters to surfaces is questionable. Indeed, he thinks using scale-invariant parameters as a replacement for traditional parameters (statistical parameters) is philosophically wrong. He warns that fractals in manufacture are not necessarily the best means of characterization. Because manufacturing characteristics are severely scale limited, surface parameters that are deliberately scale dependent should be used.

Real engineering surfaces might be multifractal or be fractal on certain length scales and nonfractal on others. It is difficult to treat the fractal parameters D and G as absolutely scale independent on all scale ranges [23]. However, as these parameters are scale independent on certain length scales (i.e., relatively scale independent), we might still assume that the fractal model is applicable.

Concerning the effect of the sampling interval on the fractal parameters, experimental results from He and Zhu [23] show that changes in sampling spacing (lateral resolution) did not influence the scale independence of fractal parameters and that if the sampling length is long enough, it will have no influence on the fractal parameter.

2.4.5 Extracting the Fractal Character of Surfaces

Fractals have many dimensions such as Hausdorff dimension, compass dimension, box dimension, mass dimension, and area perimeter dimension, and there are

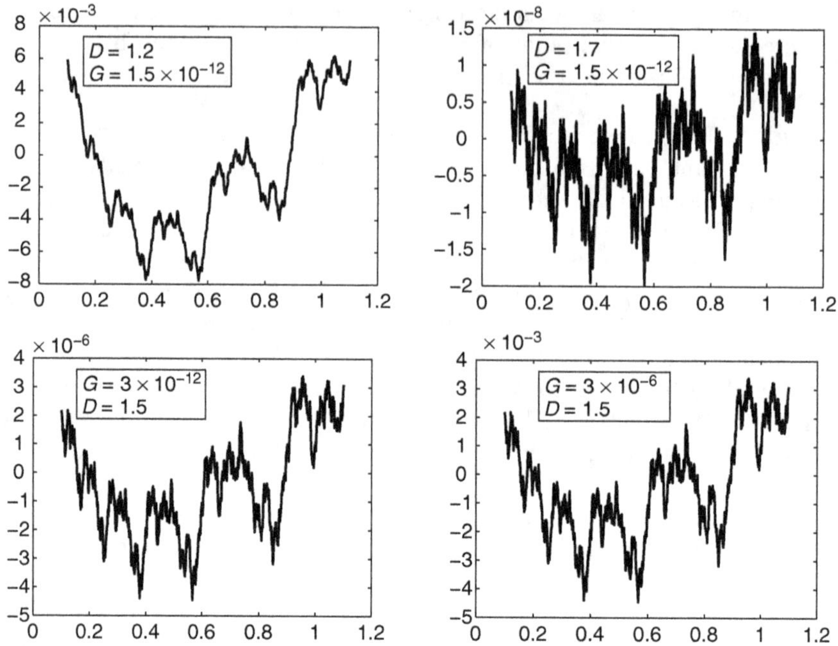

Figure 2.19. Influence of parameters D and G on a 1-μm-long profile; axis are in meters.

several methods for computing each of these dimensions [67]. In what follows, we detail two methods—power spectrum and structure function methods—in order to determine the parameters D and G to be used in Eq. [2.77].

2.4.5.1 Power Spectrum Method

The variation of roughness amplitude with spatial frequency ω of the Weierstrass–Mandelbrot function can be represented by the averaged power spectral density function $P(\omega)$. It is calculated using the real surface profile [46]:

$$P(\omega) = \frac{2\pi\ \Delta x}{N} |F(\omega)|^2 \tag{2.78}$$

where Δx is the spacing between two points, ω is the spatial frequency, N is the number of points sampled, and $F(\omega)$ is the Fourier transform of height data $z(x)$.

The power spectral density of the Weierstass–Mandelbrot function follows a power law and is defined by

$$P(\omega) = \frac{G^{2(D-1)}}{2\ \ln \gamma} \omega^{-(5-2D)} \tag{2.79}$$

Investigating the relationship between the power spectrum log $P(\omega)$ of the discrete data and log ω [22], the fractal dimensions can be deduced. The parameters G and D are found by using a *log–log* plot of the power spectrum of the fractal

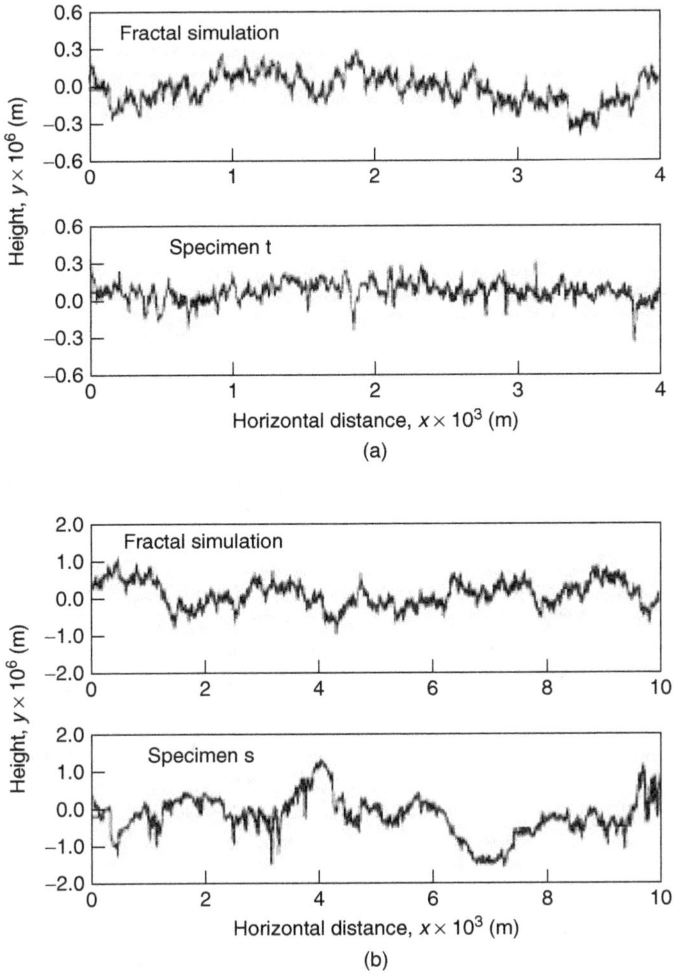

Figure 2.20. Comparison between surface profiles and fractal simulations for two different specimen of stainless steel [36].

surface. If there is a linear log–log plot, the fractal dimension can be calculated. Indeed:

$$\log(P(\omega)) = \log(G^{2(D-1)}(2\ \ln \gamma)) + (2D - 5)\log(\omega) \tag{2.80}$$

where D is thus deduced from the slope β and G is found from the intersection of the vertical axis K [41]:

$$D = \frac{\beta + 5}{2} \tag{2.81}$$

$$G = (10^K \times 2\ \ln \gamma)^{1/2(D-1)} \tag{2.82}$$

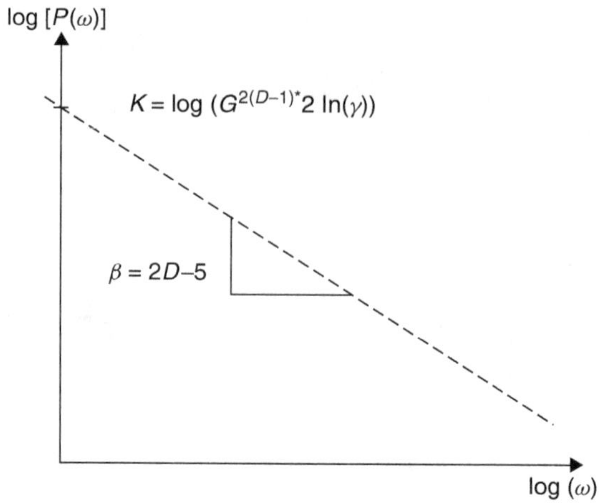

Figure 2.21. Log–log plot of the power spectral density function and deduction of fractal parameters.

By plotting the power spectrum density (Fig. 2.21), we can determine the finite range of length scales for which the surface is fractal. At the spatial frequency, which the surface deviates from the power law behavior, the limits are determined. Mandelbrot [36] characterizes the two spectral regions of stainless steel specimens by splitting the Weierstrass–Mandelbrot function into two parts.

Practically, surface characterization and extraction of fractal parameters are achieved following this process: The surface is scanned in order to get surface height information in the form of points (X, Z), the signal is treated with a fast Fourier transform resulting in the power spectral density function that can be plotted, and the fractal characteristics are extracted from this plot.

2.4.5.2 Structure Function Method

A discrete form of the structure function of a profile $z(x)$ is given by

$$S(\tau) = \frac{1}{N-k}\sum_{i=0}^{N-k}(z_{i+k} - z_i)^2 \tag{2.83}$$

where k ranges from 1 to N. Defining τ as $\tau = 2\pi(x_k - x_1)$, the fractal structure function [23] of the Weierstrass function follows a power law that is given by [18, 68]

$$S(\tau) = C\tau^{4-2D} \quad \text{where} \quad C = \frac{\Gamma(2D-3)\sin[(2D-3)\pi/2]}{(4-2D)\ln\gamma}G^{2(D-1)} \tag{2.84}$$

where Γ is the gamma function with argument $2D - 3$:

$$\Gamma(2D-3) = \int_0^\infty t(2D-2)\exp^{-t}\,dt \tag{2.85}$$

The plot of $S(\tau)$ as a function of τ (as for the power spectrum method) is a straight line in the log–log plot (Fig. 2.22):

$$\log(S(\tau)) = \log(C) + \underbrace{(4 - 2D)}_{\beta} \log(\tau) \tag{2.86}$$

If the slope of this line β satisfies $0 < \beta < 2$, the profile is fractals and the fractals are found using the slope β and the intercept with the vertical axis K:

$$D = \frac{4 - \beta}{2} \tag{2.87}$$

$$G = \left\{ 10^K \frac{(4 - 2D) \ln \gamma}{\Gamma(2D - 3) \sin[(2D - 3)\pi/2]} \right\}^{1/2D-2} \tag{2.88}$$

In practice, the principle is the same as for the so-called power spectral density (PSD) function except that in this case the structure function is plotted.

2.4.5.3 Other Proposed Methods

Other methods have been proposed:

1. The **cover method** is an original approach. Let r be the yardstick and $N(r)$ the repetitive measurement times each r. Suppose that $N(r) \times r$ is the length of the profile. For different r values, if there is a relationship between $N(r)$ and (r), the fractal dimension is expressed by [22]

$$N(r) \propto r^{-D} \tag{2.89}$$

2. The **variation method** uses the notion of δ variable to measure the variation of the profile in a δ neighborhood [17]. The order of growth of the δ variable is directly related to the fractal dimension of the profile.
3. The **reticular cell counting method** proposed by Gagnepain and Roques-Carmes [17] involves an iteration operation to an initial square whose area is supposed to be 1 and covering the entire graph. The method consists of dividing the initial square into four subsquares and then each subsquare into four subsquares, and so on. After n iterations, the initial square contains 2^{2n} subsquares. The number of subsquares containing the discrete profile are counted and the length L of the profile is approximately obrained. The calculating equation of the fractal dimension D from this method is then [22]

$$D = 1 + \frac{\log L}{n \times \log 2} \tag{2.90}$$

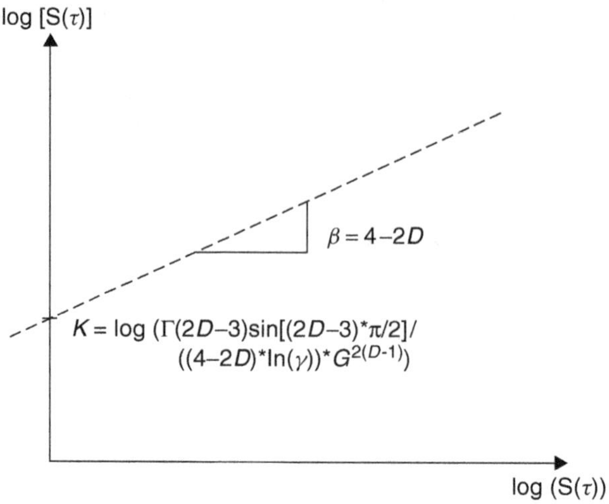

Figure 2.22. Log–log plot of the structure function and deduction of fractal parameters.

4. A **modified Gaussian fractal model** has been proposed by Zhou et al. [67] to derive equations to relate the bearing area curve with the fractal dimension D and the topothesy.
5. The **difference average law** (DAL) proposed by Jahn and Truckenbrodt [29] is based on an unusual power law for stochastically self-affine fractals. The authors state that the DAL procedure provides nearly the same relative fractal dimensions as the well-known methods (such as power spectrum) with smaller computation efforts.

2.4.5.4 Example

The idea is here to verify the feasibility of the power spectral density and structure function method:

1. A Weierstrass–Mandelbrot (WM) function is generated with fractal parameter $D = 1.55$ and amplitude parameter $G = 1e^{-12}$. Points of the profile $z(x)$ are stored and $z(x)$ is plotted in Figure 2.23(a) in dotted line.
2. The theoretical power spectral density and structure function are calculated from Eq. 2.84 and Eq. 2.84 and plotted in log–log, respectively, in Figures 2.23(b) and 2.23(c) in dotted lines.
3. From points of the profile $z(x)$, real power spectral density and structure functions are calculated and plotted in log–log, respectively, in Figures 2.23(b) and 2.23(c) in plain lines. The intersection with axis and slopes allows one to find fractal parameters

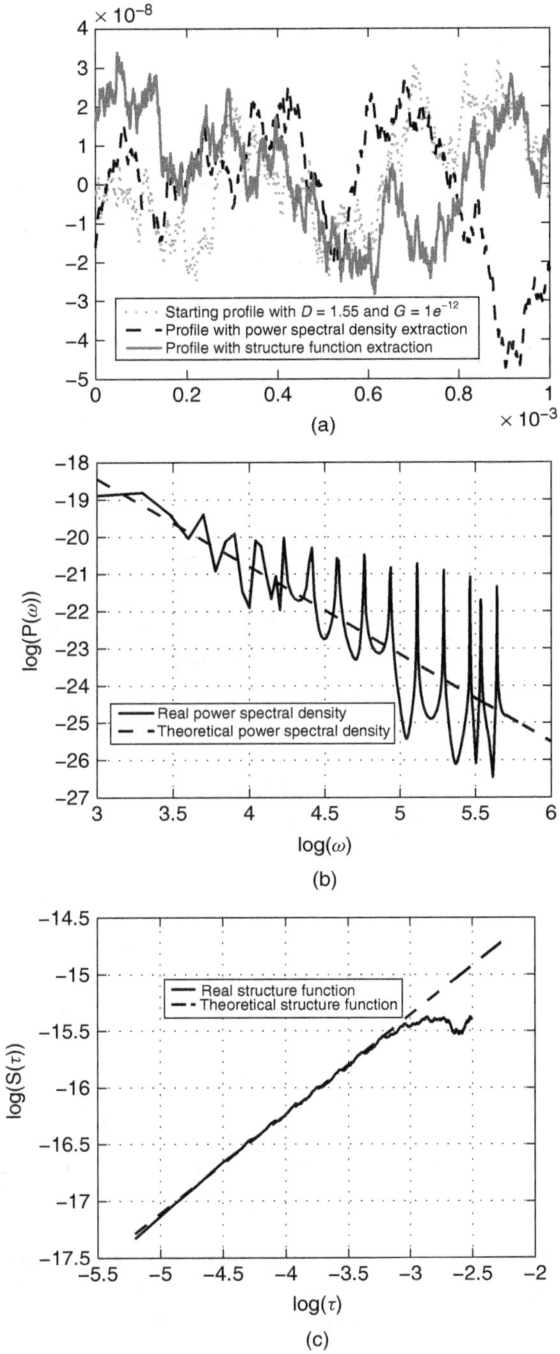

Figure 2.23. (a) Generated profile, (b) log–log plot of the power spectral density function, and (c) log–log plot of the structure function.

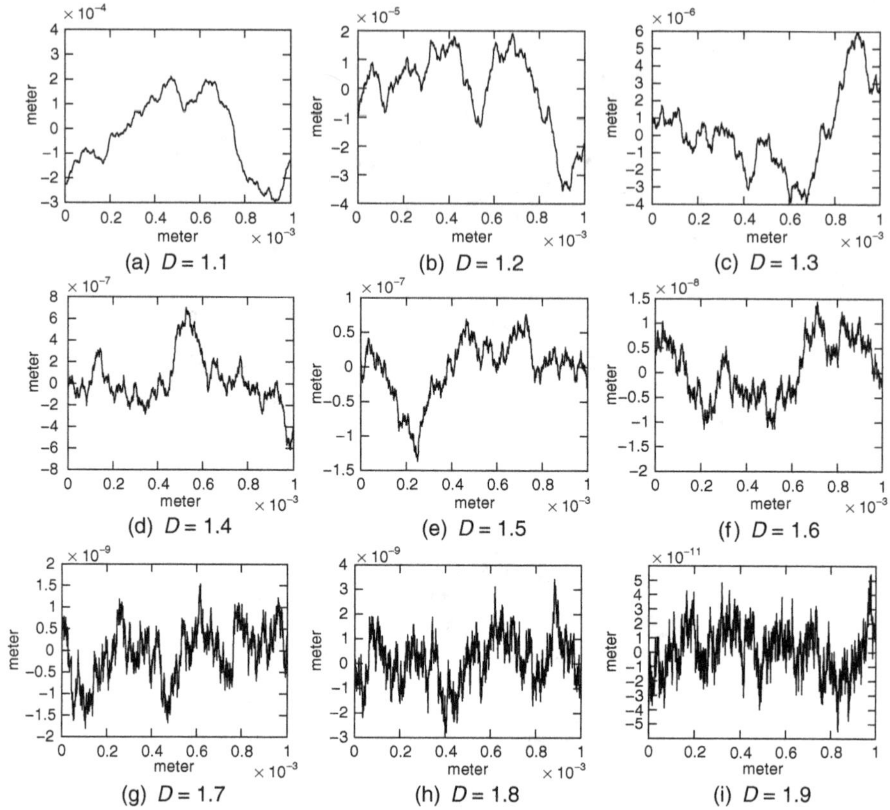

Figure 2.24. Weierstrass–Mandelbrot profile of 1 nm length, $G = 1e^{-12}$ and different values of D.

4. Finally the Weierstrass–Mandelbrot function is generated again with the new parameters and plotted in Figure 2.23(a) for comparison with the initial profile.

From the power spectral density function, the obtained parameters are $D = 1.32$ and $G = 2.39e^{-18}$. From the structure function, the obtained parameters are $D = 1.56$ and $G = 1.31e^{-12}$. This illustrates the fact that the structure function is a more accurate method for extracting fractal parameters of surfaces even if both methods are reliable.

2.4.5.5 Validity Domain for PSD and Structure Function Methods

Depending on the value of D, a method may be preferable to the other. It is the aim of this section to investigate this. A Weierstrass–Mandelbrot function is generated with a 1-mm length and $G = 1e^{-12}$ using the injected fractal parameter D (Fig. 2.24). Power spectrum and structure function are then plotted in order to extract the parameters. Extracted values are compared to injected values in order

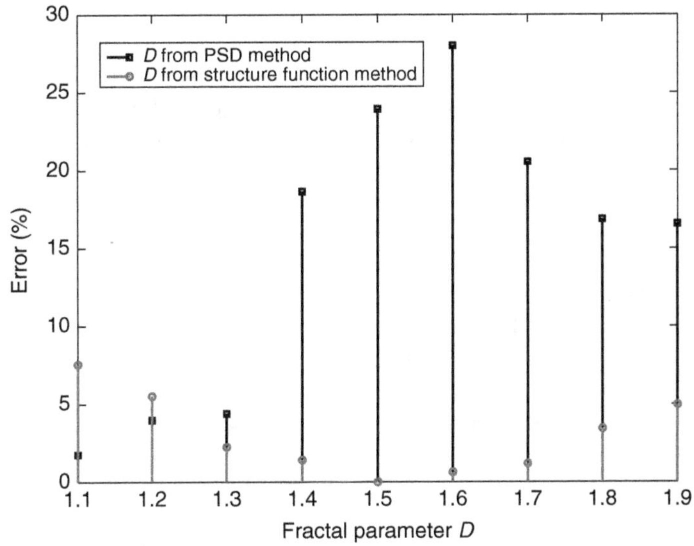

Figure 2.25. Errors on fractal parameter D evaluation depending on the extraction method.

to get the error on the extracted parameters. Results are plotted in Figure 2.25. The power spectral density method gives better results than the structure function for values of D less than 1.3 even if both methods stay reliable in this interval. For all other values the structure function method has a greater efficiency.

Power function methods and structure function method are the most widely used methods to:

1. Determine if a surface exhibits fractal properties.
2. Extract the fractal parameters from the profile.

The power spectrum function is calculated by transforming the discrete heights into the frequency domain. This results in an approximation. Structure function is calculated directly from the height information and results in a better approximation [18].

2.4.6 Conclusion

Different models have been reviewed. Most of them, however, use statistical parameters in their representation. Fractal representation has been chosen because fractal parameters do not depend on the measurement process and may be applicable to many engineered surfaces. The fractal character of surface finish for some microfabrication processes has also been demonstrated. Extraction of fractal parameters can be done with the so-called structure function method.

REFERENCES

1. Z. Adamczyk, Particle Adsorption and Deposition: Role of Electrostatic Interactions, *Adv. Colloid Interface Sci.,* 100–102:267–347, February 2003.
2. F. Arai, D. Ando, T. Fukuda, Y. Nonoda, and T. Oota, Micro Manipulation Based on Micro Physics, in Proc. of IEEE/RSJ Conf. on Intelligent Robots and Systems, Vol. 2, Pittsburgh, 1995, pp. 236–241.
3. G. M. Bell, S. Levine, and L. N. McCartney, Approximate Methods of Determining the Double-Layer Free Energy of Interaction between Two Charged Colloidal Spheres, *J. Colloid Interface Sci.,* 33:335–359, July 1970.
4. S. Bhattacharjee, M. Elimelech, and M. Borkovec, DLVO Interaction between Colloidal Particles: Beyond Derjaguin's Approximation, *Croatica Chem. Acta,* 71(4):883–903, December 1998.
5. B. Bhushan, *Handbook of Micro/Nano Tribology*, 2nd ed., CRC Press, Boca Raton, FL, 1998.
6. B. Bhushan, *Modern Tribology Handbook, Two Volume Set*, CRC Press, Boca Raton, FL, 2000.
7. B. Bhushan, Adhesion and Stiction: Mechanisms, Measurement Techniques, and Methods for Reduction, *J. Vacuum Sci. Technol. B,* 21(6):2262–2296, 2003.
8. M. S. Bobji and S. K. Biswas, Estimation of Hardness Nanoindentation of Rough Surfaces. *J. Mater. Res.,* 13(11):3227, 1998.
9. R. A. Bowling, A Theoretical Review of Particle Adhesion, in Proc. of Symposium on Particles on Surfaces 1: Detection, Adhesion and Removal, San Francisco, 1986, pp. 129–142.
10. J. A. Brant and A. E. Childress, Assessing Short-Range Membrane–Colloid Interactions Using Surface Energetics, *J. Membrane Sci.,* 203:257–273, 2002.
11. J. A. Brant and A. E. Childress, Colloidal Adhesion to Hydrophilic Membrane Surfaces, *J. Membrane Sci.,* 241:235–248, 2004.
12. G. Danuser, I. Pappas, B. Vogeli, W. Zesch, and J. Dual, Manipulation of Microscopic Objects with Nanometer Precision: Potentials and Limitations in Nano Robot Design, *Int. J. Rob. Res.,* 1998, and http://robotics.eecs.berkeley.edu/~zech/papers/IJR97.html.
13. P. G. de Gennes, F. Brochart-Wyard, and D. Quéré, *Gouttes, bulles, perles et ondes*. 2002.
14. B. V. Derjaguin, Untersuchungen über die reibung und adhäsion, IV. *Kolloid Zeitschrift,* 69(2):155–164, November 1934.
15. R. S. Fearing, Survey of Sticking Effects for Micro Parts Handling, in Proc. of IEEE/RSJ Conf. on Intelligent Robots and Systems, 5–9 August 1995, pp. 212–217.
16. F. M. Fowkes, Role of Acid-Base Interfacial Bonding in Adhesion, *J. Adh. Sci. Tech.,* 1(1):7–27, 1987.
17. J. J. Gagnepain and C. Roques-Carmes, Fractal Approach to Two-Dimensional and Three-Dimensional Surface Roughness, *Wear,* 109:119–126, 1986.
18. S. Ganti and B. Bhushan, Generalized Fractal Analysis and Its Applications to Engineering Surfaces, *Wear,* 180:17–34, 1994.
19. A. Gelb, S. Eichenlaub, and S. Beaudouin, Roughness Models for Particle Adhesion, *J. Colloid Interface Sci.,* 280:289–298, 2004.

20. R. J. Good, Theory of Adhesive Forces across Interfaces: 1. The Lifshitz–van der Waals Component of Interaction and Adhesion, in Lieng-Huan Lee, Ed., *Fundamentals of Adhesion*, Plenum, New York, 1991.
21. R. J. Good and C. J. Van Oss, *Modern Approaches to Wettability, Theories and Applications*, Plenum, New York, 1992.
22. M. Hasegawa, J. Liu, K. Okuda, and M. Nunobiki, Calculations of the Fractal Dimensions of Machined Surface Profiles, *Wear,* 192:40–45, 1996.
23. L. He and J. Zhu, The Fractal Character of Processed Metal Surfaces, *Wear,* 208:17–24, 1997.
24. M. C. Herman and K. D. Papadopoulos, Effects on the van der Waals and Electric Double Layers Interactions of Two Parallel Flay Plates, *J. Colloid Interface Sci.,* 136(2):385, 1990.
25. E. M. V. Hoek and G. K. Agarwal, Extended DLVO Interactions between Spherical Particles and Rough Surfaces, *J. Colloid Interface Sci.,* 298:50–58, June 2006.
26. R. Hogg, T. W. Healy, and D. W. Feurstenau, Mutual Coagulation of Colloidal Dispersions, *Trans. Faraday Soc.,* 62:1638–1651, 1966.
27. R. J. Hunter, *Foundations of Colloid Science*, Oxford University Press, 2001.
28. J. N. Israelachvili, *Intermolecular and Surface Forces*, 2nd ed., Academic, 1992.
29. R. Jahn and H. Truckenbrodt, A Simple Fractal Analysis Method of the Surface Roughness, *J. Mat. Proc. Tech.,* 145:40–45, 2004.
30. K. Komvopoulos, Surface Engineering and Microtribology for Microelectromechanical Systems, *Wear,* 200(1):305–327, 1996.
31. K. Komvopoulos, A Fractal Analysis of Stiction in Microelectromechanical Systems, *J. Tribol.,* 119(3):391–400, 1997.
32. M. Kostoglou and A. J. Karabelas, Effect of Roughness on Energy of Repulsion between Colloidal Surfaces, *J. Colloid Interface Sci.,* 171:187–199, 1995.
33. H. Krupp, Particle Adhesion Theory and Experiment, *Adv. Colloid Interface Sci.,* 1:111–239, 1967.
34. P. Lambert, A Contribution to Microassembly: A Study of Capillary Forces as a Gripping Principle, Ph.D. Thesis, Université libre de Bruxelles, Belgium, 2004.
35. F. F. Ling, Fractals, Engineering Surfaces and Tribology, *Wear,* 136:141–156, 1990.
36. A. Majumdar and B. Bhushan, Role of Fractal Geometry in Roughness Characterization and Contact Mechanics of Surafces, *J. Tribol.,* 112:205–216, 1990.
37. A. Majumdar and C. L. Tien, Fractal Characterization and Simulation of Rough Surfaces, *Wear,* 136:313–327, 1990.
38. B. B. Mandelbrot, *The Fractal Geometry of Nature*, W.H. Freeman, New York, 1982.
39. D. McCormack, S. L. Carnie, and D. Y. C. Chan, Calculations of Electric Double-Layer Forces and Interaction Free Energy between Dissimilar Surfaces, *J. Colloid Interface Sci.,* 169:177–196, January 1995.
40. P. Montmitonnet, F. Delamare, E. Felder, and J. Bucaille, *Cours de tribologie de la mise en forme*, Ecole des Mines de Paris, Centre de Mise en forme des matériaux, 2003.
41. C. A. Morrow, Adhesive Rough Surface Contact, Ph.D. thesis, University of Pittsburg, 2003.
42. H. Ohshima, Effective Surface Potential and Double-Layer Interaction of Colloidal Particles, *J. Colloid Interface Sci.,* 174:45–52, September 1995.

43. Q. Ouyang, K. Ishida, and K. Okada, Investigation of Micro-adhesion by Atomic Force Microscopy, *Appl. Surf. Sci.*, 169:644–648, 2001.

44. R. M. Patrikar, Modeling and Simulation of Surface Roughness, *Appl. Surf. Sci.*, 228:213–220, 2004.

45. J. Peirs, Design of Micromechatronic Systems: Scale Laws, Technologies, and Medical Applications, Ph.D. Thesis, KUL, Belgium, 2001.

46. T. Provder and B. Kunz, Application of Profilometry and Fractal Analysis to the Characterization of Coatings Surface Roughness, *Prog. Org. Coat.*, 27:219–226, 1996.

47. Y. I. Rabinovich, J. J. Adler, A. Ata, R. K. Singh, and B. M. Moudgil, Adhesion between Nanoscale Rough Surfaces I. Role of Asperity Geometry, *J. Colloid Interface Sci.*, 232(1):10–16, 2000.

48. J. E. Sader, Accurate Analytic Formulae for the Far Field Effective Potential and Surface Charge Density of a Uniformly Charged Sphere, *J. Colloid Interface Sci.*, 188:508–510, April 1997.

49. J. E. Sader, S. L. Carnie, and D. Y. C. Chan, Accurate Analytic Formulae for the Double Layer Interaction between Spheres, *J. Colloid Interface Sci.*, 171:46–54, April 1995.

50. M. Sausse Lhernould, Theoretical and Experimental Study of Electrostatic Forces Applied to Micromanipulation: Influence of Surface Topography, Ph.D. Thesis, Université libre de Bruxelles, 2008.

51. M. Sausse Lhernould, A. Delchambre, S. Régnier, and P. Lambert, Electrostatic Forces in Micromanipulations: Review of Analytical Models and Simulations Including Roughness, *Appl. Surf. Sci.*, 253:6203–6210, 2007.

52. S. Y. Shulepov and G. Frens, Surface Roughness and the Particle Size Effect of Slow Perikinetic Coagulation, *J. Colloid Interface Sci.*, 170:44–49, 1994.

53. M. Sitti. Atomic Force Microcope Probe Based Controlled Pushing for Nanotribological Characterization, *IEEE/ASME Trans. Mechatronics*, 9(2):343–349, 2004.

54. M. Sitti and H. Hashimoto, Teleoperated Touch Feedback from the Surfaces at the Nanoscale: Modelling and Experiments, *IEEE-ASME Trans. Mechatron.*, 8(1):1–12, 2003.

55. L. Suresh and J. Y. Walz, Effect of Surface Roughness on the Interaction Energy between a Colloidal Sphere and a Flat Plate, *J. Colloid Interface Sci.*, 183:199–213, 1996.

56. Surface profile parameters, available at: http://www.predev.com/smg/parameters.htm.

57. D. Tabor, Surface Forces and Surface Interactions, *J. Colloid Interface Sci.*, 58(1):2–13, 1977.

58. E. van der Wal, Bioactivity and Surface Reactivity of RF-Sputtered Calcium Phosphate Thin Films, Ph.D. Thesis, Utrecht University, 2003.

59. C. J. van Oss, *Interfacial Forces in Aqueous Media*, Taylor and Francis, Baco Raton, FL, 2006.

60. E. J. W. Vervey and J. T. G. Overbeek, *Theory of the Stability of Lyophobic Colloids*, Elsevier, New York, 1948.

61. B. Vogeli and H. von Kenel, AFM-Study of Sticking Effects for Microparts Handling, *Wear*, 238(1):20–24, 2000.

62. A. L. Weisenhorn, P. K. Hansma, T. R. Albrecht, and C. F. Quate, Forces in Atomic Force Microscopy in Air and Water, *Appl. Phys. Lett.*, 54(26):2651, 1989.

63. D. J. Whitehouse, Fractal or Fiction, *Wear,* 249(5–6):345–353, 2001.

64. D. J. Whitehouse, *Handbook of Surface Metrology*, Institute of Physics, London, 1994.

65. J.-J. Wu, Structure Function and Spectral Density of Fractal Profiles, *Chaos Solitons Fractals,* 12:2481–2492, 2001.

66. W. Wu, R. F. Giese, and C. J. van Oss, Stability versus Flocculation of Particle Suspensions in Water —Correlation with the Extended DLVO Approach for Aqueous Systems, Compared with Classical DLVO Theory, *Colloids Surfaces B: Biointerfaces,* 14:47–55, August 1999.

67. G. Zhou, M. Leu, and D. Blackmore, Fractal Geometry Modeling with Applications in Surface Characterisation and Wear Prediction, *Int. J. Mach. Tools Manufac.,* 35(2):203–209, 1995.

68. H. Zhu, S. Ge, X. Huang, D. Zhang, and J. Liu, Experimental Study on the Characterization of Worn Surface Topography with Characteristic Roughness Parameter, *Wear,* 255:309–314, 2003.

PART II

HANDLING STRATEGIES

CHAPTER 3

UNIFIED VIEW OF ROBOTIC MICROHANDLING AND SELF-ASSEMBLY

QUAN ZHOU and VEIKKO SARIOLA

3.1 BACKGROUND

Microhandling is a class of techniques for the operation of microscopic objects, which could be either artificial ones such as microfabricated parts or natural objects such as biological cells. Those techniques can include positioning, dissection, injection, aspiration, and the like. In the context of microassembly, microhandling refers to methods that manipulate the microscopic objects such that the objects will go to the destination in a desired manner, for example, placed at a target position either temporarily or permanently fixed. Robotic microassembly using microhandling techniques has been a topic of research for nearly two decades. An impressive set of techniques has been developed during the course of research, from specifically designed operation tools to various innovative handling strategies.

One lasting problem, which is still an essential problem to be tackled in today's microhandling methods, is the sticking between the operation tool and the parts under operation due to scaling down. This problem is often also referred to as the scaling effect, where the adhesion forces, consisting of van der Waals, electrostatic, and capillary forces, are dominant in comparison to the gravity and inertia. In macroscale handling, the releasing process can be taken as granted due to gravity and inertia; however, this is not the case in microscale.

Robotic Microassembly, edited by Michaël Gauthier and Stéphane Régnier

To perform microhandling successfully, such sticking problems have to be tackled, which is one of the reasons that a huge variety of microhandling techniques has been invented. This is in contrast to mini- or macroscale handling methods, where the study is largely focused on dexterity and efficiency. For robotic microhandling, efficiency is, of course, one of the primary goals, but many times the first objective is to make the process work reliably.

Despite this sticking problem, robotic microhandling does have many favorable properties, which is the fundamental reason that it has attracted so much attention. The main advantage of robotic microhandling is the capability and flexibility of robotics—if a robotic system is designed well, it can adapt to different tasks rather easily, and little or no modification to the mechatronics is required. Thanks to the advanced online and offline programming methods and control algorithms, the whole operation process, including the trajectory and the speed of both the tools and the objects, can often be reprogrammed.

By looking at the history of microassembly, it is obvious that robotic microhandling techniques have one competing technology—self-assembly. Self-assembly is a natural process for molecular structures and many examples can be found in nature. It is under intensive study in material science, physics, and biology. Self-assembly of microscopic parts is very similar to those natural processes in the sense that they obey the same principle of minimum potential energy; they are bottom up and massively parallel.

The properties of self-assembly almost contrast those of robotic microhandling. Self-assembly is driven by microforces that are a gradient of potential energy, with which robotic microhandling usually has to fight. The self-assembly process is massively parallel, in contrast to the serial nature of robotic microhandling. The process is working by design, and there is little or no reprogramming during the process, where robotic microhandling is reprogrammable by its nature. The self-assembly process is often stochastic, and in robotic microhandling the operation is fundamentally deterministic.

It seems that those two very different technologies of microassembly have little in common, except they share the same goal of positioning and/or assembling the microparts to well-defined locations. However, if we examine the processes of robotic microhandling and self-assembly, it is obvious that they still share the different phases of assembly process. In robotic microhandling, the microparts are fed by a feeding device, picked up by a tool such as a microgripper, and then moved to a desired position in a relatively high speed where the parts are then placed at a target site using various releasing techniques. In self-assembly, the microparts still need to be fed to the assembly process. Then the microparts are driven toward the receptor sites using various agitation methods and finally positioned and aligned to the receptor sites due to the principle of minimum potential energy.

It is very interesting to see how the phases of feeding, positioning, releasing, and fixing can be compared side by side for both robotic microhandling and self-assembly even though those phases might be very different in nature. Moreover, the rationales and philosophy behind the techniques of both are also very interesting and worth careful investigation. Besides those phases, many process parameters can also be discussed under the same framework. For example, what is the medium or the environment of the operations and what are the influences of the environment parameters to the processes?

In this chapter, both robotic microhandling and self-assembly will be analyzed. In the next section, various robotic microhandling techniques will be discussed. Section 3.3 will briefly discuss self-assembly techniques for microassembly. The components of both technological branches will be discussed and summarized in Section 3.4. Section 3.5 discusses possibilities to combine the components of the two branches to generate hybrid microhandling techniques, including a case study.

3.2 ROBOTIC MICROHANDLING

Robotic microhandling was under active research during the past two decades by research teams all over the world. The applications of robotic microhandling, as mentioned earlier, are mainly in two areas, biological applications and microassembly.

3.2.1 Microhandling System

The research in robotic microhandling started with the development of mechanical micromanipulators [20, 29]. One of the significant achievements in the early stage of robotic micromanipulation is the 6 degrees-of-freedom (DOF) parallel micromanipulator developed at that time by Mechanical Engineering Laboratory (MEL) [now the Agency of Industrial Science and Technology (AIST)] of Japan [52]. This micromanipulator is based on a Stewart platform driven by six piezoelectric actuators. Two-fingered operations using two such micromanipulators were implemented to achieve dexterous operation. Due to the use of piezoelectric actuators, the precision of this micromanipulator in teleoperation mode is very good even comparable to today's state-of-the-art teleoperated micromanipulators. Following that work, many other manipulator devices have been developed, such as a piezohydraulic micromanipulator [32] and a compact version of the two-handed micromanipulator from MEL [44].

In microassembly, it appears that microhandling systems based on precision positioning systems and microgrippers are more practical, where the working principle of those microhandling systems shares the basic design of macroscale robotic systems using motorized stages and a handling tool. Many commercial positioning systems can be applied in robotic microhandling, usually composed of precision mechanical rails and actuators based on various principles, such as direct-current (DC) motors, piezoelectric actuators, and piezoelectric motors. For applications that need only a small range, for example, tens of microns, compliant mechanics are often used to improve precision and reduce nonlinearity of actuators. DOF microtweezers or vacuum grippers are widely used as handling tools. In applications where dexterous manipulation is needed, multi-DOF

Figure 3.1. Typical microhandling system.

micromanipulators or microgrippers can be used as the handling tool, such as the two-handed micromanipulator, 4-DOF microgrippers [1] or 6-DOF microgrippers [59]. An example of a microhandling platform is shown in Figure 3.1, which includes a microgripper, two microscopes, and a positioning system.

To achieve automated microhandling, techniques such as motion control, visual servoing, and force control need to be applied. Some of those techniques can be borrowed directly from their macroscopic counterparts. However, many techniques need careful adaption to tackle the challenges in microassembly due to down-scaling, including the limitation in depth of view and the contradiction between resolution and field of view of optical microscopes, hysteresis and creeping properties of piezoelectric actuators, and the difficulties in implementing force control due to performance of the sensors and the limited space in microgrippers. Innovative engineering and advances in microfabrication and computing hardware have mostly overcome those problems and limitations. However, the basic problem of how to effectively perform microhandling in the context of adhesion forces remains. To discuss this, one has to check the state-of-the-art of microhandling techniques in the context of microassembly.

3.2.2 Microhandling Strategies

To achieve reliable microhandling, not mentioning efficiency, robotic microhandling needs good tools and/or strategies, such as the ones illustrated in Figure 3.2. A microhandling system composed of precision positioning stages, handling grippers, vision system, automatic control software only provides a platform on which microhandling can be done. However, such a platform cannot guarantee that microhandling can successfully be carried out without a properly designed handling tool or a handling strategy for a particular application.

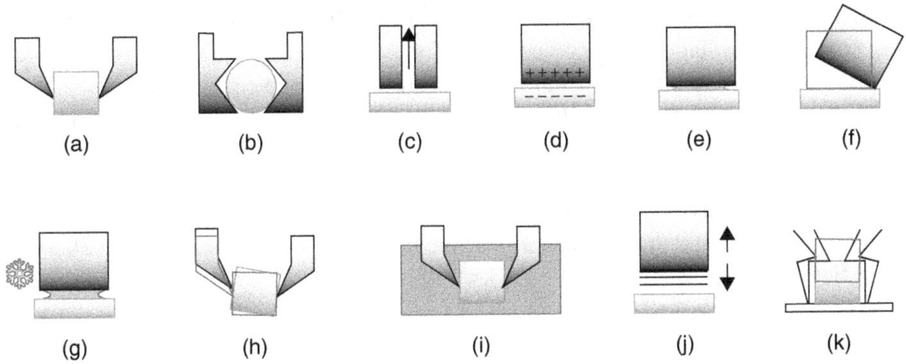

Figure 3.2. Different micromanipulation techniques: (a) Contact microgripper, (b) form closure microgripper, (c) vacuum gripper, (d) electrostatic gripper, (e) capillary gripper, (f) van der Waals gripper, (g) ice gripper, (h) collaborative manipulation, (i) submerged micromanipulation, (j) vibration release, (k) snap-locking fixing.

The problem of adhesion forces in robotic microhandling has been identified in multiple scientific papers, notable ones being by Fearing [16] and Arai et al. [3]. The understanding of those adhesion forces has greatly improved during the past decade, and a comprehensive description of this effect is discussed in Part I of this book. Those works provide a solid foundation for modeling and analysis of microhandling for many typical microhandling processes. The requirement to use those methods is that the tools and the parts under operation are of simple geometric primitives or can be approximated by simple geometric primitives. When the handling process involves more complicated scenarios, numerical algorithms have to be applied [49].

To tackle the adhesion forces, especially during the releasing process in microhandling, different microhandling strategies have been investigated. One commonly applied strategy is to fix the micropart under operation during the releasing process, such that the adhesion force between the tool and the part cannot impair the precise location of the part. However, applying this method limits the throughput of the system to the curing time of the adhesive, which may be from a few seconds to even a few minutes, depending on the adhesives and required precision. Ingenious designs have been developed to achieve reliable releasing where the adhesion forces between the part and the site is more significant than the force between the tool and part. The first work in microassembly applying this technique is probably by Feddema et al. [17], who alter the relative angle between a rod-shaped tool and a spherical part under operation, such that the effective contact area between the tool and the part is larger during picking and smaller during releasing. Consequently, the adhesion force between the tool and the spherical part is larger than the adhesion force between the part and the site when picking and is smaller than when releasing. The working principle is built upon the very range-sensitive property of van der Waals forces—when the angle between the tool and the part varies, the adhesion force between the tool and the part also varies while the force between the part and the site is constant. Saito et al. has developed a similar strategy, where the force in action is an electrostatic force in his particular setup in a scanning electron microscope (SEM) [45].

The idea of altering forces between the tool and the part has also been applied by many other researchers, even though the exact process may vary greatly. Many now so-called phase changing microgrippers [37] are based on this philosophy, beginning probably with the ice gripper [33], which freezes a small amount liquid between the tool and the part when picking and thaws the ice bond when releasing (see Section 5.3).

The force relations of tool–part and part–site can also be solved by various other methods. For example, a microhandling strategy using a capillary gripper, which picks parts by capillary force, has been developed [2]. For such a gripper, releasing can be achieved with a creative combination of capillary forces and inertia [35]. When the gripper picks the micropart, it accelerates slowly to keep the part attached to the gripper; when releasing, after the part is in contact with the surface, the gripper will retract quickly so that the sum of the inertia and the adhesion force of the micropart are stronger than the capillary force, and consequently the meniscus will break. Another example is a microfabricated gripper that uses controllable electrostatic charges for picking and releasing microparts [27].

Ambient environment conditions can have a huge impact on the force relations of the tool–part and the part–substrate interfaces. Temperature and humidity, especially, can significantly impact the force relation because van der Waals forces, electrostatic forces, and capillary forces are all functions of temperature and humidity [58, 61]. For a particular setup, the force relation of a microhandling process can be tuned by changing ambient environment parameters. This method will be more effective if the material properties and surface geometric properties are chosen such that the forces can be altered in a greater range.

Instead of changing the ambient air environment, the microassembly can be carried out in a different environment such as in water. In a water environment, the behavior of adhesion forces changes dramatically, where the van der Waals, electrostatic and capillary forces all decrease substantially (see Chapter 5). Consequently, the releasing problem caused by tool–part adhesion is reduced. However, the density and the viscosity of water are much larger than those of air. This affects the dynamics of operation and has to be taken into account.

Another slightly different alternative is the vibration release technique, which vibrates the tool at a relatively high frequency, so that the inertia of the micropart is greater than the adhesion force between the tool and the part (see Section 4.2.3). However, if the location of the releasing is above the target surface, the final location of the part after the vibration releasing process is stochastic. Therefore, vibration releasing methods where the releasing process is carried out when the part and the surface are in contact have also been proposed: By using the so-called squeeze effect [53], the randomness of the releasing process can be reduced.

Besides glue bonding and pure adhesion, other alternatives for fixing have also been invented. Snap-locking is one of the alternatives that uses a spring-guided locking mechanism to aid releasing [13, 40], where the mechanism is a microfabricated silicon structure (see Chapters 6 and 7). The assembled part will be snap-locked to the release site and the process becomes rather deterministic. This is a very effective strategy, provided that the releasing sites and the assembled parts can be designed to use such a snap-locking mechanism.

Many other locking mechanisms based on local microforces have also been investigated, for example, using capillary force [60], electrostatic force [27] or geometric constraints to aid the releasing. Those mechanisms can effectively solve the problems in releasing and will be investigated further in the latter part of this chapter.

There are many other microhandling techniques tackling adhesion forces using different physical principles. For example, surface acoustic waves can trap microparts into certain patterns [41], optical tweezers can use optical pressure to trap and move microparts in liquid [6], magnetic field can move micropart by either controlling the strength of the magnetic field of different coils [23] or resonate the micropart using alternative fields [18, 19], dielectrophoresis can transport microparts based on the nonhomogenous electric field strength [5] (also see Section 5.2). In addition to their application in microhandling, those methods can be used in the feeding process of microassembly. However, feeding techniques will not be pursued in depth in this chapter, where the emphasis is placed on the handling processes that are directly related to the assembly of the micropart.

In summary, robotic microhandling techniques today can already tackle the sticking problem quite well. That does not mean the problems brought by the adhesion forces in microhandling are gone. In fact, one of the principal problems that still persists in robotic microhandling is the trade-off between efficiency, reliability, and precision. Industrial electronic assembly machine can already assembly submillimeter microparts in tens of milliseconds with high-speed robotic positioning system and vacuum grippers. However, if the requirement of the precision of the assembly is relatively high, for example, in a few micrometers or better, the speed has to be regulated to guarantee the reliability and precision of the process. Moreover, when the size of the micropart goes even smaller, the effect of adhesion force will further impact this trade-off and make the robotic microhandling economically not viable except for research and development and for products with very high added value.

3.3 SELF-ASSEMBLY

Self-assembly has a long history in material science, chemistry, and biology. However, the application of self-assembly in microassembly starts around the beginning of 1990s. Examples of early work include using curved surfaces and geometric shape recognition aided by gravity and vibration excitation for assembly of miniaturized parts [11] and the so-called fluidic self-assembly that uses geometric shape recognition aided by gravity and fluidic agitation to assemble GaAs

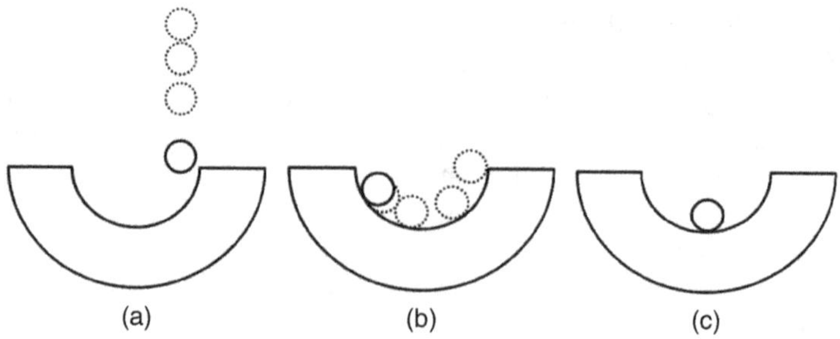

Figure 3.3. Illustration of self-assembly: a ball and a cup. (a) A ball is released to a cup and potential energy converts into kinetic energy. (b) The ball is attracted toward the energy minimum, which is in the bottom of the cup. (c) Slowly, the kinetic energy is dissipated as heat and the ball settles to the center of the cup.

parts on substrates [56]. Those initial works had only limited success, mainly because the yield was not very high. However, self-assembly techniques were refined during the past two decades. Today, the principles of fluidic agitation, vibration excitation, and pattern matching have become widely accepted concepts in self-assembly of microparts. Many self-assembly techniques for microparts have been developed, for example, two-dimensional self-assembly using capillary force[1] based on the hydrophilic-hydrophobic patterned surface [50], multibatch fluidic self-assembly [55], out-of-plane solder self-assembly [25], and three-dimensional (3D) self-assembly [10].

3.3.1 Working Principle

The fundamental working principle of microassembly using self-assembly techniques is the principle of minimal potential energy. An example of the effect of the principle in nature is that water on Earth flows downstream toward lower altitudes and forms flat surfaces in lakes and seas due to gravity of Earth. The principle of minimum potential energy can also be illustrated using the example of a ball falling into a cup (see Fig. 3.3).

In the case of self-assembly of microparts, the gravity potential can still work to a certain extent, but many other potentials, especially surface energy, are used due to the fact that gravity becomes relatively insignificant when the parts are scaled down.

The principle of minimum potential energy states that an object such as a micropart should move toward states where the total potential energy of the system is smaller. This is easy to understand in the case of gravity. In self-assembly of microparts, the surface energy is much more important than in the assembly of macroscopic parts. Therefore, in the process of the minimization of total potential energy of the system, it is often the surface energy that dominates the whole process. For example, in the case of droplet self-assembly in air (see

[1]Capillary forces are detailed in Section 1.2.2.

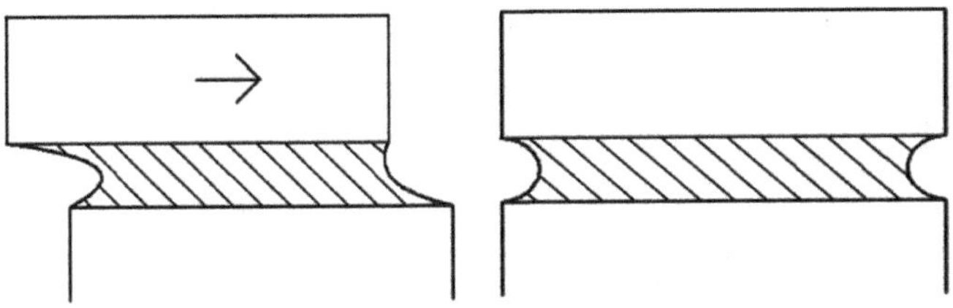

Figure 3.4. Illustration of capillary self-alignment. A meniscus is formed between a micropart and a receptor site. The energy minimum is found when the part is aligned to the receptor site.

Fig. 3.4), minimization of the total air–liquid surface and maximization of the liquid–solid surface of the capillary meniscus between a micropart and a receptor site is the principle driving force when those surfaces are hydrophilic. In fluidic phase self-assembly in water, the case is similar but the surfaces with matching pattern should be hydrophobic and a hydrophobic droplet, for example, adhesive, should be applied at the receptor site [50].

One of the challenges to be solved in all self-assembly is that the microparts might not go to the desired position due to friction and intermediate energy states—in other words, the system is stuck at a local minimum instead of reaching the global minimum of the system. To overcome such local minima, excitation such as vibration or stirring of the fluid (in the case of fluid-phase self-assembly) are applied to help the microparts to overcome those "bumps" and go forward to the global minimum of the system. Overview of the stochastic self-assembly process is illustrated in Figure 3.5.

To carry out the self-assembly of microparts, first the process should be carefully designed such that the potential energy when the parts mate the receptor sites should be smaller than the potential energy when parts are in other locations, such as in fluid or mating with other surfaces of the process. Then the parts should be fed into the system and moved to the receptor sites in a stochastic manner, agitated by, for example, stirring or vibration. When the parts are near the receptor sites, they are driven by the local potential gradient, for example, gravity or capillary force and hopefully achieve the desired assembly.

3.3.2 Self-Assembly Strategies

Many innovative self-assembly strategies have been developed since the introduction of self-assembly techniques in microassembly. Even though gravity is relatively weak at microscale, it has been successfully applied in self-assembly of millimeter microparts using shape recognition [11]. Using shape recognition and gravity, even submillimeter microparts of different shapes can be self-assembled [56]. Selective self-assembly of silicon circuit components on to a plastic substrate has been demonstrated in a solution [51], where components of different shapes can find their complementary receptor sites.

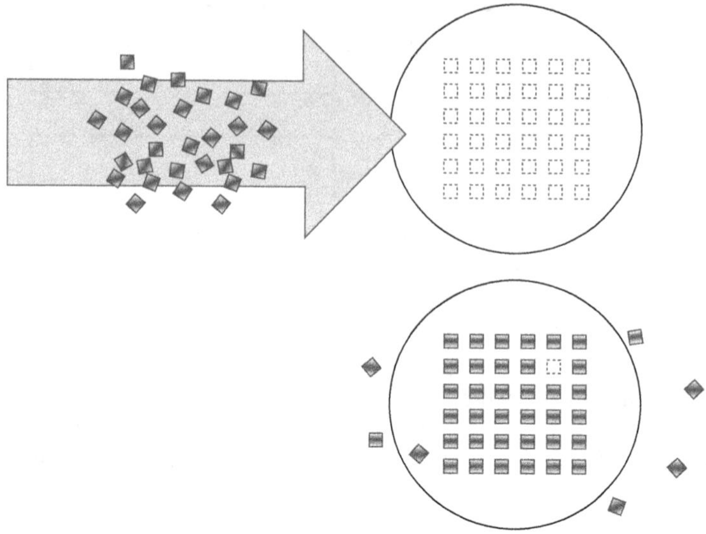

Figure 3.5. Stochastic self-assembly. Microparts are fed close to the receptor sites. After agitating the parts, most receptor sites are filled by microparts. There is always a possibility of error (e.g., one receptor site not filling, as shown in the figure) and usually more microparts than receptor sites are used.

To create a steeper gradient than the geometric shape recognition at microscale, a capillary-driven self-assembly strategy using patterned hydrophilic/hydrophobic surfaces in water has been successfully demonstrated [50]. The receptor and the bottom of the micropart are hydrophobic and the receptor sites are covered with adhesive. Therefore, the system reaches the state of minimal potential energy when the bottom of the micropart comes into contact with the receptor. Many other similar approaches have been proposed to refine the process.

The benefit of self-assembly of microparts in fluid is that the microparts can be agitated using fluid flow. However, fluidic phase self-assembly does have problems due to the pre- and/or postprocess steps required to achieve the assembly. Moreover, sometimes liquid-phase self-assembly is not compatible with the microparts to be assembled.

Self-assembly of microparts can also be carried out in air. Researchers have developed self-assembly methods that use a combination of two self-assembly techniques: geometrical shape recognition and capillary self-alignment [15]. This is a quite interesting technique because actually a two-stage positioning approach is used in this technology. First, the parts are positioned using a geometrical pattern matching technique excited by vibration. After the parts are in their rough positions, water steam is directed near the parts, so that water is condensed between the parts and substrate. Consequently, the parts are aligned by capillary self-alignment because the receptor site and the bottom of the parts are hydrophilic.

Assembly of microparts to relatively complicated structures can be done with self-assembly in multiple batches. The fundamental idea in those methods is to activate only certain interactions during each batch, so that components can be assembled into particular receptor sites. Multiple batches can then be used in different components. Xiong et al. [55] used this method in droplet self-assembly in liquid. The selectivity is achieved by electrochemically deactivating certain receptor sites in different phases. In Higuchi et al. [28], temperature is used to change the phases of various adhesives from solid to liquid or vice versa to achieve sequential capillary self-assembly. Sequential multibatch self-assembly can also be done with shape recognition and solder [57].

Most of the self-assembly of microparts are carried out on a planar surface. However, 2.5D self-assembly of microparts is also possible. Self-assembly techniques that assemble out-of-plane structures, techniques that assemble microparts on curved surfaces, and techniques that achieve stacked structures can be categorized as 2.5D. In solder ball self-assembly, capillary forces of reflown solder pull hingelike microstructures out of the plane of the substrate [25]. Silicon segments can be self-assembled on a flexible, curved support using capillary interactions and pattern matching [30]. Packaging of microcomponents, which achieves a semi-3D stacked structure, has been carried out with sequential self-assembly [57].

True 3D self-assembly has been achieved for only relatively simple structures. Examples range from crystal-like structures [54] to 3D electrical networks [10]. Reliable and reproducible 3D self-assembly of microparts is still to be concluded.

3.4 COMPONENTS OF MICROHANDLING

From the above discussion of robotic microhandling and self-assembly technologies, we can find that both branches involve various phases of the operation: After feeding, the parts need to be transported to position near their target site, and the parts should then be aligned and fixed with the receptor. Moreover, many system parameters need to be considered for both branches, such as the environment of the process, surface properties of the process, and external excitation of the process. In the following, those phases and system parameters are discussed to draw a side-by-side comparison of both technologies.

3.4.1 Feeding

Both robotic microhandling and self-assembly need to feed the parts to be assembled into the process.

3.4.1.1 Feeding in Robotic Microhandling

In robotic microhandling, the parts can be fed in a deterministic manner using, for example, trays or tapes on which the parts are placed in a certain pattern. However, parts can also be fed in a stochastic manner using a vibration feeder.

Vibration feeding can be used for parts down to the submillimeter range, after which the process becomes erratic due to the ratio of inertia of the micropart to adhesion forces being too small.

Feeding of microparts in robotic microhandling is an important aspect to be considered because the surface adhesion between the feeder and the micropart can impair the reliability of the system if not carefully designed. Otherwise, picking of microparts usually is not an issue in robotic microhandling because one can usually apply sufficient gripping force.

3.4.1.2 Feeding in Self-Assembly

In general, feeding of microparts in self-assembly is less demanding because neither the position nor the adhesion of the microparts is critical to the system. Fluidic self-assembly can be fed by transporting a droplet containing the parts with a simple pipette. Of course, the adhesion force between the microparts themselves can be an issue if the system is not designed properly. This, however, should have already been taken into account when the system is designed.

On the other hand, some self-assembly techniques require the parts to have a specific orientation before the self-assembly. For example, in the work done by Fang and Bohringer [15], the feeding is not so trivial. First, the silicon parts with hydrophobic sides are placed in water. By agitating the water, the parts are brought floating on the surface with their hydrophobic side oriented upward. After this, the parts are carefully picked with a wafer, so that they maintain their orientation when adhering to the substrate.

3.4.2 Positioning

3.4.2.1 Positioning in Robotic Microhandling

Positioning in robotic microhandling is straightforward as long as the range and the motion profile are inside the performance envelope of the positioning system (see Chapter 8). Due to the often high precision requirements of microassembly, two-stage positioning strategy is widely used where a coarse positioning subsystem can cover a large range of motion in often higher speed but lower precision, and a fine position subsystem moves the micropart in a smaller range but in a more precise manner. For example, a DC motor-driven positioner is often used for large range motions and piezoelectric actuators for precision displacements. Another example is piezoelectric motors, which can operate in stepping mode when high speed and large range are needed and in the so-called scanning mode where the motor is working in the deformation range of the piezoelectric actuator, which provides much higher resolution.

3.4.2.2 Positioning in Self-Assembly

In self-assembly, the positioning is often done in one phase after the releasing of the microparts to the process by the feeding device. The positioning of the microparts is achieved by a global force, such as fluid agitation or gravitational force. However, in some designs of self-assembly, the positioning process is actually also two stage. For example, in the work of Fang and Bohringer [15], the microparts are fine-positioned by a capillary self-assembly process after coarse geometry matching. In that example, the fine positioning is also the alignment strategy, as discussed in the following text.

3.4.3 Releasing, Alignment, and Fixing

It seems a bit tricky to compare the phase of the final stages of the microassembly process for both robotic microhandling and self-assembly. But if we check more carefully, we can see that the situation is actually very interesting.

3.4.3.1 Releasing and Alignment in Robotic Microhandling

For robotic microhandling, the final phase of the microassembly process after fine positioning is the releasing of the part. To deal with the adhesion forces, different strategies are invented to ensure reliable releasing with good precision. When throughput is not critical, releasing after fixing can be applied where the part is first bonded to the target site before the microhandling tool releases the micropart.

To aid the releasing, various releasing tools and strategies have been developed, as discussed earlier in this chapter. Those releasing strategies, including phase change, dynamic capillary gripper, vibration releasing, and electrostatic releasing do make the releasing easier. However, the positioning accuracy of the microparts is still relying completely on the fine positioning process of the robotic system. Often the releasing process makes the accuracy even worse.

One very interesting technique that aids the releasing of microparts is using traps, such as the previously mentioned snap-locking mechanism based on spring forces [13] or droplet-self alignment based on capillary force [14, 60]. Those processes actually use potential trapping based on the very same principle of minimal potential energy that is applied in the self-assembly processes. This class of releasing processes is what we refer to as the hybrid microhandling process, where self-assembly and robotic microhandling merge.

3.4.3.2 Releasing and Alignment in Self-Assembly

In self-assembly, the final phase of the microassembly process is self-alignment based on the principle of minimal potential energy. In the case of geometric pattern matching, either gravity or capillary forces will do the work. To work reliably, the shape of the potential well must be carefully designed.

Efforts to model self-assembly usually concentrate on modeling the shape of the potential well [9]. Different shapes such as circles, squares, polygons, and spirals have been investigated. Simple approximations of surface energy can be calculated by finding the overlapping area of the part and the receptor site. With such models, the potential energy can be plotted as a function of displacements and orientations and the graph can be used to find if there are local minima, which may cause undesired results in self-assembly.

Experimentally, different shapes and surfaces have been used to examine the influence on the results of self-assembly [39]. In the case of capillary self-alignment, the amount of liquid between the micropart and the receptor site also influences the precision of the assembly [14]. Moreover, the accuracy and alignment also depend on the fabrication accuracy of the microparts and receptor sites.

3.4.3.3 Fixing

Finally, in both robotic microhandling and self-assembly, the parts need to be fixed. This depends very much on the application. In some cases, the parts are already fixed before releasing or the fixing is not necessary, for example, the trapping force is sufficient or the fixing should be temporary and the part will be removed soon in applications such as micropart inspection. On the other hand, the releasing site can be coated with adhesive as in many cases of self-assembly and robotic microhandling, where successive curing or bonding can be carried out. In some applications, the postprocess curing is combined with the self-assembly process, where the self-assembly is carried out in warm liquid that keeps the adhesive in liquid phase and when the self-assembly is finished, the system is cooled down and the adhesive solidifies [51].

3.4.4 Environment

The ambient environment of microhandling can influence or determine the effectiveness of microforces and surface properties, and consequently limits what kind of strategy will be effective. Therefore, this is the basic element to be considered when a microhandling process should be designed, for both robotic microhandling and self-assembly, where the environment can be vacuum, gaseous, and liquid.

3.4.4.1 Ambient Environment for Robotic Microhandling

The most common environment for robotic microassembly is air, while biomanipulation is often carried out in liquids. Robotic microassembly has also been done submerged, where the microhandling process takes place in water even though the robotic system is mostly in air [21]. Microhandling can also be carried out in vacuum, usually in an SEM chamber. However, robotic handling in SEM is mostly for nanohandling, the cousin of microhandling, where submicron down to a few nanometer-sized components such as nanofibers or carbon nanotubes are manipulated.

In air, the ambient temperature has significant influence on both the transducers of the robotic microhandling system, as well the adhesion forces. For example, the maximum displacement of a piezoelectric actuator can change about 1.5% when humidity varies in the range of 10% relative humidity (RH) to 80%RH [61]. High humidity can also cause device failure, for example, in the case of electrostatic microgrippers [26]. As we discussed earlier, the motivation for many microhandling strategies is to tackle the adhesion forces. However, all three major components of the adhesion forces, namely van der Waals, electrostatic, and capillary forces, are functions of temperature and humidity [61]. Moreover, the ambient temperature and humidity can also greatly influence surface tribological properties [42]. Therefore, an environment control system for microassembly is beneficial [58].

In the case of water medium, the ambient environment is also very important. Due to the high thermal capacity and conductivity of the liquid, the influences will be much more direct and fast. In the case of the submerged ice microgripper, the performance is directly influenced by the water temperature. The water temperature has to be kept close to the freezing point to improve the efficiency of microhandling (see Section 5.3).

In the case of the vacuum environment, the ambient environment is a fuzzy concept. The lack of medium affects various things: (1) the capillary force of water, which has a strong contribution to adhesion in air, is nonexisting; (2) because there is no medium, van der Waals forces could be stronger; (3) because vacuum environment is often in SEM, where there will be a lot of electrostatic charges, electrostatic interactions will be more significant; (4) thermal conduction through tool and substrate is the major method of heat transportation, and radiation is usually insignificant; proper system design is important to avoid heat buildup and its consequences.

3.4.4.2 Ambient Environment for Self-Assembly

The ambient environment is also extremely important for self-assembly because the potential gradient that the self-assembly relies on is often a function of the ambient environment.

Self-assembly often takes place in water or various other liquids, such as ethylene glycol, where the local potential or the surface interaction are easier to design than in air and vacuum because more options are available [50]. One important environmental parameter is temperature, because often the curing of adhesives is triggered by changing temperature [28, 51].

However, self-assembly in vacuum and air has also been reported [8]. Recently, the research of self-assembly methods in air has been active because the relative humidity of the ambient environment is easier to vary [15]. For example, by introducing water in the form of a high-humidity airstream, capillary self-assembly can be triggered, which otherwise will not happen in a dry environment.

In other words, the ambient environment in self-assembly can be used not only to influence the process, but also as a control signal to trigger the self-assembly process, which is very interesting.

3.4.5 Surface Properties

Surface properties are a very import design aspect in microassembly, both for the robotic microhandling approach and self-assembly.

3.4.5.1 Surface Properties in Robotic Microhandling

In robotic microhandling, the surface properties between the tool–part interface and the part–receptor interface are critical design parameters. Many microhandling strategies are established on the basis of carefully determined surface properties. The tool angle variation pick-and-place technique by Feddema et al. [17] is based on the fact that van der Waals forces[2] are very distance sensitive, so when the effective contact area between the tool and the part is smaller, the van der Waals force between the two will decrease as well.

Surface roughness can be used to reduce van der Waals forces. The effective range of these forces is around 100 nm, which means that if the surface roughness is large enough, the van der Waals forces will be greatly reduced. Consequently, many microgripper designs choose large surface roughness for the tool tips to reduce the sticking problems between the tool and the part [4]. In some applications, materials with a low Hamaker number, and consequently smaller van der Waals forces, are used [12].

To reduce electrostatic effects, conductive surfaces are often used for the tips of the handling tool [4]. To reduce the effect of capillary force from water condensation during microhandling, a hydrophobic surface can also be applied [12].

It is also important that the surface properties between the part and the target site are considered together with the surface properties of the tool and the part. The adhesion force between the part and the target site should be stronger than the tool–part adhesion, unless fixing is applied before releasing. The adhesion between the picking site and the part should also be reasonable to avoid the requirement of excessive picking force, which could make the process unreliable or even damage the part.

One practical example, where surface properties are actively controlled, is the commercially available gel surface with vacuum release tray [22]. In this mechanism, an adhesive gel surface is located on a fine wire mesh. By applying vacuum under the gel surface, the surface will take the shape of the mesh and become rougher, which in turn reduces the total adhesion force considerably.

3.4.5.2 Surface Properties in Self-Assembly

In self-assembly, the surface properties are the fundamental design parameters of the system. When the self-assembly is carried out in water or in another fluid, the pattern and the hydrophilic and hydrophobic properties of the receptors as well as the parts to be assembled are the critical parameters of the process. Moreover, if the orientation of the parts after assembly should be defined, the surface properties of the microparts have to be different on different facets. The surface properties on the facet that will mate to the receptor site should be patterned such that the potential gradient will lead to a desired self-assembly. On the other hand, the surface properties of the parts should be designed to make the chance of part–part adhesion minimal. The surface properties when combined with shape recognition can also lead to orientation-controlled self-assembly [51]. Similarly,

[2] Van der Waals forces are detailed in Section 1.2.1.

self-assembly in air can combine capillary self-alignment with shape recognition [15].

Surface roughness is another parameter that should be taken into account in self-assembly. Rough surfaces can be used to make super hydrophobic surfaces [31]. However, a rough surface can also be undesirable because that will increase the fiction force and hinder the self-assembly to accomplish desired precision.

Selective binding of microparts by functionalized surfaces is also under active research. Electrochemical methods have been used to control hydrophilicity of the binding site, by applying voltage to specific binding sites [55]. Another possibility is to use temperature to activate specific binding sites [28]. A potentially more powerful technique to achieve selective binding is using DNA (deoxynibonucleic acid) as the recognition mechanism. DNA is especially attractive because of its specificity and the availability of tools to engineer it. There has been attempts to achieve self-assembly of microcomponents using DNA-functionalized surfaces; however, so far, this is still a work in progress [34].

3.4.6 External Disturbance and Excitation

Due to the requirements of often high precision in microassembly, external disturbances are usually unwanted. This is also related to the general consideration of ambient environmental conditions. However, it is interesting to note that external excitations are sometimes actually beneficial. This can be found in both cases of robotic microhandling and self-assembly.

3.4.6.1 External Disturbance and Excitation in Robotic Microhandling

The robotic microhandling system requires not only a relatively constant ambient environment, for example, temperature and humidity, but also reduced external disturbance from vibration, particle contamination, as well as electromagnetic noises.

The positioning precisions of robotic microhandling systems, especially the ones with serial kinematic configuration, are in general quite sensitive to vibration. Using rigid kinematic structures, such as parallel kinematics, can reduce the effect of vibration and mechanical deformation in steady state and during motion. However, the common practice to reduce vibration disturbances is to install the system on a proper vibration damping system, such as a vibration isolation table.

Robotic microhandling can also be hampered by dust particles, which not only cause problems in manipulation and positioning of the microparts, but also deteriorate the precision and life expectancy of precision positioning stages. This disturbance can be reduced by installing the microhandling system inside a clean room. Alternatively, the system can be installed inside a smaller chamber with clean air filtering, while the operator is in a normal room environment.

Other sources of disturbances in robotic microhandling include electric, electromagnetic, and optical disturbances that increase measurement noise and consequently reduce the precision of the closed-loop positioning systems. Electric and electromagnetic disturbances can directly increase measurement noise through the interference on the sensors or the cables. Light can, for example, disturb an optical encoder by giving a false signal to the reading head. To reduce those disturbances, proper cable shielding, noise reduction circuit design, Faraday cage, or dark box should be applied.

However, some forms of excitation are actually deliberately introduced in the robotic handling. For example, vibration is used in dynamic releasing strategy where the tool vibrates [24]. Also impulses can be used so that the part adhered on the tool is released when the inertia of the part is greater than the adhesion forces between the tool and the part [35].

3.4.6.2 External Disturbance and Excitation in Robotic Self-Assembly

The self-assembly process relies on the principle of minimum potential energy and external disturbances that help the process to overcome local minima to reach the desired results.

In self-assembly, dust particles and other contamination of the parts and the substrate are an obvious problem. The contamination can introduce local minima or deteriorate the driving potential gradient of the receptor site. Thus, cleaning methods are very important in self-assembly and cleaning methods from well-disciplined microfabrication processing are often used. On the other hand, the effect of other external disturbances is not much discussed in self-assembly.

As one of its key techniques, external excitation is very important in self-assembly. For fluid-phase self-assembly, the external disturbance can be fluidic agitation or vibration agitation that helps the microparts overcome the local minimum such as friction and part aggregation [50]. For both fluid-phase self-assembly and dry-phase self-assembly, vibration excitation can be used, including periodic excitation or impulse excitation [8].

3.4.7 Summary and Discussion

Even though there are great differences in robotic microhandling and self-assembly, they share many common components when applied to microassembly. In both cases the assembly process requires that similar steps and similar physical principles be applied. Both require feeding, positioning, releasing, and fixing of the microparts. Furthermore, both have to consider ambient environmental conditions, surface properties, external disturbances, and excitations due to the scale of the parts under manipulation and the precision requirements of microassembly. A summary of the components in both robotic microhandling and self-assembly in the context of microassembly is shown in Table 3.1.

In a previous publication of the authors [46], the relation between robotic microhandling and self-assembly has been analyzed in a different manner. Instead of phase analysis of the microassembly process (feeding, positioning, releasing, alignment, and fixing) in this chapter, a more technologically oriented approach is applied (potential trapping and hierarchical positioning). Meanwhile, similar aspects of design consideration are addressed (ambient environment, surface properties, disturbance rejection, and external excitation). In this chapter, a more

TABLE 3.1. Summary of Different Components of Robotic Microhandling and Self-Assembly

Component	Robotic Microhandling	Self-Assembly
Feeding	Component trays, vibration feeders	Fluid transport, preprocess palletization for unique orientation
Positioning	Precision positioning stages, micromanipulators	Fluid agitation, gravity, and vibration
Releasing and alignment	Release after bonding, phase change, dynamic capillary gripper, vibration releasing, electrostatic releasing, snap-locking, capillary forces	Shape recognition with gravity or capillary forces
Fixing	Adhesive	Adhesive
Ambient environment	Air, liquid, vacuum, temperature, and humidity effects	Air, liquid, vacuum, temperature, humidity, chemical composition of the solution, environment triggered self-assembly
Surface properties	Conductive, rough and low-adhesion tip surface; tool–part adhesion; part–target part–source adhesion; switchable surfaces	Hydrophilic/-phobic patterns; geometric pattern; selective surfaces; switchable surfaces
External disturbance and excitation	Vibration isolation; clean environment; electromagnetic shielding; vibration release	Clean environment; fluid agitation, vibration agitation, change of solution

side-by-side comparison of different phases is used in the hope of giving a more general comparison of both technological branches. However, the two previously identified technological components, namely potential trapping and hierarchical positioning, are very important for planning and designing microassembly processes.

The side-by-side comparison of robotic microhandling and self-assembly shows that both technologies share (a) similarity in process phases, (b) technologies based on similar physical principles, and (c) similar aspects to be considered in planning and designing of microassembly processes.

3.5 HYBRID MICROHANDLING

Based on the analysis, a natural question arises: Can we combine different technological components of robotic microhandling and self-assembly to achieve better results? One obvious advantage of self-assembly is the application of the principle of minimum potential energy, which can be used to solve the sticking problems in robotic microhandling. On the other hand, the flexibility of robotic microhandling is a desired feature of a microassembly process. Other nice features such as parallel operation, external excitation, and necessary consideration in environment and surface properties will help us design competitive microhandling strategy for different applications.

It is a widely used philosophy to combine good properties of different technology to achieve the so-called hybrid technology. The same is true for microhandling technology. If we check carefully the existing microhandling technologies, it should be possible to find that such combination of the components of robotic microhandling and self-assembly does exist already in the literature, for example, centering using capillary forces [7], capillary-force-assisted releasing [14], snap-locking [13, 40], and centering using electrostatic forces [27]. This is also natural because researchers are always trying to solve the technical problems using all the available means. However, the approach of creating such a hybrid technology was not clearly identified until recently (2006) [60].

There are many possibilities to carry out this combination and explore innovative technologies. However, two of them are obvious: (1) Robotic microhandling can help improve the yield of self-assembly by correcting the stochastic errors of the self-assembly process, and (2) self-assembly can help increase the reliability and performance of robotic microhandling by the introduction of specifically designed part–receptor and part–part interactions. While these possibilities are similar to either end of the spectrum with a small flavor of the competing technology, it can be foreseen that truly hybrid technologies, which cannot be classified into either branch, are possible.

To illustrate the idea hybrid microhandling, a case study of self-assembly-assisted robotic microhandling system is discussed. The feeding, positioning, and part of the releasing is achieved with robotic microhandling and the final releasing, alignment, and fixing are achieved by droplet self-alignment. The case study is analyzed with the help of the component analysis discussed in the previous sections.

3.5.1 Case Study: Hybrid Microhandling Combining Droplet Self-Alignment and Robotic Microhandling

The hybrid microhandling approach discussed here uses a conventional robotic microhandling system including precision positioning stages, piezoelectric micro-gripper, and microscopes to carry out robotic pick-and-place operations. A non-contact droplet dispenser is installed to the system. The dispenser can dispense discrete numbers of droplets, each sized approximately 300 pL. The system is illustrated in Figure 3.6.

When the part approaches the target site, the dispenser dispenses a droplet of water on the target site, after which the micropart, sized 300 μm × 300 μm × 70 μm, is brought in contact with the water. When the gripper releases the part,

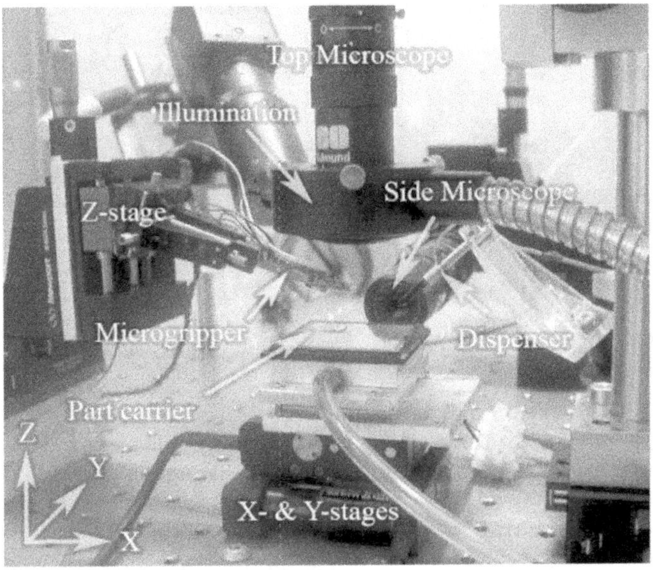

Figure 3.6. Overview of the hybrid handling platform. (Reprinted from [47].)

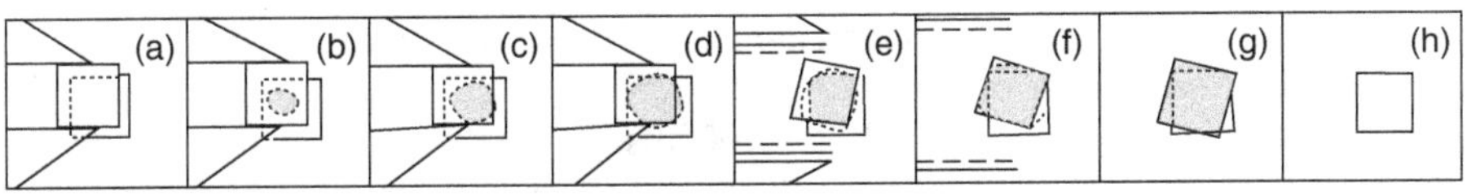

Figure 3.7. Illustration of the hybrid handling technique: (a) Microgripper approaches the release site with a part. (b) A droplet of water is dispensed between the microparts. (c) The droplet contacts with the top part and starts to wet. (d) Wetting is finished. (e) The microgripper opens, releasing the part for self-alignment. (f–g) The capillary force aligns the top part to the bottom part. (h) The water between the two parts evaporates, leaving the two parts aligned. (Reprinted from [47].)

capillary forces assist the releasing of the micropart and self-align and fix the micropart. A schematic of the basic hybrid microhandling technique is shown in Figure 3.7, and the photo of the hybrid microhandling process is shown in Figure 3.8.

In real-world experiments, the droplet self-alignment-based hybrid microhandling has quite a few nice properties when checked against the criteria that are important for microhandling, namely capability, efficiency, precision, and reliability.

3.5.1.1 Efficiency of the Hybrid Handling

Using a high-speed camera, the duration of the self-alignment of the hybrid handling process has been measured. Several repetitions were tested, each with the same release position. About 1.2 nL of water is used in each test. Here, a

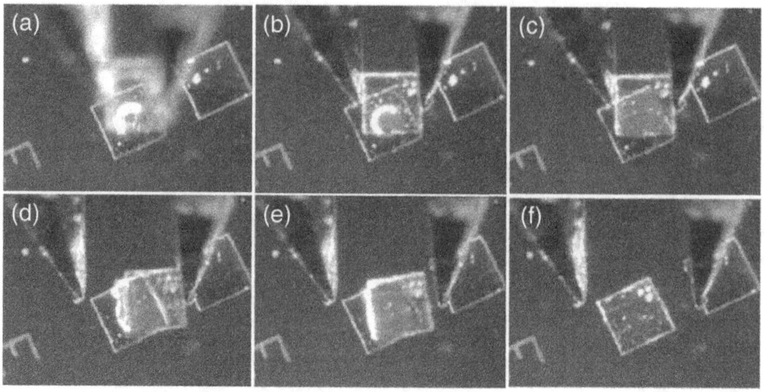

Figure 3.8. Hybrid microhandling experiment: (a) dispensing of droplet; (b) micropart contacts the droplet; (c) droplet wets the gap between the micropart and the receptor; (d) releasing starts, the adhesion between the micropart and the gripper; (e) capillary force overcomes the adhesion force between the part and the gripper; (f) self-alignment achieves fine positioning. (Reprinted from [60].)

test where the release position was approximately 200 μm in the x direction is discussed (see Fig. 3.8 for the definition of axis).

What is interesting is that there are rather large variations in the duration of the self-alignment from one test to another. The duration could vary more than one order of magnitude, from 30 to 400 ms, even if the releasing position is almost the same in each test. Small errors in position and tilting of the part and the amount and position of the liquid, combined with small imperfections of the part surfaces, can lead to very different dynamics of the self-alignment. Furthermore, if the tilting is so large that one corner or side of the top part touches the bottom part, there is large friction between the two parts, and the duration of the self-alignment will dramatically increase, if not fail completely.

This is further illustrated in Figure 3.9. In the figure, five example trajectories are shown, where the starting location is almost identical. The self-alignment is successful in all cases. However, the duration of the self-alignment varies significantly. In the extreme cases, one took 438 ms and another 38 ms.

Even with large differences between the actual process from test to test, the self-alignment seems fairly robust as all the tests here are successful, and the final outcome is very good, despite differences in process runs. The accuracy is not distinguishable from the optical microscopic image.

3.5.1.2 Accuracy of the Hybrid Handling

The accuracy of this method has been studied by the authors [14]. In these tests, ethylene glycol was used instead of water, and the accuracy was measured as a function of the liquid volume. It was observed that with only 0.4 nL of liquid, the alignment error is always high (see Fig. 3.10). The main reason for this is probably that the amount of liquid is not enough to wet the whole area.

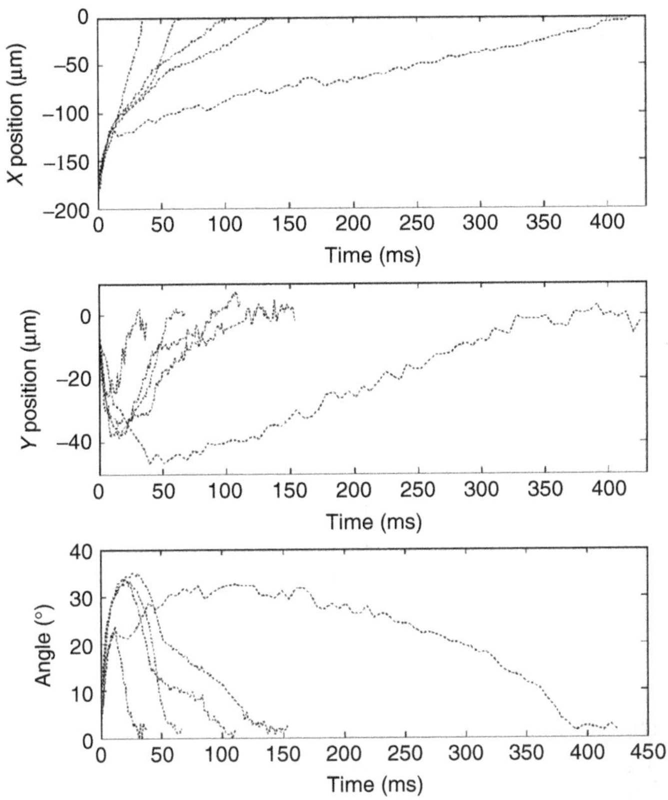

Figure 3.9. Position and angle as a function of time from five different hybrid microhandling tests with the same initial position. There is almost an order of magnitude difference in the self-alignment time between the slowest and fastest self-alignment. (Reprinted from [47].)

While it is expected that too much liquid should have a negative influence on the accuracy, this was not observed in the tests. The reason for this can be that the maximum amount of liquid is not enough to start impairing the alignment. When much larger amounts of liquid, for example, hundreds of nanoliters, is used, the alignment accuracy will be greatly reduced, even in theory, because the potential gradient will be almost flat well before the part is properly aligned.

3.5.1.3 Reliability of the Hybrid Microhandling

Experimentally, it has been observed that the self-alignment in hybrid handling is largely a binary process: It either succeeds or fails completely. Success means that the part self-aligns and the accuracy of the self-alignment is not distinguishable from the microscope image. Failure means that the part is stuck in a local minimum, typically about 50 μm from the correct position. Sometimes the

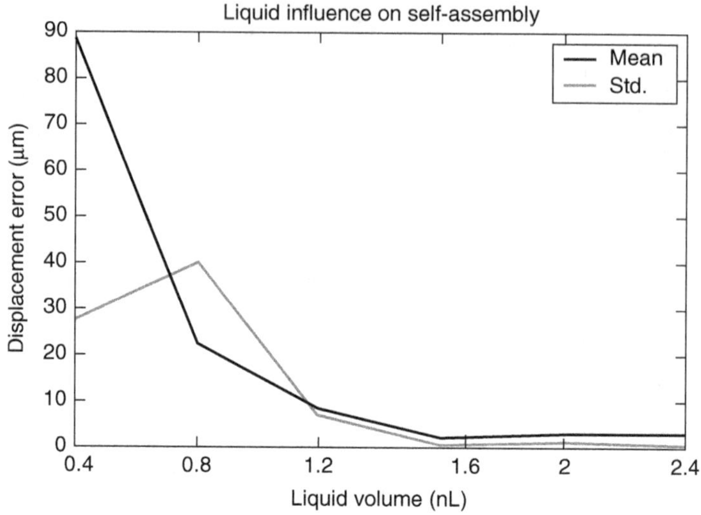

Figure 3.10. Influence of liquid volume on assembly error with SU-8 surfaces and initial alignment error of 150 μm. (Reprinted from [14].)

self-alignment never starts because the bias is so large that no meniscus forms between the parts.

The influence of four process parameters on the success rate has been measured experimentally: the release location (in x, y, and z directions) and the amount of liquid. The results of the tests are illustrated in Figure 3.11. Out of 192 tests, 69% saw successful self-assembly. By studying the figures, it is noticeable that there is no clear trend in the success rate with respect to x or y bias, even when the bias approaches 250 μm. In the case of z bias, a slightly better success rate can be observed around 30 μm. Intuitively, one would expect the success rate to be higher around zero bias; however, this is not the case as many errors are a result of the part adhering to the tool, which does not depend on the bias. Instead the decisive factors are the forming of the meniscus and if there is a dry contact between parts, which has a large enough friction force to prevent self-alignment.

However, a very clear trend in success rate can be observed when the number of droplets increases. When four or more droplets (approx. 1.2 nL) were dispensed, success rates over 80% were achieved. In Figure 3.11 it can be observed that the success rate is actually 100% with eight droplets (approx. 2.4 nL). More tests should be done for more precise estimations of the success rate; however, success rates over 95% are definitely foreseeable.

3.5.1.4 *Capabilities of the Hybrid Handling*

The capabilities of the hybrid handling methods are evaluated by using it to perform handling tasks that would be difficult using robotics or self-assembly alone. The first assembly case is to generate a 90° rotation with capillary forces. The second case tries to show that the fixing of the micropart to the receptor

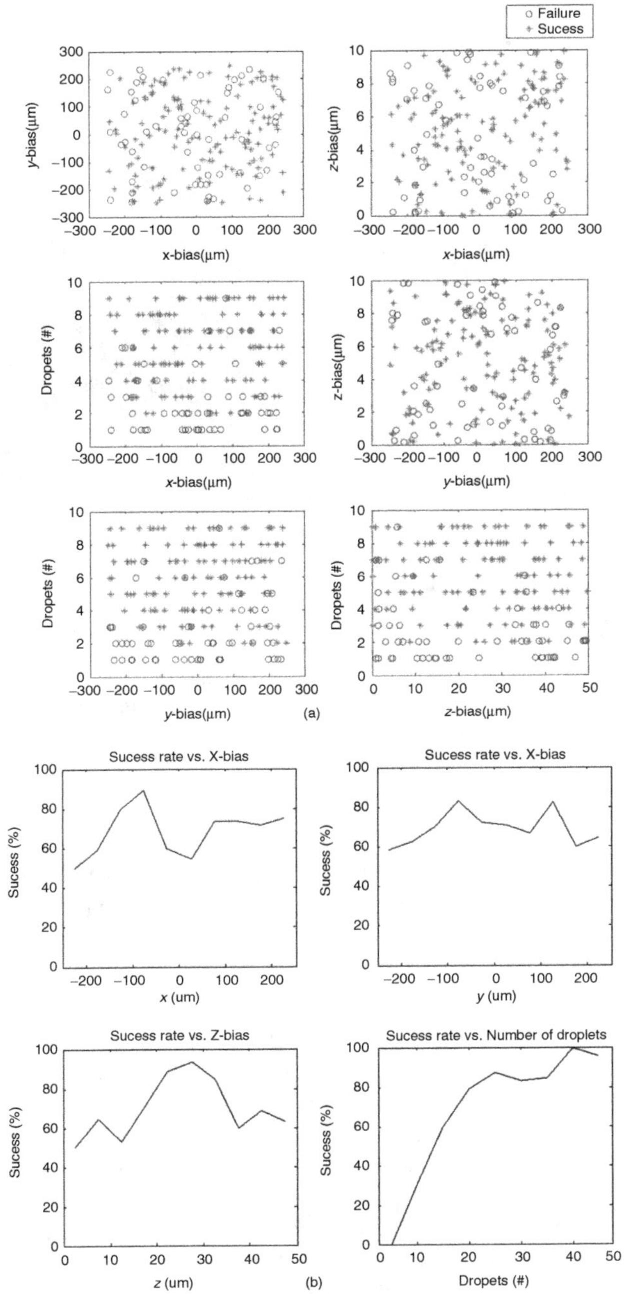

Figure 3.11. Reliability of hybrid microhandling. (a) Test results, plotted by all different parameter combinations. No clear coupling effects can be seen between parameters. (b) Success rate as a function of various process parameters. A clear effect on the success rate can be seen in the number of droplets. (Reprinted from [47].)

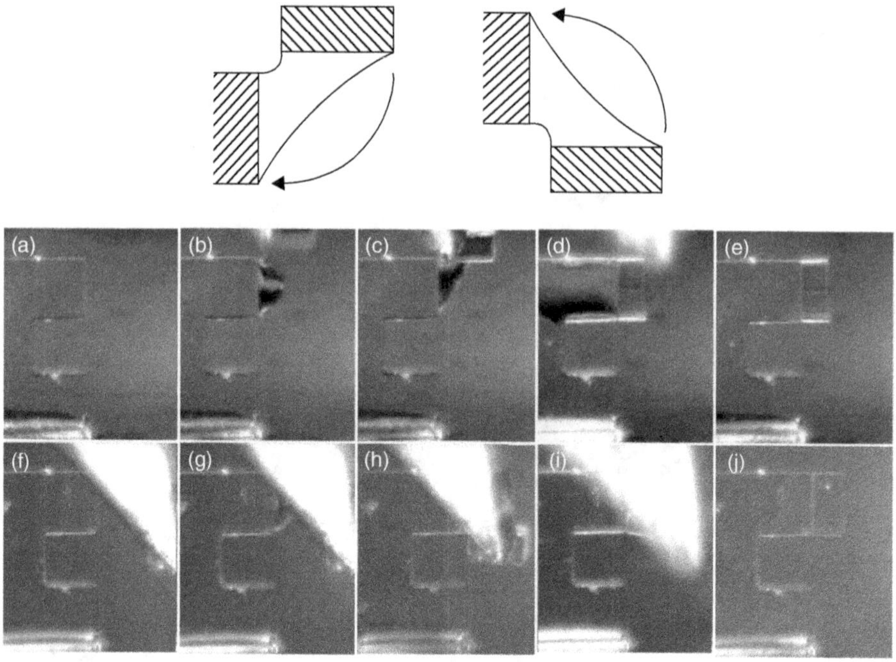

Figure 3.12. Flipping parts by hybrid handling. By approaching the receptor site from either direction, the final orientation can be chosen. (a)–(e) Part is lowered and flipped to the receptor site. (f)–(j) Part is raised and flipped to the receptor site. (Reprinted from [48].)

site is strong enough so that subsequent parts can be assembled on this part. Finally, the assembly method is used to create free-hanging structures, such as cantilevers, in the third case.

Flipping Part by Capillary Forces. Hybrid handling can be used to realize part rotation, which is achieved by choosing the release position so that capillary forces realize the rotation upon release, as illustrated in Figure 3.12. The part will deterministically assemble into one of the two final positions, depending on the side from which the part is released. This process resembles chip tombstoning in electronics assembly [43] or solder ball self-assembly of microelectromechanical systems (MEMS) structures [25].

The method relies on the fact that the wetting is constrained on one side of the parts. The droplet will wet any side that contacts with the droplet first, so that it is important to approach the receptor site from a proper direction. This is why in Figures 3.12(a)–(e) the part is lowered to the release position, while in Figures 3.12(f)–(j) the part is raised to the release position. Furthermore, in Figure 3.12(b), the dispensed droplet will hit the assembled part if the assembled part is not high enough.

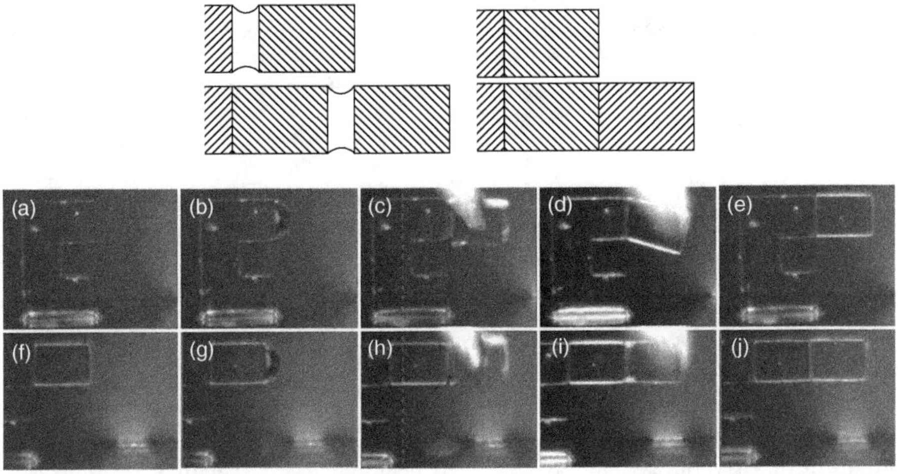

Figure 3.13. Hierarchical hybrid assembly. Dry adhesion and friction between the parts provides suitable long-term fixing, so that an assembled part can be used as the base of subsequent assembly. (Reprinted from [48].)

Actual experiment showing both cases is illustrated in Figure 3.12. The receptor site is the upper arc of a C-shaped part, which is mounted upright and glued to the surface with an ultraviolet (UV) glue. It is interesting to note that in the case of Figures 3.12 (a)–(z), the liquid spills from the receptor site upon release [Fig. 3.12(d)], but this does not prevent the self-alignment from completing.

Hierarchical Assembly. After drying of the water, a fairly stable contact forms between the parts due to dry adhesion and friction, for example, in Figure 3.12. It has been observed that this contact is much stronger than the one that would result in just placing the micropart with the microgripper alone. In fact, this contact is strong enough that subsequent assembly can be performed on top of the assembled part, as illustrated in Figure 3.13. Thus hierarchical structures can be realized.

In practice, the bond between the two parts is broken if water wets between the parts. Thus, to realize hierarchical assembly, it is important not to dispense water between the already fixed parts.

Free-Hanging Structures. If a droplet exists between the two parts with unequal dimensions, the parts will align so that the ends align to each other, provided that the releasing position is chosen so that the end of the shorter part is outside of the longer of the two parts, as shown in Figure 3.14. The technique can be used to assemble big parts on top of small ones, so that free-hanging structures, such as cantilevers, are created.

The experiment of the handling procedure is shown in Figure 3.14. The top part is 600 μm × 300 μm × 40 μm, while the bottom part is 150 μm × 300

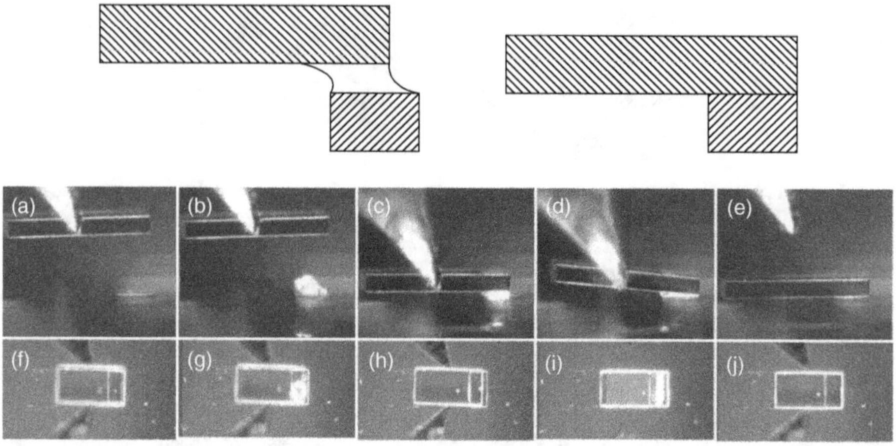

Figure 3.14. Hybrid assembly of micropart of different sizes. Starting assembly with a longer part and shorter part so that the shorter part is not inside the longer part, the ends of the parts align to each other. This can be used to realize free-hanging structures. (Reprinted from [48].)

μm × 40 μm. It can be clearly seen that capillary forces are much stronger than the gravitational forces in this scale, as the capillary forces are able to align the part. In fact, in this particular experiment, slightly uneven wetting and adhesion between the part and the gripper tip makes the part turn upward, before settling into the aligned position (Fig. 3.14). This technique can also be used to assemble smaller parts onto larger parts.

Theoretically, the final position is ambiguous, as shown in Figure 3.15. However, it is still possible to align the small part with the large part, by selecting the initial bias properly, as the authors proposed earlier [60]. The small part will usually align to the edge or the corner of the large part, when the self-alignment is started with the smaller part outside the larger part.

3.5.2 Analysis of Droplet Self-Alignment-Based Hybrid Microhandling

In the following, the components of this case of droplet self-alignment-assisted hybrid microhandling are analyzed.

3.5.2.1 Feeding

The feeding of the hybrid approach is the conventional robotic approach, where the parts are on a part carrier, the location of which can either be predefined by mechanical design or determined using machine vision system. The parts are picked by the mechanical gripper of the robotic microhandling system.

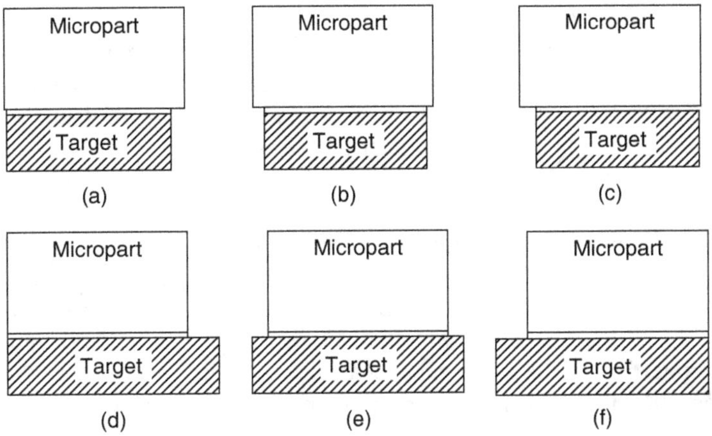

Figure 3.15. Ambiguous final position after self-alignment. (Reprinted from [60].)

3.5.2.2 Positioning

The positioning of the micropart is carried out by the positioning stages of the robotic microhandling system, where the part is held by the mechanical gripper. Therefore, any kind of robotic kinematics and automation strategies can be applied here, and dexterous manipulation and complicated operations can be achieved as in any advanced robotic microhandling system.

3.5.2.3 Releasing and Alignment

The releasing of microparts usually has to tackle the sticking problem between tool and part. In this case study, the releasing process is aided by the capillary force between the part the receptor site.

Capillary force has an almost linear scaling law. Therefore, its strength is often dominating other adhesion forces in the microscale. After the releasing, the part will automatically align to the receptor site by capillary self-alignment, which is robust and time efficient, in the meanwhile, the precision is also very good.

The capabilities of this handling technique are very promising as compared to either robotics or self-assembly alone. With only a robotic microhandling station, realizing the structures discussed in Section 3.5.1 would require rather complicated kinematic structure and dexterous manipulation. With stochastic self-assembly, the structures would also be very hard to achieve because of multiple global minima and hierarchical structures.

3.5.2.4 Fixing

After the water has evaporated, the microparts are fixed to the receptor sites by dry adhesion and friction only. Comparing to self-assembly, the fixing is similar in nature. For many applications, the fixing is sufficient. Furthermore, the water can easily be replaced by adhesive, such as UV-curable glue as experimentally validated by the authors [60]. Thus, very strong bonds can be achieved.

3.5.2.5 Ambient Environment

The hybrid microhandling takes place in air. Therefore, normal consideration of robotic microhandling and self-assembly such as cleanness should be taken care of. Due to the application of droplets, ambient vapor pressure (or relative humidity) has a very significant impact on the process.

3.5.2.6 Surface Properties

The consideration of surface properties in robotic microhandling is also relevant here, such as using rough surface and grounding in gripper is also applicable, but not critical. The reason is that the strong capillary force can largely solve those problems. On the other hand, the capillary self-alignment process relies on the surface properties of the part and the receptor, where in this case they should be hydrophilic. The surface roughness of the micropart as well as the receptor should also be small to reduce the chance of dry contact between the part and the receptor site. Moreover, the geometrical pattern or the shape of the receptor site is also very important, which sets the reference of self-alignment and affects the precision of the fine positioning process.

3.5.2.7 External Disturbance and Excitation

The hybrid microhandling process uses robots for parts transporting. Therefore, adequate disturbance reduction is still needed. However, the hybrid microhandling process has fewer requirements on vibration isolation because the fine positioning is replaced by capillary self-alignment. In contrast, vibration with small amplitude can actually help the self-alignment process to overcome local minima to reach a better precision of the final assembly.

3.5.3 Summary

The above case discussion shows that just combining robotic microhandling and capillary self-alignment, one can already achieve a highly capable, reliable, precise, and quite efficient system. Such hybrid microhandling techniques can be more reliable and efficient than the microrobotic microhandling approach in precision operations, and more capable than self-assembly and even robotic manipulation in a comparable complexity.

Therefore, hybrid microhandling is a very interesting and competing approach that probably lies between the two traditional branches of robotic microhandling and self-assembly under the name of those performance measures.

3.6 CONCLUSION

This chapter investigated the two branches of microassembly—robotic-based microhandling and self-assembly. Through component analysis, both branches of technologies are disassembled and compared side by side. The advantages of robotic microhandling are mostly on the flexibility (reprogrammability) and the capability of carrying out very sophisticated operations. On the other hand, the advantages of self-assembly is the positioning and fixing based on potential energy and the possibility of create massively parallel processes.

In some sense, robotic microassembly and self-assembly are competing technologies. Self-assembly techniques for microparts usually work in parallel in relativly large quantities, which can reach 2 million parts per hour using shape recognition in fluidic phase. This is great for industrial applications. However, they do require well-designed processes and their flexibility is not in the same scale as for robotic microhandling. A major requirement for self-assembly to take place is that the parts and receptor sites need to be specifically treated and designed, which would require redesigning and engineering of many commercial products.

Currently, the state-of-the-art industrial electronics assembly machine using high-speed robots and vacuum grippers can assemble parts from submillimeter to a few millimeters at a speed of about 200,000 parts per hour. This is remarkable, even better than the throughput of most reported self-assembly techniques. Therefore, self-assembly has yet to display a clear edge over conventional robotic assembly and its industrial penetration has been slower than expected.

On the other hand, hybrid microhandling that combines preferred components of both robotic microhandling and self-assembly may lead to innovative solutions. The components of both branches can be analyzed side by side in terms of working phase of microassembly and system parameters: feeding, positioning, releasing, alignment, fixing, environment, surface properties, and external disturbances and excitations. By combining desired components from both branches, novel hybrid microhandling strategies can be created. This idea was illustrated by a case study that uses robotic manipulation in feeding and coarse positioning, and droplet self-assembly for fine positioning and releasing/fixing. The hybrid microhandling strategy gives promising results that are highly capable, reliable, precise, and still rather efficient.

Acknowledgments

This work was supported in part by the Finnish Funding Agency for Technology and Innovation, the European Commission under grant NMP2-CT-2006-026622 Hybrid Ultra Precision Manufacturing Process based on Positional- and Self-assembly for Complex Micro-products, HYDROMEL (2006–2010) and the Graduate School in Electronics, Telecommunication and Automation of Finland.

REFERENCES

1. J. Agnus, P. De Lit, C. Clevy, and N. Chaillet, Description and Performances of a Four-Degrees-of-Freedom Piezoelectric Gripper, Proceedings of the IEEE International Symposium on Assembly and Task Planning, 10–1 1 July 2003, pp. 66–71.

2. H. Aoyama, S. Hiraiwa, F. Iwata, J. Fukaya, and A. Sasaki, Miniature Robot with Micro Capillary Capturing Probe, Proceedings of the Sixth International Symposium on Micro Machine and Human Science, MHS'95, 1995, pp. 173–178.
3. F. Arai, D. Ando, T. Fukuda, Y. Nododa, and T. Oota, Micro Manipulation Based on Micro Physics—Strategy Based on Attractive Force Reduction and Stress Measurement, IEEE/RSJ International Workshop on Intelligent Robots and Systems, IROS'95, Vol. 2, 1995, pp. 263–241.
4. F. Arai, A. Daisuke, Y. Nonoda, T. Fukuda, H. Iwata, and K. Itoigawa, Integrated Microendeffector for Micromanipulation, *IEEE/ASME Trans. Mechat.*, 3(1):17–23, 1998.
5. F. Arai, T. Kasugai, and T. Fukuda, 3D Position and Orientation Control Method of Micro Object by Dielectrophoresis, Proc. 1998 International Symposium on Micromechatronics and Human Science, 1998. MHS '98, 25–28 Nov. 1998, pp. 149–154.
6. F. Arai, T. Sakami, K. Yoshikawa, H. Maruyama, and T. Fukuda, Synchronized Laser Micromanipulation of Microtools for Assembly of Microbeads and Indirect Manipulation of Microbe, Proc. of 2003 IEEE/RSJ International Conference on Intelligent Robots and Systems, IROS 2003, Vol. 3, 27–31 Oct. 2003, pp. 2121–2126.
7. C. Bark, T. Binnenbose, G. Vogele, T. Weisener, and M. Widmann, Gripping with Low Viscosity Fluids, Proc. Eleventh Annual International Workshop on Micro Electro Mechanical Systems, MEMS'98, 1998, pp. 301–305.
8. K. F. Böhringer, K. Goldberg, M. Cohn, R. Howe, and R. A. Pisano, Parallel Microassembly with Electrostatic Force Fields, Proc. 1998 IEEE International Conference on Robotics and Automation, 1998, pp. 1204–1211.
9. K. F. Bohringer, U. Srinivasan, and R. T. Howe, Modeling of Capillary Forces and Binding Sites for Fluidic Self-Assembly, 14th IEEE International Conference on Micro Electro Mechanical Systems, MEMS 2001, 2001, pp. 369–374.
10. A. H. Cannon, Y. Hua, C. L. Henderson, and W. P. King, Self-Assembly for Three-Dimensional Integration of Functional Electrical Components, *J. Micromech. Microeng*. 15:2172, 2005.
11. M. B. Cohn, C. J. Kim, and A. P. Pisano, Self-Assembling Electrical Networks: An Application of Micromachining Technology, Proc. 6th Int. Conf. Solid-State Sensors and Actuators, 1991, pp. 490–493.
12. G. Danuser, I. Pappas, B. Voegeli, W. Zesch, and J. Dual, Manipulation of Microscopic Objects with Nanometer Precision: Potentials and Limitations in Nano-Robot Design, http://robotics.eecs.berkeley.edu/~zech/papers/IJR97.html.
13. N. Dechev, W. L. Cleghorn, and J. K. Mills, Microassembly of 3-D Microstructures Using a Compliant, Passive Microgripper, *J. Microelectromech. Syst*., 13(2):176–189, April 2004.
14. C. del Corral, Q. Zhou, A. Albut, B. Chang, S. Franssila, S. Tuomikoski, and H. N. Koivo, Droplet Based Self-Assembly of SU-8 Microparts, Proc. of 2nd VDE World Microtechnologies Congress, MICRO.tec 2003, Munich, Germany, 2003, pp. 293–298.
15. J. Fang and K. F. Bohringer, Wafer-Level Packaging Based on Uniquely Orienting Self-Assembly (the DUO-SPASS Processes), *J. Microelectromech. Syst*., 15(3):531–540, 2006.

16. R. S. Fearing, Survey of Sticking Effects for Micro Parts Handling, IEEE/RSJ International Conference on Intelligent Robots and Systems, IROS'95, Pittsburgh, 1995, pp. 212–127.

17. J. T. Feddema, P. Xavier, and R. Brown, Micro-assembly Planning with van der Waals Force, *J. Micromech.*, 1(2):139–153, 2001.

18. S. Floyd, C. Pawashe, and M. Sitti, An Untethered Magnetically Actuated Microrobot Capable of Motion on Arbitrary Surfaces, IEEE International Conference on Robotics and Automation, ICRA 2008, 19–23 May 2008, pp. 419–424.

19. D. R. Frutiger, B. E. Kratochvil, K. Vollmers, and B. J. Nelson, Small, Fast, and under Control: Wireless Resonant Magnetic Micro-agents, Proc. of the 11th International Symposium on Experimental Robotics, ISER 2008, Athens, Greece, IFRR Student Fellowship Award, July 2008.

20. T. Fukuda, M. Fujiyoshi, F. Arai, and H. Matsuura, Design and Dextrous Control of Micromanipulator with 6 DOF, Proceedings of 1991 IEEE International Conference on Robotics and Automation, ICRA 1991, Vol. 2, 9–11 Apr. 1991, pp. 1628–1633.

21. M. Gauthier, S. Regnier, P. Rougeot, and N. Chaillet, Analysis of Forces for Micromanipulations in Dry and Liquid Media, *J. Micromech.*, 3(3):389–413, 2006.

22. Gel-Pak, A division of Delphon Industries LLC, Vacuum Release Products [online]; available: http://www.gelpak.com/products/index.cfm?fuseaction=vacuum; referred: 13.11.2008.

23. C. Gosse and V. Croquette, Magnetic Tweezers: Micromanipulation and Force Measurement at the Molecular Level, *Biophys. J.*, 82:3314–3329, 2002.

24. D. S. Haliyo, Y. Rollot, and S. Regnier, Dynamical Strategies for Micromanipulation by Adhesion, in Microrobotics and Microassembly III, Bradley J. Nelson, Jean-Marc Breguet, Eds., Proceedings of SPIE vol. 4586, 2001, pp. 261–269.

25. K. F. Harsh, V. M. Bright, and Y. C. Lee, Study of Micro-scale Limits of Solder Self-assembly for MEMS, 50th Electronic Components and Technology Conference, 2000, pp. 1690–1695.

26. J. Hesselbach, S. Buettgenbach, J. Wrege, S. Buetefisch, and C. Graf, Centering Electrostatic Microgripper and Magazines for Microassembly Tasks, *Proc. SPIE*, 4568:270–277, 2001.

27. J. Hesselbach, J. Wregea, and A. Raatza, Micro Handling Devices Supported by Electrostatic Forces, *CIRP Annals—Manufacturing Technol.*, 56(1):45–48, 2007.

28. Y. Higuchi, K. Sugano, T. Tsuchiya, and O. Tabata, Temperature Controlled Capillary Driven Sequential Stacking Self-Assembly Using Two Different Adhesives, 6th International Conference on Polymers and Adhesives in Microelectronics and Photonics, Polytronic 2007, 2007, pp. 128–132.

29. I. W. Hunter, S. Lafontaine, P. M. F. Nielsen, P. J. Hunter, and J. M. Hollerbach, Manipulation and Dynamic Mechanical Testing of Microscopic Objects Using a Tele-Micro-Robot System, Proceedings of 1989 IEEE International Conference on Robotics and Automation, ICRA'89, Vol. 3, 1989, pp. 1553–1558.

30. H. O. Jacobs, A. R. Tao, A. Schwartz, D. H. Gracias, and G. M. Whitesides, Fabrication of a Cylindrical Display by Patterned Assembly, *Science*, 296(5566):323–325, 2002.

31. V. Jokinen, L. Sainiemi, and S. Franssila, Complex Droplets on Chemically Modified Silicon Nanograss, *Adv. Mat.*, 20(18):3453–3456, 2008.

32. P. Kallio, M. Lind, Q. Zhou, and H. N. Koivo, A 3-DOF Piezohydraulic Parallel Micromanipulator, Proceedings of 1998 IEEE International Conference on Robotics and Automation, ICRA 1998, vol. 2, 16–20 May 1998, pp. 1823–1828.

33. A. Kochan, European Project Develops 'Ice' Gripper for Micro-sized Components,' *Assembly Automation*, 17:114–115, 1997.

34. T. Kusakabe, T. Tanemura, Y. Higuchi, K. Sugano, T. Tsuchiya, and O. Tabata, DNA Mediated Sequential Self-assembly of Nano/Micro Components, 21st IEEE International Conference on Micro Electro Mechanical Systems, MEMS 2008, 13–17 Jan. 2008, pp. 1052–1055.

35. P. Lambert, P. Letier, and A. Delchambre, Capillary and Surface Tension Forces in the Manipulation of Small Parts, Proceedings of the IEEE International Symposium on Assembly and Task Planning, ISATP 2003, 10–11 July 2003, pp. 54–59.

36. P. Lambert and S. Régnier, Surface and Contact Forces Models within the Framework of Microassembly, *J. Micromech.*, 3:123–157, 2006.

37. D. Lang, M. Tichem Marcel, and S. Blom, The Investigation of Intermediates for Phase Changing Micro-gripping, Proceedings of 5th International Workshop on MicroFactories, IWMF, 2006.

38. B. Lopez-Walle, M. Gauthier, and N. Chaillet, Principle of a Submerged Freeze Gripper for Microassembly, *IEEE Trans. Robotics*, 24(4):897–902, 2008.

39. C. J. Morris, S. A. Stauth, and B. A. Parviz, Self-assembly for Microscale and Nanoscale Packaging: Steps toward Self-Packaging, *IEEE Trans. Adva. Packaging*, 28(4):600–611, 2005.

40. R. Murhy, A. N. Das, and D. O. Popa, High Yield Assembly of Compliant MEMS Snap Fasteners, Proceedings of ASME International Conference on Micro and Nano Systems, August 2008.

41. A. P. Neild, S. Oberti, F. Beyeler, J. Dual, and B. J. Nelson, A Micro-particle Positioning Technique Combining an Ultrasonic Manipulator and Microgripper, *J. Micromech. Microeng.*, 16(8):1562–1570, 2006.

42. S. T. Patton and B. Bhushan, Environmental Effects on the Streaming Mode Performance of Metal Evaporated and Metal Particle Tapes, *IEEE Trans. Magnetics*, 33(4):2513–2530, 1997.

43. W. J. Plumbridge, Defects in Electronics and Their Significance for Structural Integrity, *Fatigue and Fracture of Engineering Materials and Structures*, 30(5):432–442, 2007.

44. A. A. Ramadan, K. Inoue, T. Arai, T. Takubo, and I. Hatta, Micro-Nano Two-Fingered Hybrid Manipulator Hand, International Symposium on Micro-NanoMechatronics and Human Science, MHS '07, 11–14 Nov. 2007, pp. 32–37.

45. S. Saito, H. T. Miyazaki, T. Sato, K. Takahashi, and T. Onzawa, Dynamics of Micro-object Operation Considering the Adhesive Effect under an SEM, *Microrobotics and Microassembly III*, Bradley J. Nelson and Jean-Marc Breguet, Eds., Proceedings of SPIE, Vol. 4586, 2001, pp. 12–23.

46. V. Sariola, Q. Zhou, and H. N. Koivo, Hybrid Microhandling: A Unified View of Robotic Handling and Self-assembly, *J. Micro-Nanomech.*, 4(1–2):5–16, 2008.

47. V. Sariola, Q. Zhou, R. Laass, and H. N. Koivo, Experimental Study on Droplet Based Hybrid Microhandling Using High Speed Camera, Proceedings of the IEEE/RSJ 2008 International Conference on Intelligent Robots and Systems, 2008, pp. 919–924.

48. V. Sariola, Q. Zhou, and H. N. Koivo, Three Dimensional Hybrid Microassembly Combining Robotic Microhandling and Self-Assembly, submitted to 2009 IEEE International Conference on Robotics and Automation, ICRA'2009.

49. M. Savia, H. N. Koivo, and Q. Zhou, Evaluation of Adhesion Forces between Arbitrary Objects, *J. Micromech.*, 3(3–4):221–238, 2006.

50. U. Srinivasan, D. Liepmann, and R. T. Howe, Microstructure to Substrate Self-Assembly Using Capillary Forces, *J. Microelectromech. Syst.*, 10:17–24, 2001.

51. S. A. Stauth and B. A. Parviz, Self-Assembled Single Crystal Silicon Circuits on Plastic, Proc. Natl. Acad. Sci., 103(38):13922–13927, 2006.

52. T. Tanikawa, T. Arai, P. Ojala, and M. Saeki, Two-Finger Micro Hand, Proceedings of 1995 IEEE International Conference on Robotics and Automation, vol. 2, 21–27 May 1995, pp. 1674–1679.

53. T. Watanabe, N. Fujino, and Z. Jiang, Micromanipulation Using Squeeze Effect, Proceedings of 2004 IEEE/RSJ International Conference on Intelligent Robots and Systems, IROS 2004, vol. 4, 28 Sept.–2 Oct. 2004, pp. 3357–3362.

54. H. Wu, V. R. Thalladi, S. Whitesides, and G. M. Whitesides, Using Hierarchical Self-Assembly to Form Three-Dimensional Lattices of Spheres, *J. Am. Chem. Soc.* 124(48):14495–14502, 2002.

55. X. Xiang, Y. Hanein, J. Fang, Y. Wang, W. Wang, D. T. Schwartz, and K. F. Bohringer, Controlled Multibatch Self-Assembly of Microdevices, *J. Microelectromech. Syst.*, 12(2):117–127, 2003.

56. H.-J. Yeh and J. S. Smith, Fluidic Self-Assembly of Microstructures and Its Application to the Integration of GaAs on Si, Proceedings of IEEE Workshop on Micro Electro Mechanical Systems, MEMS'94, 1994, pp. 279–284.

57. W. Zheng, P. Buhlmann, and H. O. Jacobs, Sequential Shape-and-Solder-Directed Self-Assembly of Functional Microsystems, Proc. Natl. Acad. Sci., PNAS, 101:12814–12817, 2004.

58. Q. Zhou, A. Aurelian, B. Chang, C. del Corral, and H. N. Koivo, Microassembly System with Controlled Environment, *J. Micromech.*, 2:227–248, 2004.

59. Q. Zhou, P. Korhonen, B. Chang, and V. Sariola, 6 DOF Dexterous Microgripper for Inspection of Microparts, Proceedings of 2005 IEEE/ASME International Conference on Advanced Intelligent Mechatronics, 24–28 July 2005, pp. 534–539.

60. Q. Zhou and B. Chang, Microhandling Using Robotic Manipulation and Capillary Self-Alignment, Proceedings of IEEE/RSJ International Conference on Intelligent Robots and Systems, IROS 2006, Oct. 2006, pp. 5883–5888.

61. Q. Zhou, B. Chang, and H. N. Koivo, Temperature and Humidity Effects on Micro/nano Handling, *Mat. Sci. Forum*, 532–533:681–684, 2006.

CHAPTER 4

TOWARD A PRECISE MICROMANIPULATION

MÉLANIE DAFFLON and REYMOND CLAVEL

4.1 INTRODUCTION

In industrial microassemblies, reliability and performance in positioning are of the most importance to characterize a microhandling tool. Of course, the choice of one handling principle over another depends on many parameters such as cycling time, accessibility to the components (mechanically but also optically), and compatibility with the ambient conditions of manipulation. These parameters are kept in mind, but we focused here particularly on reliability and repeatability. This chapter presents theoretical models of micromanipulation tasks allowing optimizing the strategies to get precise and reliable operations and taking into account adhesion effects. Some rules and concepts to develop a precise microgripper are then extracted. We present several types of microgrippers based on a combination of principles such as grasping, adhesion, inertial release, capillary effect, and suction. Experiments were done with components of typically 50 μm in size in a gaseous environment. We investigated mainly efficiency and repeatability, but specificities inherent in each gripper family are also shown.

4.2 HANDLING PRINCIPLES AND STRATEGIES ADAPTED TO THE MICROWORLD

As discussed in Chapter 1, adhesion effects are most importance as long as the objects are getting smaller. At around 1 mm, the gravity exerted on each body

Robotic Microassembly, edited by Michaël Gauthier and Stéphane Régnier

becomes negligible compared to the adhesion forces, with the consequence that the microobjects stick to each surface they touch. This effect becomes a great problem in micromanipulation. Contrary to the macromanipulation case, opening a tweezers will no more be sufficient to release a microobject. It is then mandatory to act with new strategies in order to release properly the microobject from the gripper surface.

4.2.1 State of the Art of Micromanipulation Principles

The manipulation of microcomponents can be based on many different principles that ensue from the miniaturization of macromanipulation principles, from physical principles that become usable because of scaling laws or either from the increase of the surface forces.

Microtweezers are widely developed. They vary mainly by their actuator type as, for example, the electrothermal actuator [2, 13, 36], the electrostatic actuator [11, 39, 51], the piezoelectric actuators [1, 37, 58], the shape memory alloy component [9, 56], or pneumatic actuators [12, 47]. We could also classify them by the number and disposition of their degree(s) of freedom (DoF). For example, the simplest structure is a single 1-DOF finger [10], but other tweezers were proposed with symmetrical openings of two or more fingertips [25, 48, 57] controlled with a single actuator. At least, grippers with independent fingers that have one or several degrees of freedom [17, 59] have been proposed. Additional functionalities were also sometimes integrated as, for instance, a force sensor [11, 21, 23, 30, 41].

Among the micromanipulation tools based on a mechanical contact with the component, we can cited the principles based on the following forces: the vacuum [5, 40, 41, 55], the electrostatic force [20, 27, 33, 43], the capillary force [3, 8, 32], the adhesion force in general [4, 19, 22, 42, 49], the inertial effect [24], or even the force generated during a phase change like in cryogenics [31, 34]. However, the manipulation can also be performed without mechanical contact as with magnetic, optical, aerodynamic, or acoustical levitation methods [52]. These last ones will not be investigated in this chapter where we pointed out only on manipulation with mechanical contact between the microobject and gripper. A comparison of the micromanipulation principles is shown in the Table 4.1.

4.2.2 Adhesion Ratio at Interfaces

The efficiency of the pick-or-place operations is linked to the manner used to break one of the interfaces with the microobject in order to perform its transfer from one surface to the other. We are thus interested in the force generated by the gripper at the interface level or directly on the microobject. The final position of the object on the substrate is a function of both relative positions between the gripper and the substrate and between the object and the gripper. The first positioning is directly linked to the precision of the relative movement gripper/substrate applied during the transfer of the microobject. The second

TABLE 4.1. Comparison Between Different Micro Manipulation Principles

Principle	Microobject Material	Microobject Shape, size	Operation: Pick/Place	Environment (RH: relative humidity)	Holding force
Microtweezers	No limitation	Limited by the fingertips size	OK/adhesion to fingertips	No constraint, control of RH to limit the adhesion	Not limited
Electrostatic	Conductive material, depends on dielectric characteristic, not for microelectronic component	Optimal for flat surface	OK/perturbed by residual electric charges	Not efficient for RH>65%	Limited by sparkling
Vacuum	Not porous	Need to fit nozzle/object size	OK/adhesion to nozzle	Control of RH to limit the adhesion	According to nozzle size and pressure
Adhesion	No limitation (but need good fitting of substrate/gripper)	Optimal with size decrease	Transfer strategy	Very sensitive	According to RH, material, geometry
Inertial	No limitation	Acceleration increases when the size decreases	—/OK	Acc. increases with adhesion, so with RH	—
Capillary	Hydrophilic tendency	Maximal force for flat surfaces	OK/need another strategy	Not in liquid	According to liquid, material and geometry

positioning, usually called *referencing*, is a function of the position of the contact areas between the object and the gripper.

The study of the manipulation of a microobject consists then in considering the next four points:

- *The Forces at Interfaces.* The adhesion force and every other force that can increase or decrease the holding effect at the interface such as a suction effect or a capillary force generated by the local presence of the liquid drop.
- *The Contact Characteristics.* We consider the geometry and disposition of the contact areas around the microobject. This aspect depends also on any mechanical feature used for the referencing of the object.
- *The "Gripper/Substrate" Relative Movement.* The release movement will generate the effects of sliding, rolling, pivoting, and of course separation of the interface according to its orientation and to the disposal of the contact areas.
- *The External Forces Applied to the Microobject.* For example, an inertial force, a magnetic field, an acoustic wave, or an airflow that could be used as a manipulation principle. These forces act directly on the object and not solely within the contact areas.

The expressions of the adhesion forces and their origin and sensitivity were explained in Chapter 1. Practically, it is far from simple to evaluate precisely the adhesion force for a real configuration such as in the microassembly case because of the complex geometry when taking into account the real shape and roughness of the contact area as well as the material and coating that can be quite heterogeneous or recovered by some oxidation layer, for instance. In an industrial solution, that means for a large amount of pieces and operations over the time, one of the questions is how to know, and then keep constant, the adhesion effects in a controlled and repeatable way in order to reduce their perturbation on the operations.

We introduce here a representation of the adhesion effect in a micromanipulation case that allows us to identify the possible strategies of pick or place and then investigate their main characteristics. This representation is used to compare the amplitude of the adhesion effect at each interface without having to evaluate them precisely. The interest is to predict what happens when modifying them and which configuration would be optimal.

This representation is based on the *adhesion ratio at interfaces* (Γ), which corresponds to

$$\Gamma \equiv \left(\frac{A_G}{A_S}\right)^2 \tag{4.1}$$

where A_G and A_S are the adhesion forces at the "gripper–object" and "substrate–object" interfaces, respectively. Intuitively, the placing operation will be favored for $\Gamma < 1$ regarding Eq. 4.1, whereas picking a microobject corresponds to configurations with $\Gamma > 1$. We will see later that some other parameters influence the success of the operations, but on this basis it seems obvious that any modification of Γ during the manipulation will favor the transition between pick-and-place abilities.

The adhesion effects are sensitive to the contact geometry and the material at the interface as well as to the ambient environmental conditions and particularly to the relative humidity. The ratio Γ is thus as sensitive to these same parameters. Therefore, a few methods can be developed to modify Γ, in particular:

- *Material at the Interfaces:* When the contacts are similar on both sides of the object and considering only the pull-off forces, the ratio Γ can be expressed independently of the picked object by the ratio of the surface energies (γ_G and γ_S) as

$$\sqrt{\Gamma} \equiv \frac{A_G}{A_S} = \frac{W_G}{W_S} = \frac{2\sqrt{\gamma_G \gamma_O}}{2\sqrt{\gamma_S \gamma_O}} = \sqrt{\frac{\gamma_G}{\gamma_S}} \tag{4.2}$$

  Changing the material has a direct effect on the ratio Γ. Expression 4.2 is realistic at low relative humidity, when the capillary effects are not predominant (until 40%). For a higher relative humidity, the adhesion force will depend on the wettability of the surfaces. The use of hydrophilic and hydrophobic coatings will therefore allow modifying locally the surfaces and notably the gripping areas.
- *Roughness:* The adhesion effects are maximal for perfectly smooth surfaces and decrease rapidly with the roughness [45, 53], except in the presence of a meniscus on a hydrophilic surface [54]. It can be interesting to move the microobject from a smooth surface to a rough surface to improve its transfer, for example, by sliding or rolling it there [42].
- *Contact Area:* The contact between two flat surfaces induces an adhesion force greater than the one between two spheres [28]. The change of the gripper geometry between the picking and the placing operations or the reduction of the contact area allow modifying mechanically the adhesion effects (Fig. 4.1). For example, the transfer of a sphere of radius R_O between two flat surfaces gives expression 4.2. For a flat substrate and a spherical gripper of radius R_G the expression of the ratio Γ with the equivalent radius $R_{S/O}$ at the substrate interface and $R_{G/O}$ at the gripper one becomes

$$\sqrt{\Gamma} = \frac{A_G}{A_S} \approx \frac{R_{G/O} W_G}{R_{S/O} W_S} = \sqrt{\frac{\gamma_G}{\gamma_S}} \frac{R_G}{R_G + R_O} \tag{4.3}$$

  with $R_{S/O} = R_O$ and $R_{G/O} = R_G R_O / R_G + R_O$.
- *Presence of a Meniscus at the Interface:* The capillary force is generally the greater contribution of the adhesion force. The use of a liquid only at one of the interfaces increases considerably the adhesion force on this side solely.
- *Electrostatic Effect:* As for the capillary force, an electrostatic force exerted only at one of the interfaces can modify the ratio Γ. This force can be developed by a difference of voltage between the surfaces or by using the capacitance of structured electrodes.

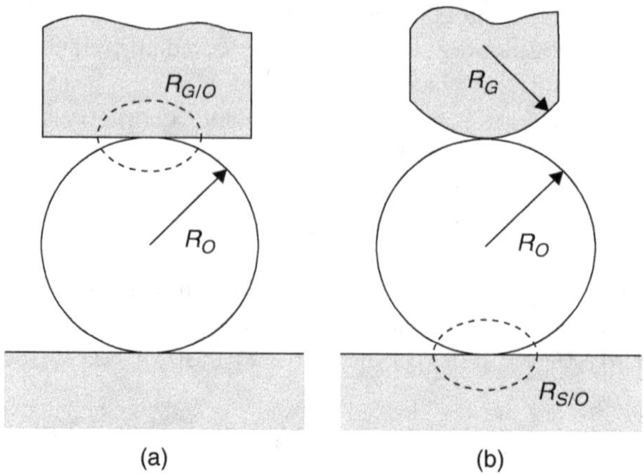

Figure 4.1. Geometries at contact: (a) transfer plane/plane and (b) transfer sphere/plane.

- *Suction Effect:* The pressure gradient between the contact area and the atmospheric pressure induces a suction effect. This is thus another way to unbalance the forces between both interfaces.
- *Blowing Effect:* An overpressure, contrary to the suction effect induced by an underpressure, allows decreasing the force at the interface until it provokes a repulsive effect.

Seeing all these ways to modify the configuration, the adhesion ratio Γ can be expressed in a general manner as

$$\sqrt{\Gamma} \equiv \frac{A_G + F_G}{A_S + F_S} \tag{4.4}$$

with A_G and A_S the adhesion forces at the gripper–object and substrate–object interfaces and F_G and F_S the gripping forces that act at these same interfaces but are controlled by the gripper or the substrate. Adhesion ratio Γ depends on many parameters, by taking into account the characteristics of the microobject, of the gripper, of the substrate, and of the environment as well as of every additional force that occurs during the operation at the interfaces level:

$$\Gamma \equiv \Gamma(\text{object, gripper, substrate, environment, \ldots}) \tag{4.5}$$

Furthermore, the gripper efficiency will depend on the ability to change from a picking configuration to a placing configuration. The efficiency can thus be qualified by the variation of the ratio Γ between these two operations.

4.2.3 Adhesion-Based Micromanipulation

We are interested in predicting how the microobject will act (supposed to be spherical) during its transfer between the substrate and gripper surfaces depending on the release movement, the disposal of the contact areas, and the adhesion forces at both interfaces gripper–object (interface G) and substrate–object (interface S). Of course, the first objective is to know which side the object will stay on at the end of the process. The best configuration in terms of positioning would be that the release movement induces no displacement at all on the receiver side, neither sliding, nor rolling, or even pivoting of the microobject. This study allows extracting the best strategy to pick or place the characteristics of both interfaces gripper–object and substrate–object. The conditions of separation, sliding, and rolling at each interface are first investigated following previous studies proposed in the literature [18, 42, 50]. The object's behavior is then analyzed among different configurations.

4.2.3.1 General Expressions of the Constraints at the Interfaces

Figure 4.2 summaries the disposal of the forces and moments at each interface gripper–object and substrate–object. The manipulator typically induces the external force F during the release. With this force F and the adhesion forces A_G and A_S at each interface, the sphere can be in a sliding, rolling, or pivoting movement before being separated from one of the interfaces. The forces T_{SX}, T_{SY} (T_{GX}, T_{GY}) and N_S (N_G) are the tangential and normal reactions at the substrate–object interface (gripper–object, respectively). The moments M_{SX}, M_{SY} and M_{SZ} (M_{GX}, M_{GY}, and M_{GZ}, respectively) are generated by the tangential strains and can induce rolling or pivoting movements of the sphere. The substrate plane defines the system of reference $\Re_{OS}$. The force F acting during the release can be written as

$$\vec{F} = \begin{pmatrix} F_x \\ F_y \\ F_z \end{pmatrix} = F \begin{pmatrix} \sin\psi \cos\phi \\ \cos\psi \\ \sin\psi \sin\phi \end{pmatrix} \tag{4.6}$$

Considering a static configuration, the forces on the object as shown in Figure 4.2 allow to write the following equations at equilibrium:

$$\begin{aligned} 0 &= T_{SX} - A_G \sin\alpha + N_G \sin\alpha - T_{GX} \cos\alpha \\ 0 &= T_{SX} - T_{GY} \\ 0 &= N_S - A_S - N_G \cos\alpha + A_G \cos\alpha - T_{GX} \sin\alpha \end{aligned} \tag{4.7}$$

Considering the subset gripper–object [Fig. 4.2(c)], the following expression gives the normal and tangential reactions in function of the force F:

$$\begin{aligned} T_{GX} &= F_X \cos\alpha - F_Z \sin\alpha = F \sin\psi \cos(\alpha + \phi) \\ T_{GY} &= F_Y = F \cos\psi \\ N_G &= A_G - F_X \sin\alpha - F_Z \cos\alpha = A_G - F \sin\psi \sin(\alpha + \phi) \end{aligned} \tag{4.8}$$

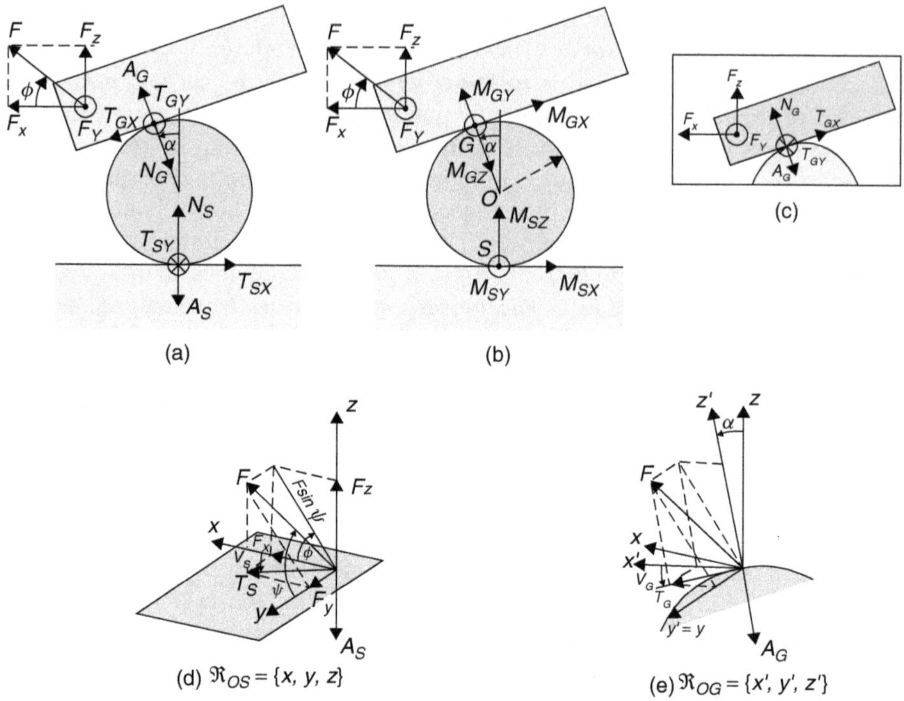

Figure 4.2. Arrangement of (a) the forces (T_x,T_y, N) and (b) the moments (M_x, M_y, M_z) acting on the microobject at each interface during the transfer with $F = (F_x, F_y, F_z)$ the force applied during the release and the adhesive forces A_G and A_S. (c) View from the gripper side. Systems of reference (d) $\Re_{OS}$ linked to the substrate, and (e) $\Re_{OG}$ inclined of an angle α and linked to the gripper.

They allow expressing the reactions on the substrate side as

$$
\begin{aligned}
T_{SX} &= F_X = F \sin\psi \cos\phi \\
T_{SY} &= F_Y = F \cos\psi \\
N_S &= A_S - F_Z = A_S - F \sin\psi \sin\phi
\end{aligned}
\tag{4.9}
$$

4.2.3.2 Separation Threshold

The minimal force necessary for the separation of the object from the surface of the gripper or the substrate is by the projection of the adhesion and friction forces onto the force F. The separation threshold at the gripper–object interface is then

$$
F > A_G \sin\psi \sin(\alpha+\phi) + \mu_G N_G\, |\cos\nu_G \sin\psi \cos(\alpha+\phi) + \sin\nu_G \cos\psi| \tag{4.10}
$$

The friction force is directly opposed to the tangential reactions induced by F and oriented by an angle ν_G in the plane tangent to the interface [Fig. 4.2(e)]:

$$\tan \nu_G = \frac{\cos \psi}{\sin \psi \cos(\alpha + \phi)} \tag{4.11}$$

By introducing Eq. 4.8 in to 4.10, the separation threshold becomes

$$F > \frac{A_G \sin \psi \sin(\alpha + \phi) + \mu_G |\cos \nu_G \sin \psi \cos(\alpha + \phi) + \sin \nu_G \cos \psi|}{1 + \mu_G \sin \psi \sin(\alpha + \phi) \cdot |\cos \nu_G \sin \psi \cos(\alpha + \phi) + \sin \nu_G \cos \psi|} \equiv \Sigma_G \tag{4.12}$$

The same approach gives the separation threshold at the substrate–object interface where the angle $\alpha = 0$:

$$F > \frac{A_S \sin \psi \sin \phi + \mu_S |\cos \nu_S \sin \psi \cos \phi + \sin \nu_S \cos \psi|}{1 + \mu_S \sin \psi \sin \phi \cdot |\cos \nu_S \sin \psi \cos \phi + \sin \nu_S \cos \psi|} \equiv \Sigma_S \tag{4.13}$$

with $\tan \nu_S = \cos \psi \ (\sin \psi \cos \phi)$. A release condition could be expressed by $\Sigma_G < \Sigma_S$. In other words the release will be effective if the interface gripper–object separates itself first. It goes, of course, inversely for the picking condition, thus: $\Sigma_G > \Sigma_S$. Thus, we can define a "separation limit" on the adhesion ratio $\Gamma = (A_G/A_S)^2$ from Eq. (4.1) as a function of the release movement and the friction coefficients. This limit can be written as

$$\Gamma_{\text{separation}} = \left(\frac{\sin \phi + \mu_S |\cos \phi|}{\sin(\alpha + \phi) + \mu_G |\cos(\alpha + \phi)|} \cdot \frac{1 + \mu_G \sin(\alpha + \phi) |\cos(\alpha + \phi)|}{1 + \mu_S \sin \phi |\cos \phi|} \right)^2 \tag{4.14}$$

4.2.3.3 Sliding Threshold

The friction force depends on the friction coefficient μ similarly to the macroscale with the adhesion force added to the normal load. The sliding thresholds at both interfaces are obtained with

$$|T_S| > \mu_S N_S \quad \text{and} \quad |T_G| > \mu_G N_G \tag{4.15}$$

Considering the inequalities 4.15 and the expressions 4.8 and 4.9, the sliding thresholds can be defined as

$$\begin{aligned} |T_S| &= \sqrt{T_{SX}^2 + T_{SY}^2} \\ &= F\sqrt{(\sin \psi \cos \phi)^2 + (\cos \psi)^2} > \mu_S (A_S - F \sin \psi \sin \phi) \end{aligned} \tag{4.16}$$

$$|T_G| = F\sqrt{[\sin \psi \cos(\alpha + \phi)]^2 + (\cos \psi)^2} > \mu_G [A_G - F \sin \psi \sin(\alpha + \phi)] \tag{4.17}$$

The minimal forces inducing the sliding at each interface become then

$$\text{At interface } S \quad F > \frac{\mu_S A_S}{\sqrt{(\sin\psi\,\cos\phi)^2 + \mu_S \sin\psi\,\sin\phi}} \tag{4.18}$$

$$\text{At interface } G \quad F > \frac{\mu_G A_G}{\sqrt{[\sin\psi\,\cos(\alpha+\phi)]^2 + \mu_G \sin\psi\,\sin(\alpha+\phi)}} \tag{4.19}$$

Similarly, to the separation limit a sliding limit can be defined. It determines the interface where sliding occurs first and can be written as

$$\Gamma_{\text{sliding}} = \left[\frac{\mu_S}{\mu_G} \cdot \frac{|\cos(\alpha+\phi)| + \mu_G \sin(\alpha+\phi)}{|\cos\phi| + \mu_S \sin\phi}\right]^2 \tag{4.20}$$

Once the sliding is induced on one of the interfaces the reaction forces at this interface can then be written as:

$$\begin{aligned} T_X &= \mu N \cos\nu \\ T_Y &= \mu N \sin\nu \text{ with} \quad \tan\nu = \frac{\cos\psi}{\sin\psi\,\cos(\alpha_0+\phi)} \\ N &= A - F\sin\psi\,\sin(\alpha_0+\phi) \end{aligned} \tag{4.21}$$

where $\alpha_0 = \alpha$ if the sliding occurs first at the gripper–object interface and $\alpha_0 = 0$ if it appears first at the substrate–object interface. The reactions on the opposite interface can so be established with Eqs. 4.7 and 4.21 and allow determining the corresponding sliding threshold.

4.2.3.4 Rolling and Pivoting Thresholds

A rolling movement on one of the interfaces is obligatory, accompanied by a pivoting of the microobject on the other interface. Then the object can pivot on the gripper side and roll on the substrate [Fig. 4.3 (a)] or the object can pivot on the substrate and roll on the gripper [Fig. 4.3 (b)].

Rolling and pivoting of the microobject are possible if the generated moments M^R and M^P at each contact area are bigger than the maximum resisting moment $M^{R_{\max}}$ and $M^{P_{\max}}$ of the interface. The rolling condition can be expressed as [42]

$$\left|M^R\right| > M^{R_{\max}} \quad \text{with} \quad M^{R_{\max}} = c_r a W \tag{4.22}$$

where c_r is the rolling resistant coefficient, a the radius of the contact area, and W the work of adhesion at the interface (see Chapter 1). The condition of pivoting depends on the friction coefficient at the specific interface and is written as

$$\left|M^P\right| > M^{P_{\max}} \quad \text{with} \quad M^{P_{\max}} = 2a\mu N \tag{4.23}$$

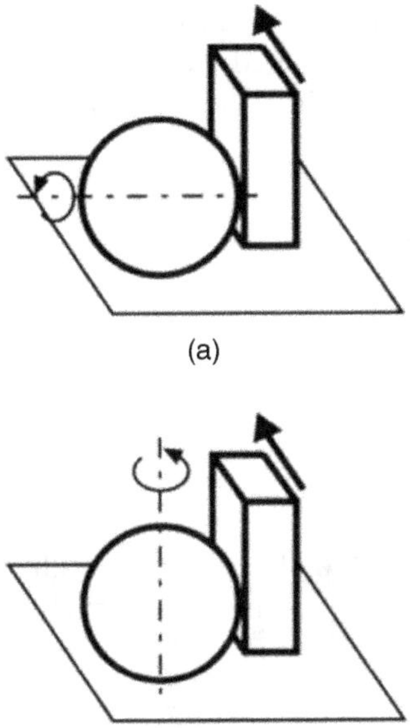

(a)

(b)

Figure 4.3. Types of pivoting movement: (a) on G and (b) on S.

with μ the friction coefficient and N the normal load at the interface. The rolling and pivoting moments can be expressed by the tangential and normal components for each interface as described in Figure 4.2(b). The rolling movement is induced by the tangential components. So the rolling moments are expressed by $M_G^R = \sqrt{M_{GX}^2 + M_{GY}^2}$ and $M_S^R = \sqrt{M_{SX}^2 + M_{SY}^2}$. The components normal to the interfaces contribute only to the pivoting moment, thus $M_G^P = M_{GZ}$ and $M_S^P = M_{SZ}$. Still in a static equilibrium, the sum of the moments acting on the object of radius R is obtained by the tangential forces at each interface as

$$\sum \vec{M} = \overrightarrow{OG} \otimes \vec{T_G} + \overrightarrow{OS} \otimes \vec{T_S}$$
$$= \begin{pmatrix} R \sin \alpha \\ 0 \\ R \cos \alpha \end{pmatrix} \otimes \begin{pmatrix} T_{GX} \cos \alpha \\ T_{GY} \\ -T_{GX} \sin \alpha \end{pmatrix} + \begin{pmatrix} 0 \\ 0 \\ -R \end{pmatrix} \otimes \begin{pmatrix} -T_{SX} \\ -T_{SY} \\ 0 \end{pmatrix} \tag{4.24}$$

where the operator $\otimes$ is the outer product.

We can thus write the following expressions:

$$\begin{aligned} R(T_{SY} + T_{GY} \cos \alpha) &= M_{SX} + M_{GX} \cos \alpha + M_{GZ} \sin \alpha \\ R\,(T_{SX} + T_{GX}) &= M_{SY} + M_{GY} \\ RT_{GY} \sin \alpha &= M_{SZ} + M_{GX} \sin \alpha - M_{GZ} \cos \alpha \end{aligned} \tag{4.25}$$

The maximum resisting moments are proportional to the contact radius (a_G and a_S). We can assume that the generated moments follow the same rules [42]. With respect to Eq. 4.25, at each interface, the moments can be written in the following ways:

$$\begin{aligned}
M_{GX} &= \frac{a_G}{a_S + a_G} R\,(T_{GY} + T_{SY}\ \cos\alpha) & M_{SX} &= \frac{a_S}{a_S + a_G} R\,(T_{SY} + T_{GY}\ \cos\alpha) \\
M_{GY} &= \frac{a_G}{a_S + a_G} R\,(T_{SX} + T_{GX}) & M_{SY} &= \frac{a_S}{a_S + a_G} R\,(T_{SX} + T_{GX}) \\
M_{GZ} &= \frac{a_G}{a_S + a_G} R(T_{SY}\ \sin\alpha) & M_{SZ} &= \frac{a_S}{a_S + a_G} R(T_{GY}\ \sin\alpha)
\end{aligned} \tag{4.26}$$

By using these expressions in condition 4.22 with the reactions explained in Eqs. 4.8 and 4.9, the thresholds of the rolling movement without any sliding become, at both interfaces:

$$\frac{a_G}{a_S + a_G} R \cdot F \sqrt{[\cos\psi\ (1 + \cos\alpha)]^2 + \{\sin\psi\ [\cos(\alpha+\phi) + \cos\phi]\}^2} > a_G c_{rG} W_G \tag{4.27}$$

$$\frac{a_S}{a_S + a_G} R \cdot F \sqrt{[\cos\psi\ (1 + \cos\alpha)]^2 + \{\sin\psi\ [\cos(\alpha+\phi) + \cos\phi]\}^2} > a_S c_{rS} W_S \tag{4.28}$$

Both these conditions have to be simultaneously fulfilled to allow the object to roll. The relation on the force F inducing the object to *roll without any sliding* becomes

$$\begin{aligned}
&F\sqrt{[\cos\ \psi\ \ (1 + \cos\ \alpha)]^2 + \{\sin\psi\ [\cos(\alpha+\phi) + \cos\phi]\}^2} \\
&\quad > \frac{a_G + a_S}{R} \max(c_{rG} W_G, c_{rS} W_S)
\end{aligned} \tag{4.29}$$

Moreover, a rolling movement of the microobject can follow the sliding generated at one of the interfaces. The minimal force inducing the sliding on one of the interfaces and the rolling of the object is obtained by introducing expressions 4.7 and 4.21 into 4.26. This threshold allows determining which movement the object would get when it slides at one of the interfaces.

To get a movement of *pivoting*, the conditions on the force are issued from Eqs. 4.23 and 4.26 and can be written as

$$\text{At the substrate–object interface}\ |T_{GY}\ \sin\alpha| > \frac{a_G + a_S}{R} 2\mu_S N_S \tag{4.30}$$

$$\text{At the gripper–object interface}\ |T_{SY}\ \sin\alpha| > \frac{a_G + a_S}{R} 2\mu_G N_G \tag{4.31}$$

In conclusion, when the object is pivoting on the gripper side and rolling on the substrate [Fig. 4.3(a)], the pivoting threshold is done by the combination of Eqs. 4.28 and 4.31. Similarly, when the object is pivoting on the substrate and rolling on the gripper [Fig. 4.3(b)], Eqs. 4.27 and 4.30) define the pivoting threshold. In the case of sliding, expressions 4.7 and 4.21 have to be used to define the tangential strain before introducing it in 4.30 and 4.31.

4.2.3.5 Behavior of the Microobject

By comparison of the separation and sliding limits according to the orientation of the gripper and the release direction, it appears that in some areas sliding and separation do not occur first on the same interface (Fig. 4.4). The placing of these areas varies in function of the friction coefficients. Generally, a great difference between the forces at the interfaces (e.g., Γ near 100 or 0.01, respectively) allows getting operations that are clearly in a picking state (or a releasing state, respectively). In reality, it is not so obvious to get such conditions of Γ. Also, a ratio near 1 allows operating pick and release by changing only the orientation of the gripper and/or the release direction. The positioning is not optimized by this configuration because mainly of the induced rolling movement. By choosing wisely the parameters, at least one of the operations can be improved. It is thus important to know most precisely the configurations that limit these areas of uncertainty. For this purpose, the thresholds of separation, sliding, and then rolling or pivoting need to be investigated. These force thresholds are computed based on the expressions of adhesion forces presented in Chapter 1 where the forces A_G and A_S are considered as the pull-off forces. As shown in Figure 4.5

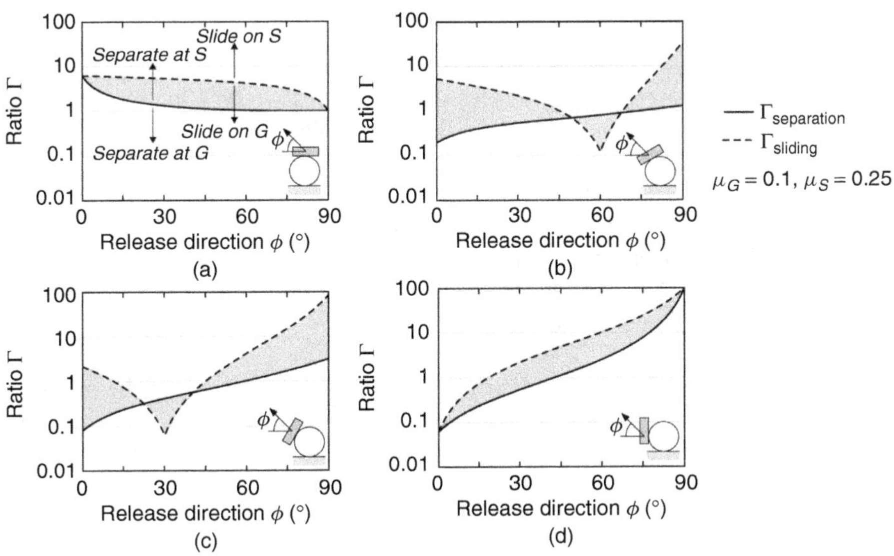

Figure 4.4. Behavior of limits of separation and sliding in function of release direction for gripper oriented by (a) 0°, (b) 30°, (c) 60°, and (d) 90°.

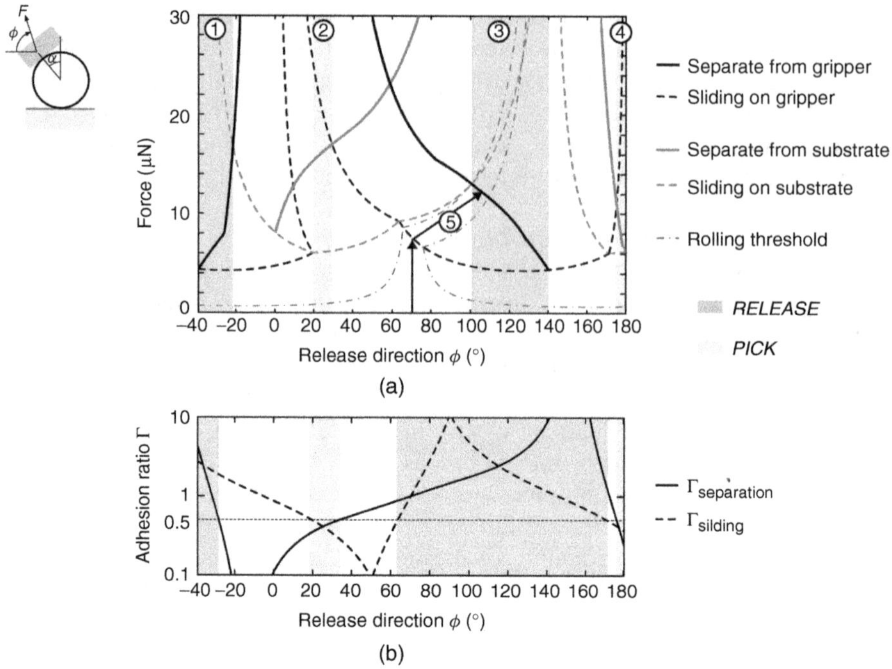

Figure 4.5. (a) Behavior of sphere made of polystyrene of 50 μm in diameter between two flat surfaces for $\Gamma = 0.5$, $\mu^G = \mu^S = 0.25$ and gripper orientation $\alpha = 40^\circ$ in function of the release direction (considering the release direction is included in the x-z plane: $\psi = 90^\circ$ as described in Fig. 4.2) areas of pick and release are shown; (b) separation and sliding limits for the same configuration.

the study of the force thresholds define more precisely the different areas. The uncertainty areas correspond to the fact that the object would slide on both interfaces before achieving any separation threshold. No way to know safely what could happen then. However, for sure the positioning of the microobject in this situation is no more controlled.

Figure 4.5 shows the situation of a gripper oriented by $\alpha = 40^\circ$. Four areas of pick or release appear to function in the release direction ϕ. Between each of them, a sliding behavior is induced successively on both interfaces. The transfer strategy varies from one to the other:

- Both extreme areas (1 and 4) represent a transfer by shearing the interface. The movement is tangent to the interface to break. In this case the transfer can only be effective when the entire gripper or substrate surface will slide on the object: The surface of contact has to be mechanically eliminated. One of the interfaces is in a splitting case but the other one is ensured or even forced in contact.
- Areas 2 and 3 are, on the contrary, characterized by the splitting of one interface opposed directly to the adhesion force on the other side. Each

of these areas implies a rolling movement of the microobject. A release without rolling could be obtained only with a precise and fine measured and controlled release force by following area 5 in Figure 4.5.

The micromanipulation based only on the adhesion forces at the two interfaces needs the following characteristics in order to have robust and repeatable operations:

- A maximal difference of adhesion forces between both interfaces, but with a ratio that can be inverted to go from a picking configuration to a placing one.
- An effective contact area as large as possible to increase the adhesion force on the gripper side, but with a minimal size in order to not perturb every other microobject on the substrate (optimization of the grasping step).
- The control of the contact orientation to use the most efficient release strategy. The correction of the alignment allows also achieving a transfer area without rolling.
- A limitation of the force applied on the microobject: first to prevent any undesired movement and then to prevent (plastic) deformation of the microobject. The measure of the force exerted on the microobject allows generating only minimal perturbations. To limit deformation and movement the sensor needs a resolution lower than 0.1 μN. The force limitation can also be achieved with a passive system.

4.2.4 Grasping — A Special Case of Adhesion Handling

The use of tweezers is very common even in micromanipulation, although the release operation becomes very perturbed by the presence of the surface forces. Grasping a microobject is usually not so problematic. Some issues concerning the limitation of the grasping force as well as the size and disposal of the fingertips are despite everything quite sensitive points notably concerning the positioning performances. The release operation with tweezers is in fact an adhesion-based micromanipulation problem. Actually, during the release with tweezers the microobject is finally in contact only with one of the fingertips, which has a preferred orientation of $\alpha = 90^\circ$ according to Figure 4.2.

4.2.4.1 Grasping a Microobject

One of the advantages of microtweezers is their ability to generate high holding force. This can be very advantageous when the component has, for example, to be detached from a wafer. The minimal grasping force depends then on how much the substrate hold it. The maximal grasping force is defined by the plastic deformation of the microobject. It is interesting to note that by increasing the number and size of the contact area on the gripper side, the adhesion effect at the gripper–object interface will increase, and so the grasping force can be decreased as well. There becomes a limit where the grasping force is only necessary to place the object at its right position inside the gripper jaws, but the adhesion would be then sufficient to hold it.

4.2.4.2 The Release Operation

The release operation is first a question of ensuring the contact between the object and the substrate. Without this connection, the object will for sure stick on one of the fingertips, ending in a failed release. Once the contact is effective with the substrate, the release direction will, of course, depend on the disposal of the contact areas, particularly if some mechanical referencing features were placed on the fingertips. In a general case, we can assume that the three directions of the space are available. A first investigation about the separation limit in Eq. 4.14 for the specific case of tweezers ($\alpha = 90^\circ$) shows immediately the difference between the release directions as illustrated in Figure 4.6:

1. For a vertical release ($\psi = 90^\circ$) the separation condition is simply expressed by $\mu_G A_G < A_S$. The separation limit becomes

$$\Gamma = \frac{\Gamma_G}{\Gamma_S} < \frac{1}{\mu_G^2} \tag{4.32}$$

2. For a tangential release ($\psi = 0^\circ$) the separation condition is easily obtained by $\mu_G A_G < \mu_S A_S$. It gives the following separation limit:

$$\Gamma < \frac{\mu_S^2}{\mu_G^2} \tag{4.33}$$

3. Finally a lateral release responds to the condition $A_G < \mu_S A_S$, which determines the limit:

$$\Gamma < \mu_S^2 \tag{4.34}$$

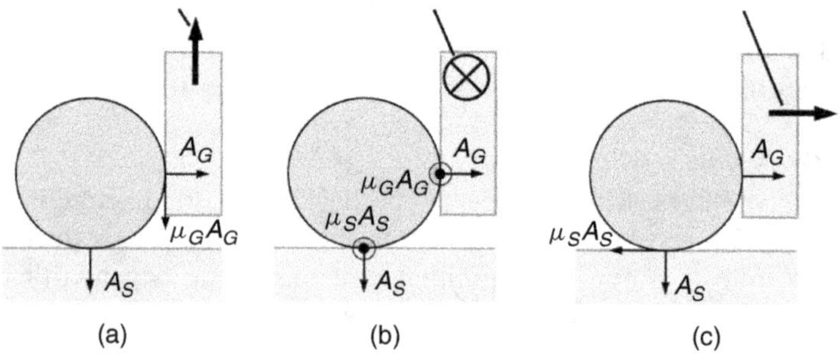

Figure 4.6. Schemes of the three different release directions: (a) vertical, (b) tangential, and (c) lateral.

As friction coefficients are smaller than 1, at least in the ambient air, the lateral release represents the most restrictive situation. In this case the adhesion ratio Γ would have to be strictly smaller than 1 to be able to operate a successful release. Inversely the vertical release brings the widest freedom: Even with an adhesion ratio Γ greater than 1, so intuitively for a catching configuration, the release would be possible. Finally, the tangential release is less restrictive in term of adhesion ratio than a lateral release, but according to the geometry, it can generate some pivoting movement.

The comparison of these three separation limits gives important clues about the design of microtweezers. The lateral release is the less efficient direction. Nevertheless, microtweezers are often designed with two mobile fingertips: Once in contact with the substrate and at the opening of the tweezers, the object is without doubt subjected to a lateral release. The object will thus stick to one of the fingertips during the opening of the tweezers. Any previous effort of positioning will be affected by this operation and a recentering step is then needed. The same problem occurs in the case of smooth fingertips when the grasping force deforms it. In consequence, the design of tweezers with one stiff mechanical reference (fixed fingertip) and only one mobile finger should be preferred to optimize the positioning performance at the opening and release steps.

Finally applying a vertical movement in the direction of the substrate before operating the release will initiate a break at the gripper–object interface independently of the adhesion ratio Γ (Fig. 4.7). The other main advantage is to ensure the contact of the microobjet and the substrate before applying any release direction. This strategy allows to decrease the adhesion effect at the gripper–object interface and to better take advantage of the adhesion at the substrate–object interface. In this way the release is possible even with high adhesion ratio Γ. Figure 4.8 shows the relation between the adhesion ratio Γ that limits the successful release, the orientation of the movement over the substrate (ϕ), and the friction coefficient with the substrate (μ_S). In conclusion, manipulations of microobjects with tweezers require taking into account not only the characteristics at both interfaces but also the structure of the tweezers that plays an important role during the opening and release as well as the geometry of the fingertips. Complex fingertips with a referencing feature will, of course, allow a well-positioned object inside the jaws. Indeed the adhesion effect may be increased by this multicontact interface

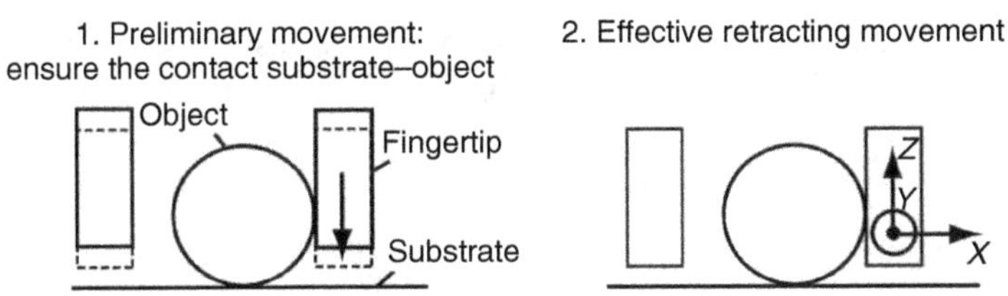

Figure 4.7. Release strategy for microtweezers: (1) get the contact with the substrate and simultaneously a reduction of the adhesion to the gripper side, and (2) release movement.

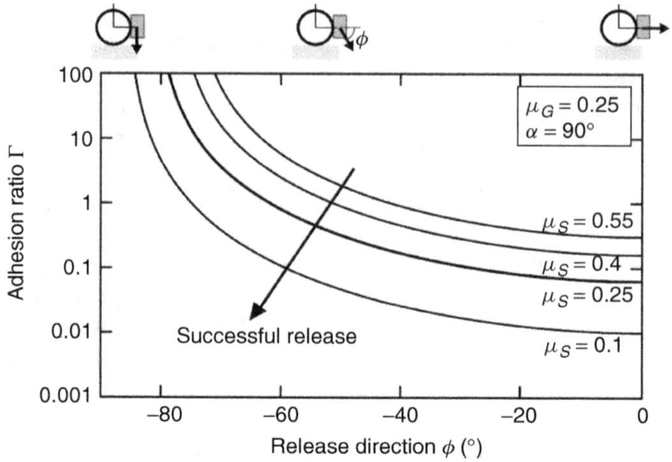

Figure 4.8. Separation limit (Γ) in function of release direction for different friction coefficients on the substrate ($\mu_G = 0.25$, $\alpha = 90°$).

requiring a high adhesion to the substrate or limiting the orientation of the release movement.

4.2.5 Case of an Additional Force Acting at the Interface

This case is typically the situation of a vacuum handling and may be pointed out to characterize a capillary manipulation. In this situation, the development of the gripper needs a way to evaluate the amplitude of the force to apply at the interface. We consider Figure 4.9, which illustrates the different forces and their directions. The force F_G will, of course, depend on the adhesion effect at both interfaces and comes directly from expressions 4.4 and 4.14:

$$F_G = A_S \frac{\sin\phi + \mu_S |\cos\phi|}{\sin(\alpha+\phi) + \mu_G |\cos(\alpha+\phi)|} \frac{1 + \mu_G \sin(\alpha+\phi) |\cos(\alpha+\phi)|}{1 + \mu_S \sin\phi |\cos\phi|} - A_G \tag{4.35}$$

The same evaluation can, of course, be made in case of an additional force on the substrate–object interface. One interesting behavior can occur when the force at interface applied an attractive effect at a certain distance. In this case, the microobject can be picked without having to ensure the contact with the gripper if

$$F_G > A_S \frac{\cos\alpha + \mu_S |\sin\alpha|}{1 + \mu_S |\sin\alpha| \cos\alpha} \tag{4.36}$$

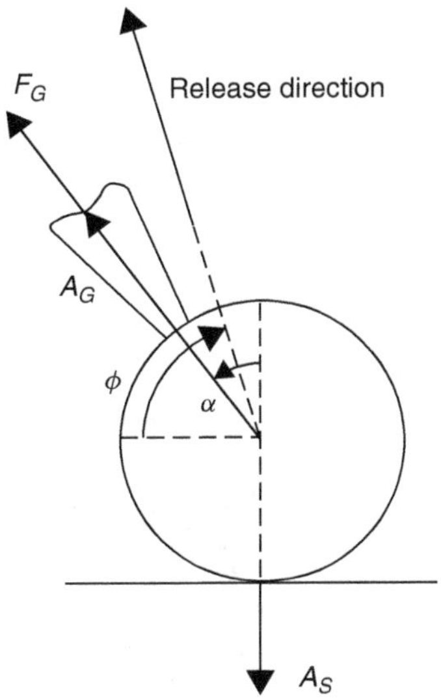

Figure 4.9. Scheme of forces and their orientations.

When the force is not sufficient to counterbalance the adhesion to the substrate side, it allows sometimes sliding the object on the substrate surface if

$$F_G > \frac{\mu_S A_S}{|\sin \alpha| + \mu_S \cos \alpha} \tag{4.37}$$

The contact with the gripper is thus facilitated but the picking operation follows expression 4.35.

4.2.6 Case of an External Force Acting on the Component

Until now, we took into consideration only the forces applied inside the contact area. However, it can happen that another force constrains the microobject and usually on the gravity center. In case of the gravity force, we already show its too small effect compared to the adhesion forces. Nevertheless, for a greater acceleration the inertial force exerted becomes no more negligible. We speak then about an inertial release. Let us now consider the orientations given by Figure 4.10. As in the previous case, we are interested in the evaluation of the force (inertial force) to operate a transfer of the microobject.

There are two situations. The first one corresponds to a microobject in contact only with the gripper. The minimal force allowing to counterbalance the adhesion at the gripper–object interface is expressed as

$$F_{\text{In}} > \frac{A_G\,[\sin(\phi - \alpha) + \mu_G\,|\cos(\phi - \alpha)|]}{1 + \mu_G\,\sin(\phi - \alpha)\,|\cos(\phi - \alpha)|} \tag{4.38}$$

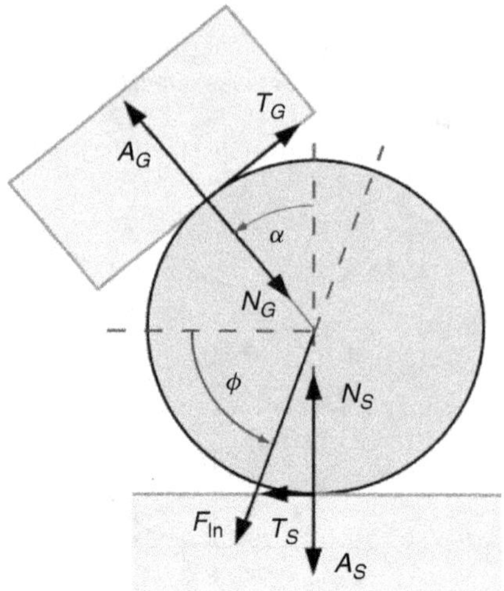

Figure 4.10. Scheme of forces and their orientations.

For the second case, the microobject is in contact with the gripper and the substrate. The adhesion force at the substrate–object interface will thus reduce the necessary inertial force. This one becomes

$$F_{\text{In}} > \frac{A_G\,[\sin(\phi-\alpha)+\mu_G\,|\cos(\phi-\alpha)|] - A_S\,(\sin\phi+\mu_S\,|\cos\phi|)}{1+\mu_G\,\sin(\phi-\alpha)\,|\cos(\phi-\alpha)|+\mu_S\,\sin\phi\,\cos\phi} \tag{4.39}$$

The direction of the external force will differentiate between the pick-and-place operations. The use of an external force can allow being independent of the substrate to generate the release. The positioning performance will then depend, of course, on the control of the acceleration direction but also on the ability of the substrate to stop the object.

4.3 MICROMANIPULATION SETUP

The micromanipulation setup used in this study was developed in order to make comparative tests of microgrippers based on different principles. The setup has sufficient adaptability and standardization to be able to be interfaced with the different configurations of microgrippers. As well, the vision system and the integration of all elements were designed to give the opportunity to work with components between 5 and 500 μm in size. The measurements concern the reliability of the pick-and-place operations as well as the positioning repeatability for components of 50 μm size. All of the experiments are performed in a gaseous environment. The setup is shown in Figure 4.11 and the main elements are illustrated in Figure 4.12 [14, 16].

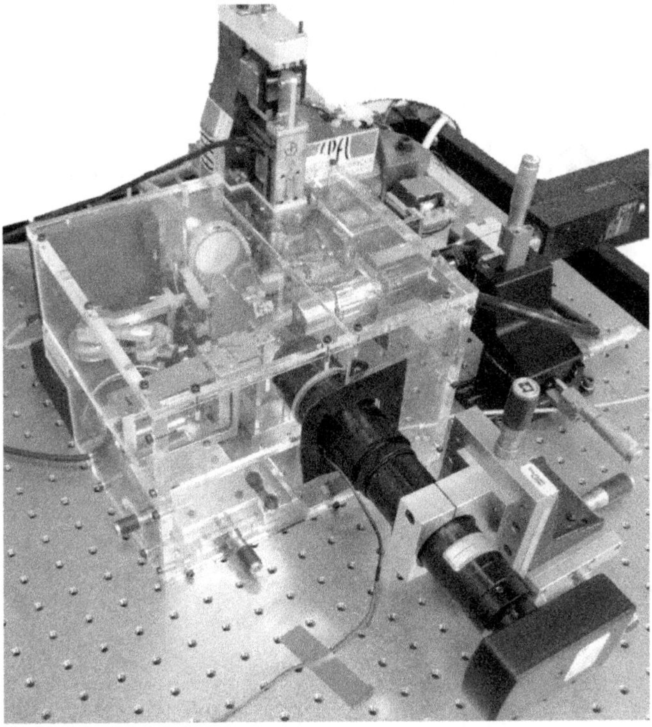

Figure 4.11. View of micromanipulation setup.

The manipulator is a robot Delta^3 based on a parallel kinematics of 3 degrees of freedom. It allows strokes of ±2 mm and a repeatability of ±10 nm. It takes up in a cube of 210 mm side. Its design is based on flexible hinges with noncontact actuators (moving magnet) and sensors. Thus no friction at all occurs in the actuator and measurement loops [7]. The different microgrippers have been conceived to be fixed to the Delta^3 by a unique pneumatic gripper closed by default. Each microgripper is then designed based on a standardized interface. Thanks to mechanical referencing features under the form of three balls contacting three V-grooves, the tip of the gripper is always brought in the field of view of the microscope with a repeatability of 0.3 μm. In this way a tool changer has been integrated in the setup. Its original hypocycloidal kinematics has the advantage of a very small occupied volume and allows an elliptical trajectory with only one actuator. An automated change of the tool can be performed without any perturbation of the working space, notably without having to open the closed surroundings. Different microgrippers can be tested in exactly the same environmental conditions. The substrate is fixed on a motorized z axis with a stroke of 300 μm and a resolution of 0.1 μm. Two manual tables allow placing the microobject in the field of view of the microscopes before the experimentations. The substrate can be rigidly fixed either to the moving axis or through an elastic

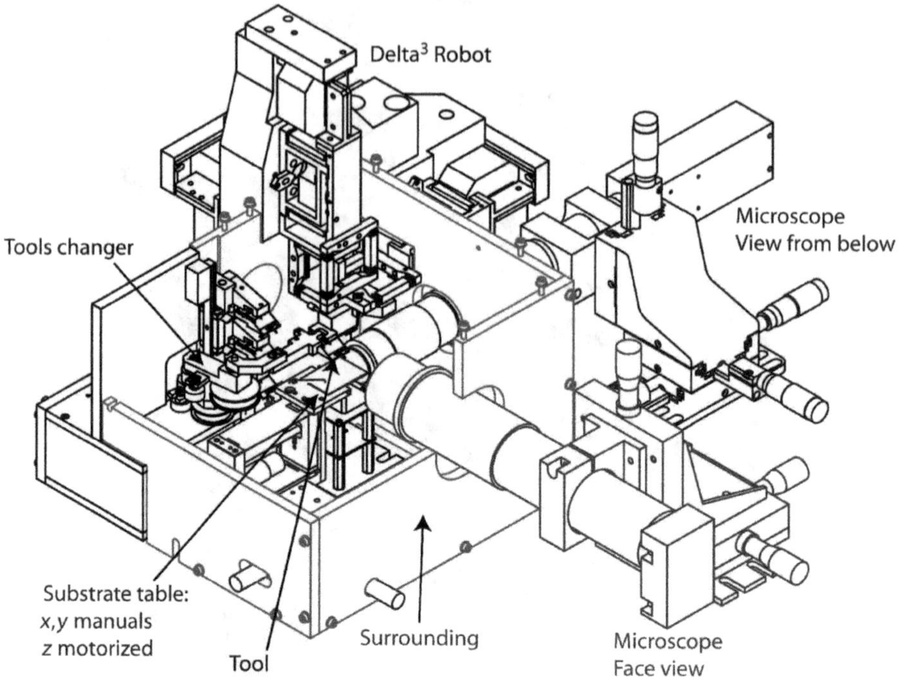

Figure 4.12. Micromanipulation setup with different functionalities.

link in order to limit the force applied on the microobject by the gripper during the operations. The stiffness is then only 10 μN/μm.

The working space is placed in a box in which the relative humidity is monitored and controlled by injecting a nitrogen flow. The surrounding is not sealed but allows keeping out of the box the elements that generate dust, keeping inside the most sensitive parts. In this way and because of the overpressure created inside, the setup does not need to be placed in a clean room but just with standard laboratory conditions. Finally, the supervision of the operations is made with two microscopes that show back and side views of the manipulation scene. The back view is used for the positioning measurement as well as for the detection and tracking of the microobjects and microgripper. The side view gives the vertical information data. Automated procedures have been implemented in order to decrease the influence of the operator on the result of a micromanipulation task.

4.4 EXPERIMENTATIONS

As we have seen above, some configurations or principles can be more adapted to only one of the operations of pick or place. The grippers presented here are consequently generally a combination of principles. They were developed to manipulate

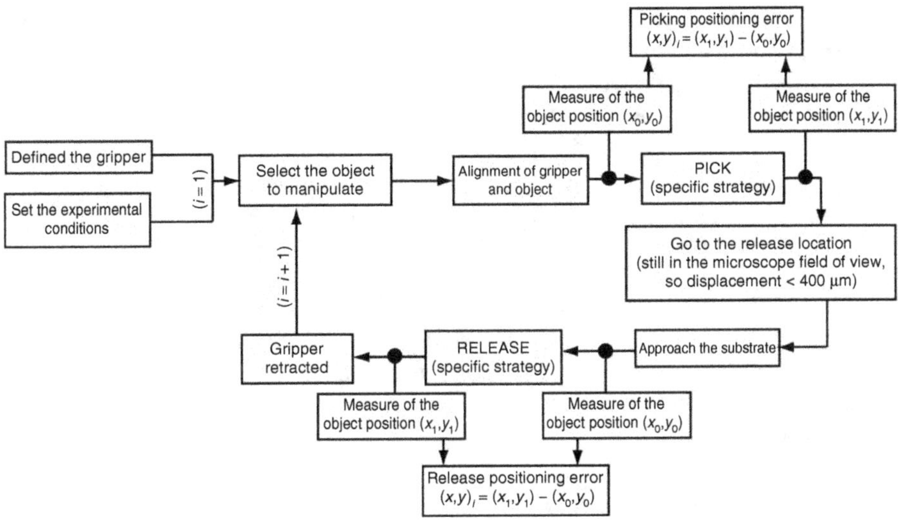

Figure 4.13. General operating mode for characterization of pick-and-place operations.

components of 50 μm in size and the objects were typically polystyrene spheres of 50 μm in diameter. The gripper families discussed here are the following:

1. Microtweezers for which two main configurations are presented: They vary by their guidance structure and the integration and type of actuation.
2. Inertial microgripper based on adhesion for the picking operation.
3. Thermodynamic microgripper that uses the cycle condensation/evaporation of the ambient relative humidity for capillary manipulations.
4. Vacuum gripper assisted by vibration for the release operation.

The experimental study is based on the success rate measurement for both operations of pick and release as well as on the positioning performances represented by the positioning repeatability. The influences of the material at interfaces, the relative humidity, the release direction, or the holding force were investigated in this way. Each micromanipulation principle requires, of course, some specific steps: Thus the pick-and-place strategies are particular to each gripper family. Figure 4.13 shows the general operating mode common to all grippers.

The characterization is based on the success or failure of the operation and on the positioning error measurement. Success rate and positioning repeatability are then extracted from these data. These parameters are defined as:

- *Positioning Error.* It is induced by the transfer operation itself (usually the release operation). The positioning error (x,y) is the difference between the final position of the microobject (x_1,y_1) and its original position (x_0,y_0) before the transfer.

- *Success Rate.* This is the ratio between the number of successful operations (pick or release) and the total number of these operations. An operation is considered as successful if the transfer was made correctly and with a positioning error of less than 20 μm.
- *Positioning Repeatability.* It quantifies the operation variability in terms of positioning and depends on the standard deviation σ. The repeatability is then considered as two times σ. Each measure corresponds to at least 30 successful operations under similar conditions.

No binding feature was used to fix the microobject to the substrate but the adhesion effects. So this is solely pick-and-place operations and not microassembly. This last one would depend on the constraints and influences of the binding feature itself and thus would represent another large field of research that will not be discussed here.

4.4.1 Microtweezer Family

This study focuses on the precision in micromanipulation. We saw in Section 4.2.4.2 that having a fixed and stiff fingertip optimizes the release operation in that sense. Two types of microtweezers were considered:

- *Modular Microtweezers.* This allows a large flexibility especially for the tip shape, size, and material. The actuator is independent of the guidance structure of the fingertips. This structure is a low-cost tool with the advantage of an easy adaptability to each microobject.
- *Monolithic Microtweezers.* Contrary to the first one, here everything is integrated: guidance structure, fingertips, and actuator. There is no modularity after fabrication, but a higher precision.

4.4.1.1 Modular Microtweezers: Pneumatic Microtweezers

The possibility of using fingertips with material and shape adapted to the microcomponent and a unique actuator has motivated the development of such pneumatic microtweezers (Fig. 4.14). The gripper is then composed of a monolithic guidance structure based on flexible hinges obtained by laser cutting from a stainless steel plate. The tips are then fixed on it. In order to guaranty their alignment, the tips are cut as a unique piece and detached from each other after assembly. Two types of fingertips were used:

- Stainless steel tips of 50 μm in thickness cut by microelectro-discharge machining [Fig. 4.16(a)].
- Silicon tips of 12 μm in thickness fabricated by deep reactive ion etching (DRIE) [26] [Fig. 4.16(b)].

The pneumatic actuator is made of a bellow that simply pushes the mobile finger of the tweezers. The bellow is fabricated by a process of nickel electrodeposition and coated with gold (Servometer Precision Manufacturing Group

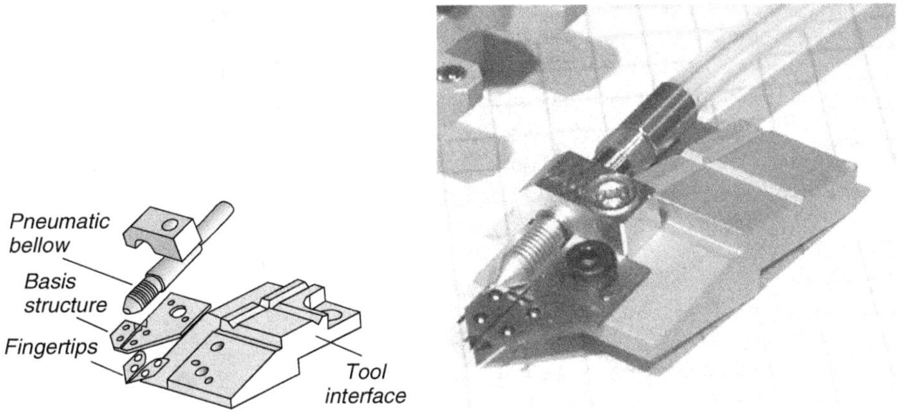

Figure 4.14. Pneumatic microtweezers.

LLC: external diameter of 2.44 mm for a length of 5.9 mm). The choice of this pneumatic actuator was made because of its small volume and the easy way to interface it with the guiding structure.

The actuation of the structure needs a force of several hundreds of millinewtons. Such microtweezers can thus apply high grasping force, for example, to detach the component from its support. Nevertheless, it needs a careful use in order to not achieve any plastic deformation of the component or of the fingertips. A pressure sensor has been added to the setup, allowing measuring the grasping force with a resolution of 51 μN for the stainless steel tips (respectively, 40 μN for the silicon tips because of the length difference).

The main disadvantage of such gripper is the lack of tip alignment due to the assembly itself. Effectively even a small difference in the glue layer thickness can induce a problematic misalignment. We have proposed an in situ correction process based on microelectro-discharge machining to correct and even shape the gripper tips [35]. Another solution is to add a degree of freedom to one of the tips to get an actively controlled alignment, for example, by integrating a piezoelectric bimorph actuator to one of the fingertips. Such microtweezers were proposed in the literature [1, 16, 58].

4.4.1.2 Monolithic Microtweezers — MEMS Tweezers

The use of the silicon lithographic process allows integrating fine flexible guiding structures, actuator, and sensor on a same device. The monolithic microtweezers experimented during this study were designed and fabricated at the Institute of Robotics and Intelligent Systems (IRIS) at Swiss Institute of Technology in Zürich. The microtweezers have an electrostatic actuator based on a comb drive structure and a capacitive force sensor that measures in fact the displacement of the fingertips when grasping an object. The device is fabricated by DRIE from a silicon on insulator (SOI) wafer. Figure 4.15 shows this microtweezers. The details of the conception and fabrication can be found in Beyeler et al. [11].

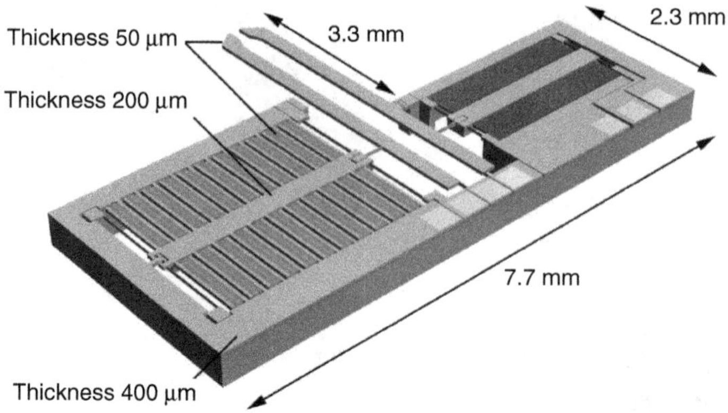

Figure 4.15. MEMS tweezers with comb drive actuator and capacitive force sensor fabricated by DRIE [11].

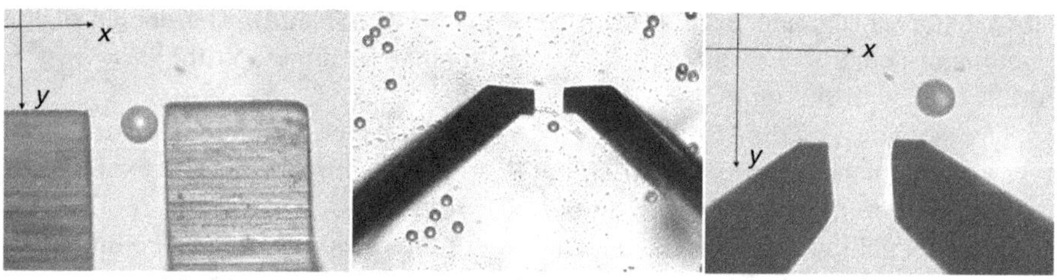

Figure 4.16. Views of fingertips of different configurations of microtweezers: (a) stainless steel tips of 50 μm in thickness and (b) silicon tips of 12 μm in thickness of the pneumatic microtweezers; (c) silicon tips of 50 μm in thickness of the MEMS microtweezers.

The displacement of the fingertip is 100 μm under 150 V. Two configurations were developed for the force sensor: the first one with a range of ±2800 μN and a resolution of 520 nN and the second one with a range of ±360 μN and a resolution of 70 nN. This microtweezers was designed for fine force-controlled operations. The goal here is not to achieve high grasping force but inversely to allow applying minimal forces and even to be able to quantify simultaneously the adhesive effects. These high performances are counterbalanced by the fragility of the structure as well as by the sensitivity to dust. Interfacing such tools is then more constraining and with high risks of breaking the device.

4.4.1.3 Experimentations on Microtweezers

Figure 4.16 shows the different fingertips of the microtweezers as well as the related x-y directions.

Grasping Operation. Picking polystyrene spheres of 50 μm in diameter does not oppose any problem but electrostatic charges that sometimes make the objects

TABLE 4.2. Influence of Release Direction on Positioning Error of Polystyrene Sphere of 50 μm in Diameter.[a]

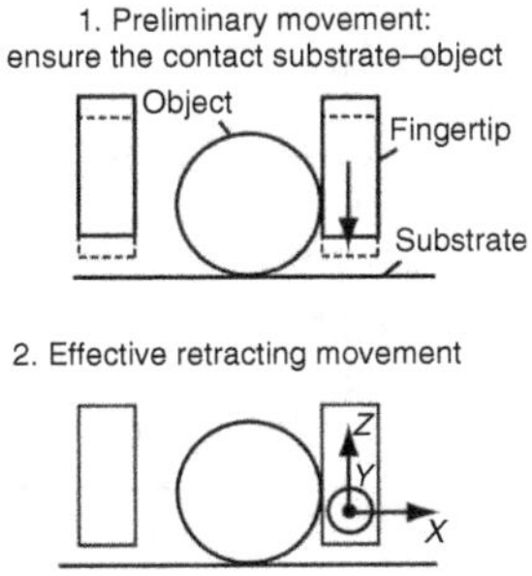

Release Direction	Repeatability x (μm)	Repeatability y (μm)
X : lateral	**4.20**	2.20
Y : tangential	2.34	**12.54**
Z : vertical	**1.40**	**1.76**

[a] The manipulations were practiced with the MEMS tweezers. A hydrophobic coating (perfluoroalkylchlorosilane, PFS) was deposed on the fingertips; the substrate is in glass + 10 nm chromium.

jump. As object and substrate are of insulated material, their electrical connection is not possible. Anyway, we tested ionic nozzles to balance the charges in the working space. Its influence could not be demonstrated in a repeatable way because also charging effects were not repeatable. However, the spheres just jump on the fingertips of the gripper as the ionic nozzle is switched on. This effect could be used as a picking principle based on an external electrostatic charge generator acting on a conductive gripper plate as shown also in Hesselbach et al. [27].

Release Direction. To get a successful release of a microobject, this one has to touch the substrate first in order to use its adhesion force for maintaining it during the retracting movement. The direction of this movement will then influence the positioning of the object. The strategies shown in Table 4.2 start all, after the opening of the tweezers, by a first movement normal to the substrate and in the direction to this last one (10 μm). The complete retracting of the gripper is then realized. The positioning performances will, of course, depend on the quality of the orientation of the gripper, the substrate, and the release direction. Without any mechanical referencing feature on the substrate, any angular error will be reproduced at each step on the positioning error. The lateral release is the less sensitive to any misalignment but at the same time the most influenced by the adhesion ratio. The tangential release is subjected to create pivoting movement of the object. The large difference between the positioning error in x and y shows well this influence (Table 4.2).

The vertical release gives the best positioning results as expected in Section 4.2.4.2. The perturbation is minimal with this strategy and compatible with a high adhesion ratio. All related operations that follow will use this strategy.

Influence of the Relative Humidity. The increase in relative humidity provokes the spontaneous condensation of the liquid on the surfaces and, as well, increases

TABLE 4.3. Positioning Performances for Different Relative Humidity Levels.[a]

Substrate	Relative Humidity (%)	Repeatability x (µm)	Repeatability y (µm)	Success rate (%)
Glass	3±1	7.98	8.94	76
	21±1	5.08	8.18	89
	44±3	3.44	4.76	91
Glass + hydrophobic coating	3±1	11.9	2.62	95
	21±1	6.14	2.34	89
	44±3	13.88	3.14	93

[a]The pneumatic microtweezers with silicon fingertips was used.

the capillary effects at the interfaces. Note that an interface will be less sensitive to this variation if there is a hydrophobic coating on at least one of the surfaces. Table 4.3 shows the results of the release operations on a glass substrate with different humidity levels. We can see an improvement of the positioning with the increase of the humidity as well as of the success rate. As the substrate has a hydrophilic tendency, the capillary effect will increase with the humidity and so the ability to fix the object on the substrate. The operations were then pursued with the same gripper but on a hydrophobic substrate. Results in Table 4.3 show that there is a clear influence of the finger tip movement during the placing operation as the repeatability is deteriorated mainley in the x-direction. At low humidity, the positioning was more perturbed by electrostatic chargers. At higher level the capillary effect may increase only on the gripper side, including larger uncontrolled release positions. In consequence the distribution in x and y is not as homogeneous as with the glass substrate. The release operations are effectively very dependent on the substrate adhesion characteristics. The increase of the relative humidity has a positive effect under the condition that the substrate has a hydrophilic tendency whatever the gripper characteristic.

Influence of the Materials. Intuitively, the decrease of the adhesion effects on the gripper side should make easier the release operations. The measures shown in Table 4.4 confirm this fact. The success rate is larger in case of a hydrophobic coating on the gripper surface. Nevertheless, this evolution is not so important and is even not observed for the positioning repeatability. On the contrary, and as just seen above, a hydrophobic coating on the substrate will perturb the positioning. Considering a fine chromium[1] deposition (10 nm) on the glass substrate, we remark a decrease of the success rates (Table 4.4). The difference is greater without using any hydrophobic coating on the gripper surface. However, the positioning repeatability is less affected. The configurations "silicon on gripper" and "glass + chromium on substrate" represent thus the highest adhesion ratio and so the worst case for a release operation.

[1]The 10-nm chromium layer increases, in fact, the roughness of the substrate. Its very small thickness makes an inhomogeneous layer that does not allow ensuring any electrical connection.

TABLE 4.4. Results of Positioning of Polystyrene Spheres of 50 μm in diameter with MEMS Tweezers with/without Hydrophobic Coating on Fingertips

Substrate	Hydrophobic Coating on Gripper	Repeatability x (μm)	Repeatability y (μm)	Success Rate (%)
Glass	No	1.34	1.48	86
	Yes	1.68	1.84	93
Glass + 10 nm chromium	No	1.32	2.20	71
	Yes	1.40	1.76	89

Discussion. Both major parameters for a successful manipulation with microtweezers (as well as for any adhesion transfer) concern the ability to induce an effective contact object–substrate and the quality of the adhesion on the substrate side.

Without being in contact with the substrate, the object will stick to one of the fingertips and stay on it at release time. Once the contact is made, the substrate adhesion allows limiting the object movement. Of course, the less the gripper perturbs the object, the more precise will be the positioning: This shows the importance of the release direction as well as of a low adhesive gripper surface. The micro electromechanical systems (MEMS) tweezer is the best adapted for getting high precision. This is due to its monolithic structure, to the very well defined contact areas, and to the high resolution of the actuator. But we have to be careful about the disposal of the force sensor as generally it is based on the measure of the deformation given by the grasping force itself. In our case, the sensor is placed on the opposite finger to the actuated one. A small displacement occurs when grasping and then the fingertip comes back at the resting position during the opening. The amplitude is small (lower than 1 μm) but is still a cause of positioning error when high performances are anticipated. For that reason, it would be better to integrate actuator and sensor on the same side in order to get a fix and stiff fingertip. The stiffness of the fingertips plays an important role in the positioning. When too smooth, they are deformed by the grasping force and will come back to their resting position at the opening. The optimum would be to get stiff fingertips in every direction.

In conclusion the microtweezers were the best controlled operations that we considered. Their use can then be limited by the great space they need to have access to the microcomponent.

4.4.2 Inertial Microgripper Based on Adhesion

The inertial microgripper combines the use of adhesion effects for the picking operation with, as its name implies, an inertial release. The orientation of the contacts and the characteristics of both interfaces influence the success of the operations for both these manipulation principles. For getting reliable and precise

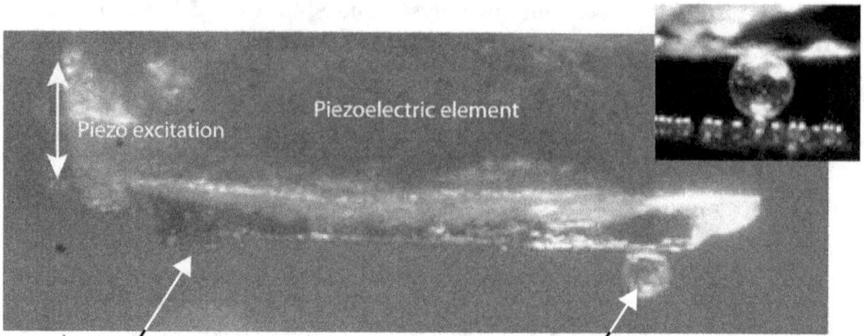

Figure 4.17. View of piezoelectric element and of available contact surfaces: a flat silicon piece and a glass sphere of 200 μm in diameter compared to the 50 μm in diameter polystyrene spheres.

operations, the direction of the release acceleration was chosen normal to both interfaces. In this way, the accessibility to the microobjects is also optimized.

4.4.2.1 Conception of the Inertial Microgripper

The controlled deformation of a piezoelectric element will generate the high release acceleration. Figure 4.17 shows the construction of such a microgripper. The acceleration generated by a piezoelectric actuator working in the d_{33} mode and excited with a sinusoidal signal can be expressed by

$$a* = \delta\omega^2 = \eta V d_{33}(2\pi f)^2 \tag{4.40}$$

with δ the displacement, V the applied voltage, f the frequency, ω the pulsation, and d_{33} the piezoelectric coefficient (450×10^{-12} m/V for PIC 151 from Physik Instrument GmbH). The coefficient η represents the attenuation factor of the piezoelectric displacement [15]. The attenuation is mainly due to the reaction of the support of the piezoelectric element. It can be evaluated by making the equilibrium of the inertial forces acting on the structure and on two parts of the piezoelectric element. The first one participates to the structure acceleration and only the second one generates the desired acceleration. A first attenuation factor corresponds to our design to $\eta_1 = 0.65$.

A second coefficient comes from the limited bandwidth of the high-voltage amplifier that was used. For the frequency range of 150 – 350 kHz, the attenuation varies from 0 to 4%, giving an average value of $\eta_2 = 0.98$. At higher frequency, until 600 kHz the attenuation achieves 10%. The estimated effective displacement generated by the piezoelectric actuator at the placing of the microobject is finally

$$\delta = \eta_1\eta_2 d_{33} V \approx 0.65(0.98)450 \times 10^{-12}\ \text{m/V}(200)\ \text{V} = 57\ \text{nm} \tag{4.41}$$

TABLE 4.5. Success Rates for Picking Operation with Glass Sphere Tip on Glass Substrate with Hydrophobic Coating

Relative humidity	40%	20%	2%
Success rate	83%	71%	54%

The frequency range of 150 – 350 kHz, for a voltage of 200 V, corresponds to accelerations of 5.1×10^4 to 2.8×10^5 m/s^2. The mass of a polystyrene sphere of 50 μm in diameter is 6.9×10^{-11}kg. These accelerations allow generating inertial forces of 3.5–19.3 μN

4.4.2.2 Experimentations on Inertial Gripper

Picking by Adhesion. For a configuration where both interfaces are facing each other, one of the main parameters during the transfer by adhesion is the force exerted on the microobject. Without its limitation, the deformation at contact achieves quickly the plastic domain. For that reason we did the next operations with the substrate fixed on a compliant table (0.10 μN/μm). To improve the picking operation, the glass substrate was recovered with a hydrophobic coating. The picking by adhesion with the silicon flat surface of the gripper allowed success rates between 70 and 80% even with 2% relative humidity. The strategy was simply to come vertically into contact and then to leave by the same way. The same operations with the glass sphere tip were really less efficient: The adhesion ratio Γ is in practice just above 1. For that reason the strategy was the following: A rolling movement of the object allows reducing the adhesion force at the interface before applying a vertical release. The success rates achieve then more than 70%, with a reduction at low relative humidity (Table 4.5). A force between 100 and 200 μN was applied to pick the object. The rolling distance was 2 μm. It can be evaluated as the diameter of the deformed contact area (see Chapter 1).

Inertial Release. Two types of excitation signals (Fig. 4.18) were used to generate the necessary acceleration. They are in both cases sinusoidal signals with an amplitude of 200 V and a variable frequency. The mode *single* means a unique sine pulse and the mode *multiple* is a pack of 10 sinusoidal pulses. These signals are sent every 20 ms to the piezoelectric actuator. The minimal release frequency is reached by increasing the frequency in steps of 10 kHz to between 60 and 500 kHz. An operation is then considered as successful if the release is done at a frequency lower than 500 kHz.

Minimal Frequency of Release. Table 4.6 shows the results of the measures of the minimal frequency of release. The adhesion forces were estimated based on a spherical gripping surface of 150 μm in diameter with an adhesion energy of $\gamma_G = 0.03$ J/m^2. The pull-off force acting at the interface with a polystyrene sphere can then be evaluated as 5.6 μN. An acceleration of 8.1×10^4 m/s^2 would be necessary to release the micro-object. It corresponds to a frequency of 190 kHz.

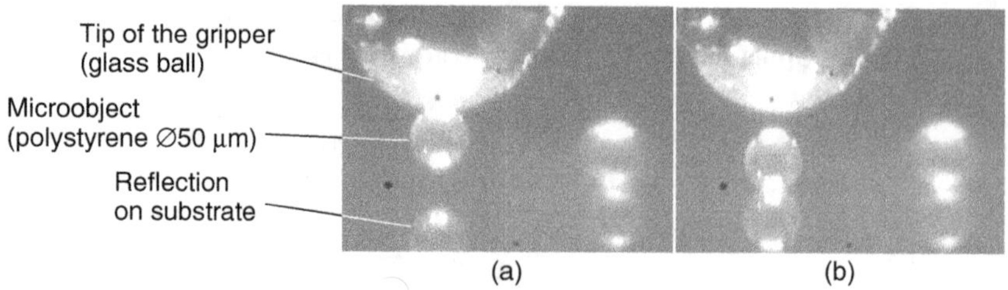

Figure 4.18. Inertial release of polystyrene sphere of 50 μm in diameter: (a) hold by adhesion and then (b) ejected under the high acceleration issued from the piezoelectric actuator (not visible on the picture).

TABLE 4.6. Measure of Minimal Frequencies to Get the Release, Depending on Excitation Mode and Relative Humidity.[a]

			Frequency (kHz)		Force (μN)		
RH (%)	Mode	Success Rate (%)	Average	Standard Deviation	Average	Min	Max
2	Single	84	213	44.1	7.6	3.5	15.0
	Multiple	91	252	52.9	10.3	4.0	22.3
20	Single	77	239	43.5	9.2	4.0	16.8
	Multiple	92	228	41.0	8.3	3.5	17.9
40	Single	81	253	55.8	10.4	3.0	24.7
	Multiple	92	243	65.4	9.8	2.6	37.1

[a] We give here also the equivalent inertial force computed from the minimal frequency.

When the capillary force dominates, the force would be of 22.6 μN and thus a frequency of 382 kHz. We remark an increase of the frequency with the relative humidity denoting an increase of the adhesion force due to the capillary effect. This tendency can be really seen in the single mode, but the minimal frequency in the multiple mode is quite constant. With relative humidities of 20 and 40%, the frequencies are lower in the multiple mode than in the single. Mode it is thus possible that in reality the release would need more than one pulse. This fact was already demonstrated by Haliyo et al. [24] in the case of an acceleration close to the release threshold. The object would be ejected only after a few pulses and not already after the first one. The multiple mode would therefore be more efficient. The success rates confirm this tendency. The dispersion of the measures is also quite large. From one operation to the next one, the conditions at the interface are thus not stable. By analyzing the equivalent forces, we remark that the maximal values for each configuration would confirm the presence of a meniscus at the interface. At low relative humidity this effect is less important but still not negligible. But the minimal values are lower than the evaluated pull-off force. That would confirm that the quality of the surface influences greatly the adhesion effect.

TABLE 4.7. Positioning Performances at Minimal Frequency and with Threshold of 350 kHz

Frequency	RH (%)	Mode	Repeatability x (μm)	Repeatability y (μm)	Success Rate (%)
Minimal	2	Single	8.8	6.8	87
		Multiple	3.8	4.0	91
	20	Single	11.0	2.0	77
		Multiple	6.9	4.9	92
	40	Single	7.4	3.1	81
		Multiple	7.9	5.2	92
350 kHz	2	Single	7.8	3.7	48
		Multiple	5.6	1.8	97
	20	Single	9.5	1.9	70
		Multiple	11.8	4.4	93
	40	Single	8.7	5.3	89
		Multiple	7.8	4.9	90

Positioning Performances. The positioning measurements were done at the minimal frequency as well as by applying a frequency threshold. This threshold was set to 350 kHz or an equivalent inertial force of 19 μN. This frequency is higher than 98% of the measured minimal frequencies. The acceleration is applied after placing the object 5 μm above the substrate. Seeing the success rates in Table 4.7, the results confirm the interest to apply a pack of pulses instead of a unique one. The use of a high threshold that would generate a higher inertial force than the one just necessary does not improve the success rate or the positioning. The best configuration looks to be a minimal but repetitive signal. It seems also that the positioning repeatability is better at low relative humidity. It could be explained by a less homogeneous break of the meniscus interface compared to a dry contact. The measured positioning error does not inform us about the relative position of the gripper–object before release and thus on the centering error. The error in x is bigger than in y, but the x direction is also the rolling direction used to pick the object. The repeatability along y would show here the expected performances for a centered microobject.

4.4.3 Vacuum Nozzle Assisted by Vibration

The use of suction in micromanipulation is limited by the adhesion effect. Below a size of around 20 μm the microspheres can no more be manipulated by this principle because the suction force generated by the pressure difference becomes lower than the adhesion force.

A vacuum tool in micromanipulation will use either the adhesion or an over-pressure for the release. This last solution cannot be applied to high positioning requirements because the object is simply ejected from the tool. We propose here to combine the suction effect for the picking operation with a vibration of the

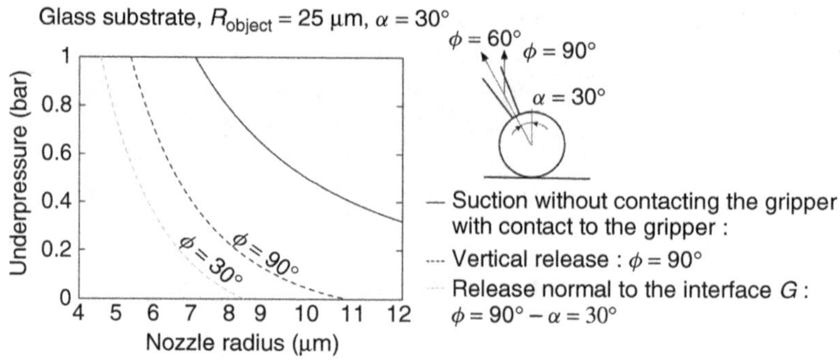

Figure 4.19. Evaluation of the nozzle radius for manipulating a polystyrene sphere of 50 µm in diameter and with an gripper tip orientation of 30°.

gripper to help the placing operation by adhesion. We could suppose that if we need a suction force to pick a microobject we should be able to depose it just by adhesion. However, the difficulty to control the effectiveness of the contact with the object appears quite often. By applying a vibration to the gripper, the interface gripper–object should be damaged as soon as the contact is established and without having to induce any releasing movement.

4.4.3.1 Conception

Figure 4.19 allows to evaluate the size of the vacuum nozzle for different release directions and pressures. Picking the object without an effective contact with the gripper tip is limited to a radius of more than 7 µm for a polystyrene sphere of a radius of 25 µm. Glass pipettes in borosilicate allow obtaining such small nozzles and keeping a compact volume. This will be comfortable for the user. The repeatability of the operations will depend on the repeatability of the size and shape of the nozzles, which can be easily modified (Fig. 4.20). They are fragile but also quite inexpensive. In that sense they could be integrated as consumed and changed often, or at least be adapted to the objects they manipulate. The gripper orientation is fixed to 30° from the vertical. A piezoelectric actuator is inserted between the pipette fixture and the gripper support: The whole system will then vibrate. The actuator is axially polarized to induce a shearing effect at the contact area gripper–object (Fig. 4.21).

4.4.3.2 Experimentations on Vacuum Gripper

As vibration will be continuous during the release step, it is interesting to know its influence on the picking operation. For a small influence, this vibration could always be applied to the gripper independently of which operation is processed. The influence of the vibration on the positioning and reliability of both operations were considered. A further study would need to investigate the best type of signal and amplitude. The manipulations related here were held on a glass substrate with a relative humidity of 42% ±3%.

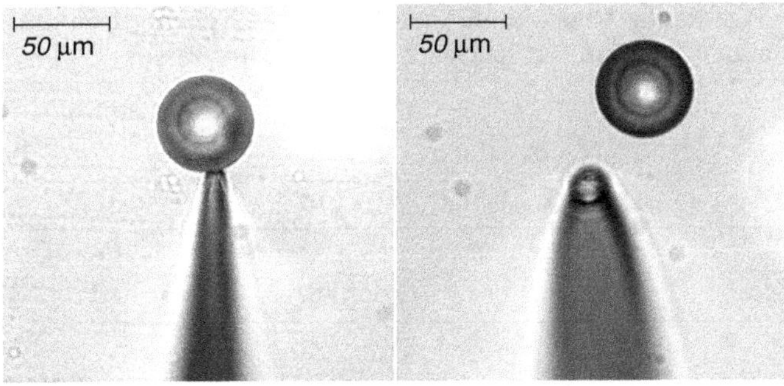

Figure 4.20. Pipette tips with variation of size compared to polystyrene sphere of 50 μm in diameter.

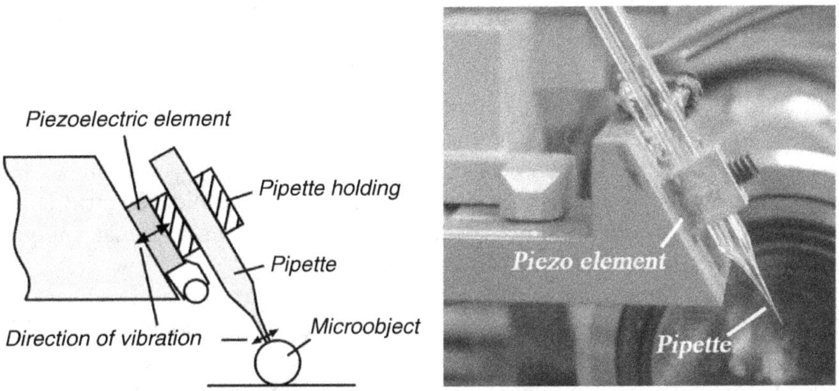

Figure 4.21. Vacuum gripper with piezoelectric actuator for vibrating plate.

Picking with Vacuum. With objects of 50 μm in diameter, a nozzle diameter smaller than 8 μm does not allow picking the object, whatever is the pressure (success rate lower than 20%). For a larger diameter, an effective contact gripper–object was not necessary in 78% of the cases. The spheres were even attracted, sometimes at a distance of 12 μm, but generally at a distance of 1–3 μm between the nozzle tip and the object. The vibration does not have an effect on the picking success rate, but perturbs in some way the position of the object on the nozzle.

Releasing Operation. The manipulation without vibration is only based on a transfer by adhesion, as stopping the vacuum is not sufficient to make the sphere fall down. The contact has thus to be made effective with the substrate before shearing or breaking the interface with the tool. Most of the time, these operations needed successive trials to get a successful transfer. With vibration, the release occurs when contacting the substrate and without having any other movement but the vertical touching and then a vertical release. Table 4.8 gives the position repeatability results for the same configuration but with and without the vibration.

TABLE 4.8. Positioning Repeatability of Polystyrene Sphere of 50 μm in Diameter with and without Vibration of Pipette on Glass Substrate

	Positioning Repeatability			
	Pick		Place	
Release Type	x (μm)	y (μm)	x (μm)	y (μm)
With vibration	2.8	5.8	1.6	3.2
Without vibration	2.2	3.8	3.3	16.4

The pipette diameter was 13 μm (±0.2 μm). For picking, the values of repeatability are certainly affected by any misalignment between the pipette and the object, but they show anyway the small influence of the vibration on the picking step. For placing, the vibration allows a well-improved positioning performance. Without vibration, the positioning is more influenced in the y direction, which corresponds to the pipette direction (Table 4.8). The excitation of the piezoelectric actuator is a square signal of 300 mV at a frequency of 2 kHz. The amplitude varies between 300 mV and 1 V for some manipulation where the sphere was more adhering to the pipette. This is the fact, for instance, when using a larger diameter pipette. The amplitude of the movement of the piezoelectric actuator is smaller than 1 nm at such a voltage. Finally, we should note that applying a sinusoidal signal is less efficient on the release operation.

The reason a small movement of the pipette could reduce significantly the adhesion effect at the interface could be explained by the models of Mindlin et al. [38], Savkoor [46], and finally Johnson [29]. They studied the behavior at the contact before getting a complete sliding. Their model assumes that a lateral force lower than the friction force induces a displacement inside the contact area. Thus, a local sliding occurs and reduces greatly the adhesion force until it affects the whole area where the complete sliding is produced (Fig. 4.22). The use of such a type of vacuum gripper has the great advantage of the best accessibility among all the microgrippers presented here. The second advantage is the very easy and intuitive way to use such a tool that allows a user to be quickly autonomous.

4.4.4 Thermodynamic Microgripper

The manipulation based on adhesion can be improved by using the capillary forces as they represent the main contribution of the surface forces. Even if there is quite often a liquid adsorbed layer on the surfaces, this last one is usually not sufficient to ensure the complete capillary force. Below 70% relative humidity, the contribution of the capillary forces to the adhesion effects is not total. Thus it is generally preferred to use a controlled volume of liquid to make the manipulations. Instead of picking a drop of water with the gripper, we propose here to use the effects of condensation/evaporation of the ambient relative humidity as the supply of water. This principle gives the name to the thermodynamic microgripper.

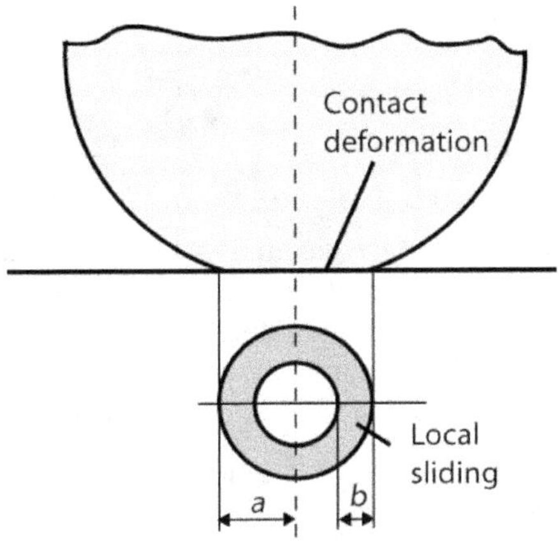

Figure 4.22. Scheme of deformation at contact area and of local sliding area. When $b = a$, the sliding is complete.

4.4.4.1 Conception

The temperature of condensation depends on the ambient temperature and on the relative humidity (Fig. 4.23). The expression of the local relative humidity and the temperature of condensation are obtained by the Clapeyron equation at the liquid–gas transition considering the law of the perfect gas. Assuming that the enthalpy of vaporization ($\Delta_{vap}H$) is independent of the temperature, after

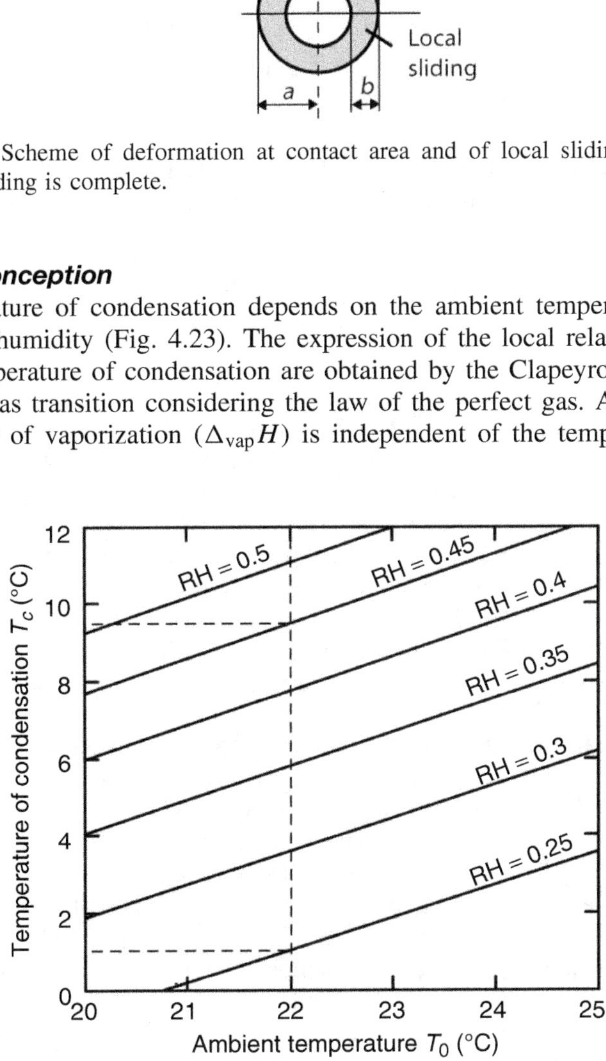

Figure 4.23. Condensation temperature is function of ambient temperature and relative humidity.

integration we get the following expression for the pressure of saturation p_s at temperature T [6]:

$$p_s(T) = p_s(T_0)e^{-\chi} \quad \text{with} \quad \chi = \frac{\Delta_{\text{vap}}H}{R}\left(\frac{1}{T} - \frac{1}{T_0}\right) \tag{4.42}$$

with R the constant of the perfect gas and $p_s(T_0)$ the pressure of saturation at temperature T_0. The relative humidity is the ratio between the ambient pressure and the pressure of saturation at temperature T [$\text{RH} = p/p_s(T)$]. By modifying locally the temperature and assuming the pressure stays uniform in the whole volume, the local relative humidity can be expressed by

$$\text{RH}_{\text{local}} = \frac{p}{p_s(T_{\text{local}})} = \frac{\text{RH}[p_s(T_0)]}{p_s(T_{\text{local}})} = \text{RH}(e^{\chi}) \tag{4.43}$$

When the relative humidity achieves 100%, the Kelvin radius becomes infinite. We can then determine the local condensation temperature T_C, for $\text{RH}_{\text{local}} = 1$, as

$$T_C = \frac{T_0\, \Delta_{\text{vap}}H}{\Delta_{\text{vap}}H - \ln(\text{RH})RT_0} \tag{4.44}$$

The environment, where the setup is located, has between 25 and 45% relative humidity for a temperature of 22°C. The minimal temperature to achieve will be between 1 and 9.5°C. For the evaporation, a temperature of 60°C would allow decreasing the local relative humidity between 4 and 7%. The conception of such a gripper is based on a Peltier element (PE-008-03-09 by Supercool AB) that will allow to heat and cool down alternatively the tip of the gripper. It is recovered by a steel plate of 50 μm in thickness, which were cut by laser to have a tip as small as the microobject. The tip was then polished. A temperature sensor (type J thermocouple) is fixed on this plate and finally the whole device is insulated. Figure 4.24 shows a closeup view of the active part of this gripper.

4.4.4.2 Experimentations on Thermodynamic Microgripper

The adopted strategy to pick and place a microobject is the following: For the picking operation, the gripper is cooled down to 3°C to induce the condensation on the gripping area; the gripper comes then in contact with the microobject and a meniscus is created at the interface; finally a vertical release of the gripper is applied. The placing operation starts with the gripper placed 80 μm above the substrate; the gripper temperature is increased until 30°C, and the object encounters the substrate; the gripper temperature is again cool down to 3°C in order to condensate water on the substrate. The temperature is finally increased again to 30°C and the gripper is retracted. Finally, the liquid on the substrate evaporates naturally.

Figure 4.25 shows the different steps of the release process of a silicon cube of 50 μm side. The conditions were of 40% ±4% for the relative humidity

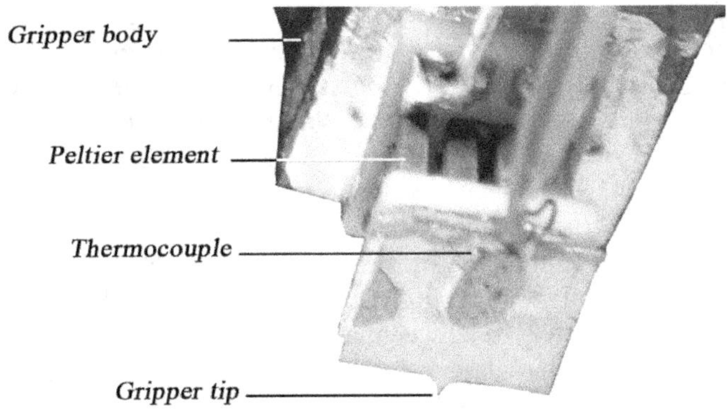

Figure 4.24. Thermodynamic microgripper.

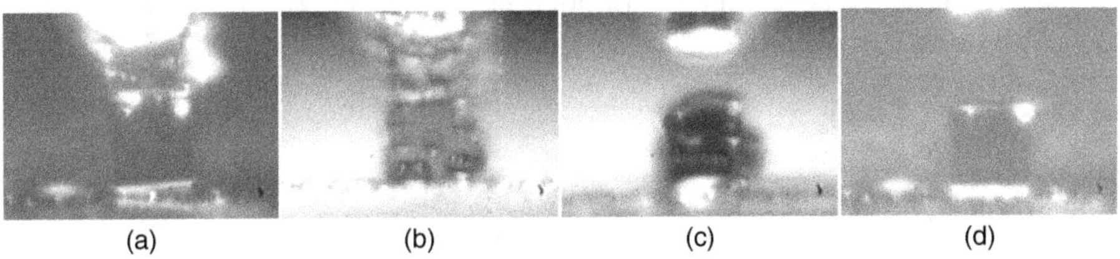

Figure 4.25. Release of silicon cube of 50 μm side with the thermodynamic gripper: (a) approach and increase of the gripper temperature at 30°C, (b) cooling at 3°C, (c) heating at 30°C and retracting of the gripper, and (d) evaporation.

TABLE 4.9. Positioning Performances for Pick-and-Place Operations with Thermodynamic Microgripper

		Repeatability		
Microobject	Operation	x (μm)	y (μm)	Success Rate (%)
50-μm side silicon cube	Pick	3.4	2.5	97
	Place	4.0	6.4	83
50-μm polystyrene in diameter sphere	Pick	8.1	8.9	75
	Place	9.8	6.8	79

and a temperature of 23°C ±1°C. The results of pick-and-place operations with polystyrene spheres and silicon cubes are shown in Table 4.9.

The operations were easier with the silicon cubes than with the polystyrene spheres. The polystyrene is less hydrophilic than the silicon. Moreover the capillary forces are larger between two flat surfaces than between a sphere and a plane. The two types of components could indeed be manipulated.

Concerning the silicon cubes, the positioning repeatability is better in picking than in placing. As the gripper has a similar size (50 μm) and shape, it appears as a centering effect due to the meniscus. For the spheres, this effect is less marked due to the geometry of the component. During the picking operation the use of a temperature cycle “heat until ambient temperature—cool down to 3°C” will increase the meniscus size. The component can even be nearly covered by the drop. This strategy was more efficient to take the polystyrene spheres.

Finally, there is an attractive effect due to the meniscus that facilitates greatly the manipulation. It is then not necessary to have a real contact between the object and the gripper, so no need to worry about the applied force like for a “dry” manipulation. The increase of the temperature until around 60°C or more has shown experimentally to be not efficient to release the object. In fact, there will be always a water layer in between, and this layer generates a large capillary force that could not be overbalanced by the substrate adhesion. We thus decide to “transfer” the water drop from the gripper to the substrate to make the placing of the component as shown here above: Condensation is provoked on the substrate side for the release operation. Another considered solution was to reduce the size of the contact area. This strategy got good performances for the manipulation of components of 2 mm in diameter. The limit is here the fabrication challenge of such a gripper.

4.5 CONCLUSION

The choice of one gripper and mainly of the manipulation principle can depend on various parameters such as the reliability and precision of the operation. Some constraints like the accessibility or the condition of the environment can be more important for parameters that depend on the micro-object, for instance, a sensitivity to the presence of a liquid or to electrostatic charges or some specific geometry. Finally, a manipulation setup could sometimes have to be used by different users. In this case, the simplicity of use would perhaps have a great influence on this choice.

The models presented here allow evaluating the feasibility of principles and then optimizing the gripper as well as the strategy of manipulation. Adhesion effects are difficult to evaluate, but taking their ratio in account allows representing the configuration in order to make iterative steps to optimize first the model then the strategy.

Considering the conception of a micromanipulation or microassembly setup, the development of the microgripper cannot be kept out of the whole system. The difficulties to evaluate the adhesion effects in real conditions as well as the great dependence between all the interfaces in the success of the operations ask for a very strong interaction between the designers of the different elements that are the component, the receiver, and the gripper. In such a way optimal choices for the surfaces and principles of manipulation can be done.

After having defined the main trends theoretically, the difficult evaluation of the adhesion effect in real conditions causes the need to check rapidly by experimentations the feasibility of the pick-and-place operations. The whole integration is not necessary at this step, only the validation of the principles. It allows then a few iterations to optimize the gripper before developing a more industrial tool with, for instance, the integration of some sensors or other functionalities.

Manipulation with tweezers presents the advantage of the best control of the position. Going to smaller dimensions will increase the adhesion problem and will certainly combine them with some other principles such as vibration effects or by locating a picking principle on the substrate (e.g., located capillary effect).

Of course, evolution of the presented principles is still expected. For example, the cycle evaporation/condensation is not optimal on the evaporation process: It may be possible to improve the evaporation by getting an explosion process that would first decrease the meniscus but also apply an overpressure to eject the object. Another combination of principles could be an electrostatic inertial gripper that would have the advantage of not needing effective contact in both operations.

Focus on microdevice assembly and keep in mind the adhesion effects as well as scaling effects have the great advantage to stimulate the integration, or the combination, of innovative micromanipulation principles. The conception and the development of reliable and precise manipulation tools are then improved by considering the overall situation simultaneously with the evaluation of the particular case of each contact interface.

REFERENCES

1. J. Agnus, P. De Lit, C. Clevy, and N. Chaillet, Description and Performances of a Four Degrees-of-Freedom Piezoelectric Gripper, in Proc. of the IEEE International Symposium on Assembly and Task Planning, 2003, pp. 66–71.
2. K. N. Andersen, K. Carlson, D. H. Petersen, K. Molhave, V. Eichhorn, S. Fatikow, and P. Boggild, Electrothermal micro grippers for pick-and-place operations, *Microelect. Eng*., 85(5–6):1128–1130, 2008.
3. H. Aoyama, S. Hiraiwa, F. Iwata, J. Fukaya, and A. Sasaki, Miniature robot with micro capillary capturing probe, in Sixth International Symposium on Micro Machine and Human Science, 1995.
4. F. Arai, D. Ando, T. Fukuda, Y. Nonoda, and T. Oota, Micro manipulation based on micro physics-strategy based on attractive force reduction and stress measurement, in International Conference on Intelligent Robots and Systems, Human Robot Interaction and Cooperative Robots, vol. 2, 1995, pp. 236–241.
5. F. Arai, and T. Fukuda, A new pick up and release method by heating for micromanipulation, in MEMS '97, 1997, pp. 383–388.
6. P. W. Atkins, Chimie physique, De Boeck Université, 2000.
7. J.-P. Bacher, Conception de robots de très haute précision à articulations flexibles: Interaction dynamique-commande, Thèse No. 2907, EPF Lausanne, 2003.
8. C. Bark, T. Binnenbose, G. Vogele, T. A.-W. Weisener and M. A.-W. Widmann, Gripping with low viscosity fluids, in Proc. of the Eleventh Annual International Workshop on Micro Electro Mechanical Systems, 1998, pp. 301–305.
9. Y. Bellouard, T. Lehnert, J.-E. Bidaux, T. Sidler, R. Clavel, and R. Gotthardt, Local Annealing of Complex Mechanical Devices: A New Approach for Developing Monolithic Microdevices, *Mat. Sci. Eng*., A273–275:795–798, 1999.

10. Y. Bellouard, Conception de dispositifs en alliage à mémoire de forme en microtechnique, thèse no. 2308, EPF Lausanne, 2000.

11. F. Beyeler, A. Neild, S. Oberti, D. J. Bell, Y. Sun, J. Dual, and B. J. Nelson, Monolithically Fabricated Microgripper with Integrated Force Sensor for Manipulating Microobjects and Biological Cells Aligned in an Ultrasonic Field, *J. Microelectromech. Syst.*, 16:7, 2007.

12. S. Bütefisch, V. Seidemann, and S. Büttgenbach, Novel Micro-pneumatic Actuator for MEMS, *Sensors Actuators A: Phys.*, 97–98:638–645, 2002.

13. T. Chu Duc, G. K. Lau, and P. M. Sarro, Polymer Constraint Effect for Electrothermal Bimorph Microactuators, *App. Phys. Lett.*, 91:101902, 2007.

14. M. Dafflon, B. Lorent, and R. Clavel, A Micromanipulation Setup for Comparative Tests of Microgrippers, in International Symposium on Robotics (ISR), Münich, Germany, 2006.

15. M. Dafflon, B. Lorent, W. Driesen, and R. Clavel, Characterization of an Inertial Micro Gripper Based on Adhesion Forces, in 5th International Workshop on MicroFactories (IWMF), Besançon, France, 2006.

16. M. Dafflon, Prehenseurs, conditions et strategies pour une micromanipulation de precision, Thèse No. 4160, EPF Lausanne, 2008.

17. P. De Lit, J. Agnus, C. Clévy, and N. Chaillet, A Four-Degree-of-Freedom Microprehensile Microrobot on Chip, *Assembly Automation*, 24:33–42, 2004.

18. F. Dionnet, Télé-micro-manipulation par adhésion, Thesis, Université Paris 6, Pierre and Marie Curie, 2005.

19. W. Driesen, T. Varidel, S. Régnier, and J.-M. Breguet, Micro Manipulation by Adhesion with Two Collaborating Mobile Micro Robots, *J. Micromech. Microeng.*, 15, 2005.

20. E. T. Enikov, L. L. Minkov, and S. Clark, Microassembly Experiments with Transparent Electrostatic Gripper under Optical and Vision-Based Contro, *IEEE Trans. Indus. Elect.*, 52, 2005.

21. S. Fahlbusch, A. Shirinov, and S. Fatikow, AFM-based Micro Force Sensor and Haptic Interface for a Nanohandling Robot, in IEEE/RSJ International Conference on Intelligent Robots and System, 2002, pp. 1772–1777.

22. J. T. Feddema, P. Xavier, and R. Brown, Micro-assembly Planning with van der Waals Force, in International Symposium on Assembly and Task Planning, Porto (Portugal), 1999.

23. M. A. Greminger, and B. J. Nelson, A Deformable Object Tracking Algorithm Robust to Occlusions and Spurious Edges, in Proc. of the 2005 IEEE International Conference on Robotics and Automation, 2005, pp. 1264–1269.

24. D. S. Haliyo, S. Regnier, and J.-C. Guinot, [mu]mad, the Adhesion Based Dynamic Micromanipulator, *Eur. J. Mech. - A/Solids*, 22:903–916, 2003.

25. S. Henein, M. Thurner, and A. Steinecker, Flexible Micro-gripper for Micro-factory Robots, CSEM, 2003.

26. D. Heriban, J. Agnus, J.-R. Coudevylle, M. Gauthier, and N. Chaillet, Design of Silicon Finger Tips for a MOC (Microrobot on Chip) Microgripper, in TMMF 2005, Desktop MMES and Nano Factories, Tsukuba (Japan), 2005.

27. J. Hesselbach, J. Wrege, and A. Raatz, Micro Handling Devices Supported by Electrostatic Forces, *CIRP Annals—Manufact. Technol.*, 56:45–48, 2007.

28. J. Israelachvili, *Intermolecular and Surface Forces*, Academic, San Diego, 1997.

29. K. L. Johnson, Adhesion and Friction between a Smooth Elastic Spherical Asperity and a Plane Surface, *Proc. R. Soc. Lond. A*, 453:163–179, 1997.

30. D.-H. Kim, B. Kim, H. Kang, and B.-K. Ju, Development of a Piezoelectric Polymer-based Sensorized Microgripper for Microassembly and Micromanipulation, in International Conference on Intelligent Robots and Systems (IROS), 2003, pp. 1864–1869.

31. A. Kochan, European Project Develops Ice Gripper for Micro-sized Components, *Assembly Automation*, 17:114–115, 1997.

32. P. Lambert, F. Seigneur, S. Koelemeijer, and J. Jacot, A Case Study of Surface Tension Gripping: The Watch Bearing, *J. Micromech. Microengin.*, 16:1267–1276, 2006.

33. S. H. Lee, K.-C. Lee, S. S. Lee, and H.-S. Oh, Fabrication of an Electrothermally Actuated Electrostatic microgripper, in 12th International Conference on TRANSDUCERS, Solid-State Sensors, Actuators and Microsystems, 2003, pp. 552–555.

34. B. López-Walle, M. Gauthier, N. Chaillet, A Submerged Freeze Microgripper for Micromanipulations, in IEEE International Conference on Robotics and Automation, 2007, pp. 826–831.

35. B. Lorent, M. Dafflon, C. Joseph, and R. Clavel, In situ Micro Gripper Shapping by Electro Discharge Machining, International Symposium on Robotics (ISR), Münich, Germany, 2006.

36. M. Mayyas, P. Zhang, W. H. Lee, P. A.-S., P., Shiakolas, and D. A.-P., D. Popa, Design Tradeoffs for Electrothermal Microgrippers, in IEEE International Conference on Robotics and Automation, 2007, pp. 907–912.

37. A. Menciassi, A. Eisinberg, M. C. Carrozza, and P. Dario, Force Sensiong Microinstrument for Measuring Tissue Properties and Pulse in Microsurgery, *IEEE/ASME Trans. Mech.*, 8:10–17, 2003.

38. R. D. Mindlin, W. R. Mason, J. F. Osmer, and H. Deresiewicz, Effects of an Oscillating Tangential Force on the Contact Surfaces of Elastic Spheres, in Proc. of the 1st National Congress of Applied Mechanics, 1952, pp. 203–208.

39. M. Mita, H. Kawara, H. Toshiyoshi, M. A.-A. Ataka, and H. A.-F. Fujita, An Electrostatic 2-Dimensional Micro-gripper for Nano Structure, in 12th International Conference on TRANSDUCERS, Solid-State Sensors, Actuators and Microsystems, 2003, pp. 272–275.

40. M. Nienhaus, W. Ehrfeld, F. Michel, V. Graeff, and A. Wolf, Automatic Microassembly of Radar Sensors for Automotive Applications, *Proc. SPIE, Micromach. Microfabri. Process Technol. IV*, 3511, 1998.

41. D. Petrovic, G. Popovic, E. Chatzitheodoridis, O. Del Medico, A. Almansa, F. Sümecz, W. Brenner, and H. Detter, Gripping Tools for Handling and Assembly of Microcomponents, in 23rd international Conference on Microelectronics (MIEL 2002), Nis (Yugoslavia), 2002, pp. 247–250.

42. S. Saito, H. T. Miyazaki, T. Sato, K. Takahashi, and T. Onzawa, Dynamics of Micro-object Operation Considering the Adhesive Effect under an SEM, *Proc. SPIE, Microrobot. Microassembly III*, 4568:12–23, 2001.

43. S. Saito, H. Himeno, K. Takahashi, and T. Onzawa, Electrostatic Detachment of a Microobject from a Probe by Applied Voltage, in International Conference on Intelligent Robots and System, 2002, pp. 1790–1795.

44. Y. Sang, M. Dubé, and M. Grant, Thermal Effects on Atomic Friction, *Phys. Rev. Lett.*, 87, 2001.
45. M. Sausse-Lhernould, A. Delchambre, S. Régnier, and P. Lambert, Electrostatic Forces in Micromanipulations: Review of Analytical Models and Simulations Including Roughness, *Appl. Surf. Sci.*, 253:6203–6210, 2007.
46. A. R. Savkoor, *Fundamentals of Friction: Macroscopic and Microscopic Processes*, I. L. Singer and H. M. Pollock, Eds. Kluwer Academic, Dordrecht, the Netherlands, 1992, p. 111.
47. J. Schlick, and D. Zühlke, Design and Application of a Gripper for Microparts Using Flexure Hinges and Pneumatic Actuation, *Proc. SPIE, Microrobotics and Microassembly III*, 4568:1–11, 2001.
48. V. Seidemann, S. Bütefisch, and S. Büttgenbach, Fabrication and Investigation of Inplane Compliant su8 structures for MEMS and Their Application to Micro Valves and Micro Grippers, *Sensors Actuators A: Phys.*, 97–98:457–461, 2002.
49. M. Sitti, and H. Hashimoto, Two-Dimensional Fine Particle Positioning under an Optical Microscope Using a Piezoresistive Cantilever as a Manipulator, *J. Micromech.*, 1:25–48, 2000.
50. M. Sitti, Atomic Force Microscope Probe Based Controlled Pushing for Nanotribological Characterization, *IEEE Trans. Mechatronics*, 9:343–349, 2004.
51. Y. Sun, D. Piyabongkarn, A. Sezen, B. Nelson, R. Rajamani, R. Schoch, and D. P. Potasek, A Novel Dual-Axis Elecrostatic Microactuation System for Micromanipulation, in IEEE International Conference on Intelligent Robots and Systems, Lausanne (Switzerland), 2002.
52. V. Vandaele, P. Lambert, and A. Delchambre, Non-contact Handling in Microassembly: Acoustical Levitation, *Precision Eng.*, 29:491–505, 2005.
53. B. Voegeli, and H. von Kanel, AFM-Study of Sticking Effects for Microparts Handling, *Wear*, 238:20–24, 2000.
54. R. N. Wenzel, Resistance of Solid Surfaces to Wetting by Water, *Ind. Eng. Chem.*, 28:988–994, 1936.
55. W. Zesch, M. Brunner, and A. Weber, Vacuum Tool for Handling Microobjects with a Nanorobot, in International Conference on Robotics and Automation, Albuquerque (New Mexico), 1997.
56. H. Zhang, Y. Bellouard, E. Burdet, R. Clavel, A.-N. Poo, and D. W. Hutamacher, Shape Memory Alloy Microgripper for Robotic Microassembly of Tissue Engineering Scaffolds, in Proc. of IEEE International Conference on Robotics and Automation, Vol. 5, 2004, pp. 4918–4924.
57. J. W. L. Zhou, H.-Y. Chan, T. K. H. To, K. W. C. Lai, and W. J. Li, Polymer MEMS Actuators for Underwater Micromanipulation, *IEEE/ASME Trans. Mechatronics*, 9:334–342, 2004.
58. Q. Zhou, P. Korhonen, B. Chang, and V. Sariola, 6 DOF Dexterous Microgripper for Inspection of Microparts, in IEEE/ASME International Conference on Advanced Intelligent Mechatronics, Monterey (USA), 2005.
59. Q. Zhou, Strategies in Automatic Microhandling, in International Conference on Mechatronics and Automation, Harbin (China), 2007.

CHAPTER 5

MICROHANDLING STRATEGIES AND MICROASSEMBLY IN SUBMERGED MEDIUM

MICHAËL GAUTHIER

5.1 INTRODUCTION

As presented in Part I, the surface forces may affect the micromanipulation task and especially the release of the microobject. Most of microhandling tasks are done in the dry medium (air or vacuum). But, the liquid medium could have a lot of advantages in micromanipulation of manufactured objects as presented in Section 2.2. The objective of this chapter is to present different microhandling strategies used in liquid and a microassembly station.

Section 2.2 has shown that distance forces (van der Waals, electrostatic, capillary forces) and contact forces (pull-off forces) are reduced in liquid. Adhesion and electrostatic perturbations are thus reduced, which represent a major advantage of the liquid medium. Moreover, hydrodynamic forces increase in liquid and the behaviour of the microobject is thus more stable (see Fig. 2.5). Though the adhesion forces are reduced in liquid, sticking effects are usually not totally canceled [2.1] and the release task stays a critical problem. Consequently, the study of new release strategies of manufactured microobjects in liquid is a key point to perform submerged microassembly.

As current microhandling strategies of manufactured objects are performed in the air (or vacuum), new micromanipulation strategies are required to manipulate in liquid. Two ways can be explored:

Robotic Microassembly, edited by Michaël Gauthier and Stéphane Régnier

- Strategies based on principles used in biomicromanipulation. In this case, principles can be improved or modified to be able to handle manufactured objects in spite of biological objects (no more biocompatibility, more degree of freedom required, etc.).
- New strategies, currently not used in liquid because of biological constraints.

This chapter deals with four handling strategies: the dielectrophoretic gripper, the submerged freeze gripper, the chemical control of the release, and the release using adhesive substrates.

5.2 DIELECTROPHORETIC GRIPPER

In current micromanipulations, usual approaches consist in control of a repulsive physical force to overcome the pull-off force (e.g., acceleration in air: see Section 4.4.2). In liquid the repulsive dielectrophoretic force can be used to overcome pull-off force in order to induce the release of the microobjects. This principle usually used in biological cell manipulations is easily controllable by an electric field and is particularly efficient in liquid.

5.2.1 Principle of Dielectrophoresis

The electrostatic forces are usually used to manipulate microobjects without contact in an electric field. Let us consider an electric field $\overrightarrow{E(t)}$ induced by electrodes. A dielectric object in the electric field can be modeled by an electric dipole whose momentum is $\overrightarrow{m(t)}$. The force and the torque applied on the object are classically done by (see Fig. 5.1)

$$\vec{F} = (\overrightarrow{m(t)} \cdot \vec{\nabla})\overrightarrow{E(t)} \tag{5.1}$$

$$\vec{\Gamma} = \overrightarrow{m(t)} \otimes \overrightarrow{E(t)} \tag{5.2}$$

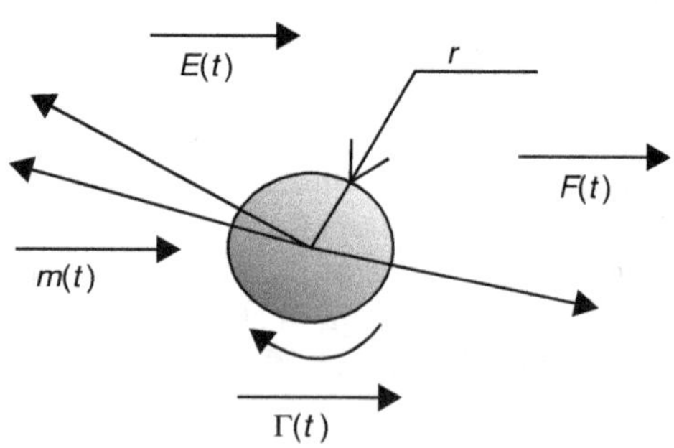

Figure 5.1. Electrostatic force and torque applied on a microobject in an electric field.

where the dipole momentum $\overrightarrow{m(t)}$ is a function of the geometry and the dielectric properties of the microobject. In case of a spherical particle, the momentum is given by

$$\overrightarrow{m(t)} = 4\pi r^3 \epsilon_3 \frac{\kappa_1 - \kappa_3}{\kappa_1 + 2\kappa_3} \overrightarrow{E(t)} \tag{5.3}$$

where r is the radius of the particle, and κ_i the complex dielectric constant of the object ($i = 1$) and the medium ($i = 3$). These complex dielectric constants κ_i depend on the pulsation w of the electric field and is defined by $\kappa_i = \epsilon_i + j\sigma_i/w$, where ϵ_i and σ_i are the dielectric constant and the electric conductivity. The complex fraction of this equation (5.3) is consequently a function of the pulsation w and is named the Clausius–Mossoti factor, $K(w)$:

$$K(w) = \frac{\kappa_1 - \kappa_3}{\kappa_1 + 2 \cdot \kappa_3} \tag{5.4}$$

These expressions enable the calculation of the electrostatic efforts applied in an object in any case. Two particular examples that represent the dielectrophoresis force and torque are presented next.

5.2.1.1 Dielectrophoresis Force

In order to build the expression of the dielectrophoresis force, we consider a stationary sinusoidal electric field $\overrightarrow{E(t)}$ (see in Fig. 5.2) [36]:

$$\overrightarrow{E(t)} = \overrightarrow{E_0} \cos(wt) \tag{5.5}$$

The dipole momentum, which represents the electric behavior of the particle, is thus defined by

$$\overrightarrow{m(t)} = 4\pi r^3 \epsilon_3 [\mathrm{Re}(K(w)) \cos(wt) + \mathrm{Im}(K(w)) \sin(wt)] \overrightarrow{E_0} \tag{5.6}$$

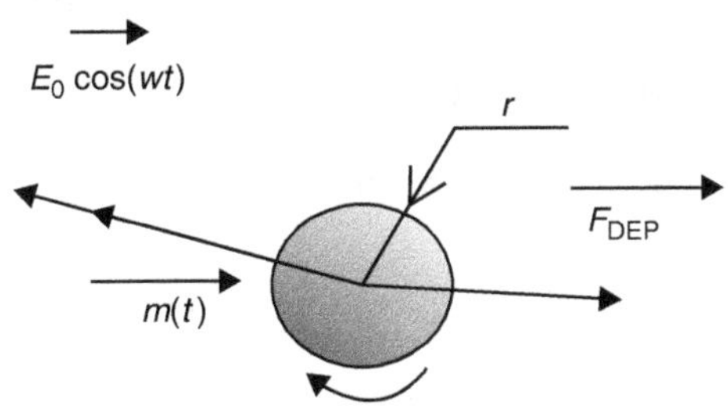

Figure 5.2. Dielectrophoresis force applied on a dielectric microobject.

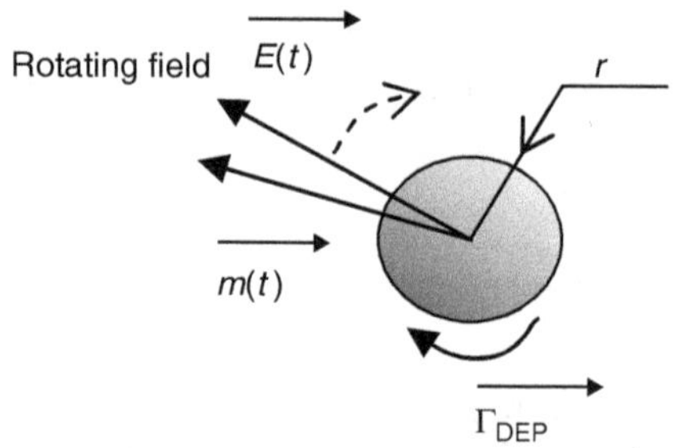

Figure 5.3. Dielectrophoresis torque applied on a microobject in a rotating field.

Consequently, the variable force $\overrightarrow{F(t)}$ at every time t is

$$\overrightarrow{F(t)} = 4\pi r^3 \epsilon_3 [\mathrm{Re}(K(w))\cos^2(wt) + \mathrm{Im}(K(w))\sin(wt)\cos(wt)](\overrightarrow{E_0}\cdot\vec{\nabla})\overrightarrow{E_0} \tag{5.7}$$

If the pulsation w of the electric field is sufficiently high, the force applied on the object can be modeled by the average force. In the previous equation, the first term in the sum, which is a function of the square of the cosine of the pulsation, has a nonzero average value in spite of the second term, which is sinusoidal. Consequently, the average force $\overrightarrow{F_{\mathrm{dep}}}$ (named the dielectrophoresis force) applied to the object is a function of the real part of the Clausius–Mossoti factor:

$$\overrightarrow{F_{\mathrm{dep}}} = \pi r^3 \epsilon_3 \mathrm{Re}(K(w))\vec{\nabla}\overrightarrow{E_0^2} \tag{5.8}$$

This equation represents the average force applied to a dielectric object placed in a sinusoidal electric field whose pulsation is w. In case the medium and the object are dielectric, this force is classically expressed by

$$\overrightarrow{F_{\mathrm{dep}}} = \pi r^3 \epsilon_3 \frac{\epsilon_1 - \epsilon_3}{\epsilon_1 + 2.\epsilon_3}\vec{\nabla}\overrightarrow{E_0^2} \tag{5.9}$$

5.2.1.2 Dielectrophoresis Torque

In order to build the expression of the dielectrophoresis torque, we consider a rotating electric field $\overrightarrow{E(t)}$ whose norm is constant (see Fig. 5.3):

$$\overrightarrow{E(t)} = E_0[\cos(wt)\vec{x} + \sin(wt)\vec{y}] \tag{5.10}$$

Consequently, the dipole momentum verifies

$$\overrightarrow{m(t)} = 4\pi r^3 \epsilon_3 E_0 \Big[(\mathrm{Re}(K(w))\cos(wt) + \mathrm{Im}(K(w))\sin(wt))\vec{x} + (\mathrm{Re}(K(w))\sin(wt) - \mathrm{Im}(K(w))\cos(wt))\vec{y} \Big] \tag{5.11}$$

Thus, the dielectrophoresis torque applied on the object is done by

$$\overrightarrow{\Gamma_{\text{DEP}}} = 4\pi r^3 \epsilon_3 \, \text{Im}(K(w)) E_0^2 \tag{5.12}$$

5.2.2 Application of the Dielectrophoresis in Micromanipulation

Two typical behaviors are observed in dielectrophoresis. In fact, the sign of the force (5.8) depends on the sign of the real part of the Clausius–Mosotti factor $K(w)$:

- If the real part of the Clausius–Mossoti factor $\text{Re}(K(w))$ parameter is positive, the microparticle tends to move to the highest electric field gradient (near to the electrode). The dielectrophoretic force is attractive and is called "positive DEP."
- In case of a negative $\text{Re}(K(w))$, the microparticle tends to move to the lowest electric field (far from the electrode). The dielectrophoresis force is repulsive and is called "negative DEP."

The dielectrophoresis (DEP) is usually used in cell micromanipulation to perform direct cell sorting [6, 42] or field-flow-fractionation (FFF-DEP) [17, 18]. In specific configurations, it enables to catch individual cells too [38]. However, the noncontact manipulation does not enable large blocking force and thus is not relevant for robotic microassembly. Moreover, dielectrophoresis is used to manipulate carbon nanotubes (CNT) in the field of nanomanipulation [40]. Although this principle is not really effective in air, recently Subramanian presents first tests on the use of DEP in manufactured objects manipulation in air [40]. In this medium, this kind of physical principle requires high voltage (e.g., 1000 V).

In robotic microassembly, the dielectrophoresis can be used in submerged robotic handling in order to induce repulsive force during release.

5.2.2.1 Robotic Microhandling Using Dielectrophoresis

Using a hybrid approach between microhandling and dielectrophoresis is an original way to perform manufactured microobject positioning. As the grasping by a gripper with two fingers enables to induce complex three-dimensional (3D) trajectories and complex microassembly task (i.e., insertion), two-fingered grippers are usually chosen. Consequently, the release task is disturbed by the adhesion force (pull-off force). The use of negative dielectrophoresis to control the microobject release has been proposed [19]. An electric field could be produced by electrodes placed on the gripper or by using a conductive microgripper. After opening the gripper, an alternating electric field is applied on the gripper electrodes and induces a repulsive force on the microobject in order to release the object.

The behavior of the microobject is composed of two phases:

- The microobject is in contact with the gripper and is immobile [Fig 5.4(a)].
- The microobject is in motion in the liquid [Fig 5.4(b)].

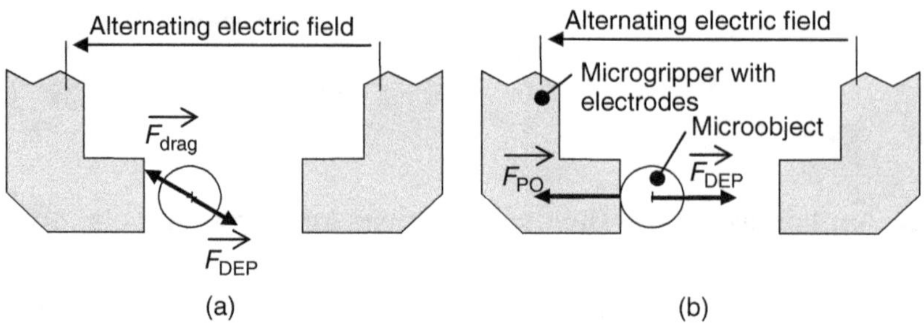

Figure 5.4. Principle of the dielectrophoretic release: (a) First step: the dielectrophoretic force F_{DEP} overcomes the pull-off force F_{PO}. (b) Second step: the dielectrophoretic force F_{DEP} is opposed to drag force F_{drag}.

Before the release, forces applied to the microobject are the adhesion force and the dielectrophoresis force. The release appears if the dielectrophoresis is greater than the pull-off force:

$$F_{DEP} > F_{PO} \tag{5.13}$$

After the release, in a very short time the microobject reaches its maximum velocity. The microobject trajectory is then defined by the equilibrium of the dielectrophoretic force and the hydrodynamic force F_{drag} induced by the liquid:

$$\overrightarrow{F}_{DEP} = -\overrightarrow{F}_{drag} \tag{5.14}$$

Consequently, from equation 2.40 the trajectory of the particle is defined by its velocity $\overrightarrow{V}$:

$$\overrightarrow{V} = \frac{1}{k\mu_3}\overrightarrow{F}_{DEP} \tag{5.15}$$

where k is a function of the geometry of the object and μ_3 the dynamic viscosity of the liquid.

The transition (acceleration of the microobject) between both cases is made in a very short time (e.g., 50 μs) because of the small inertia of the microobject. As the precise description of this acceleration phase has no specific interest in micromanipulation, the complete behavior of the microobject can be described by equations (5.13)–(5.15).

5.2.2.2 *Experimentations*

To validate this approach, experimentations were performed on glass microsphere with a diameter of 20 μm. The gripper is a 4 degree-of-freedom (DoF) piezoelectric microgripper described in [2]. Specific end-effectors in silicon were built with microfabrication technologies (DRIE) and glued on the microgripper as presented

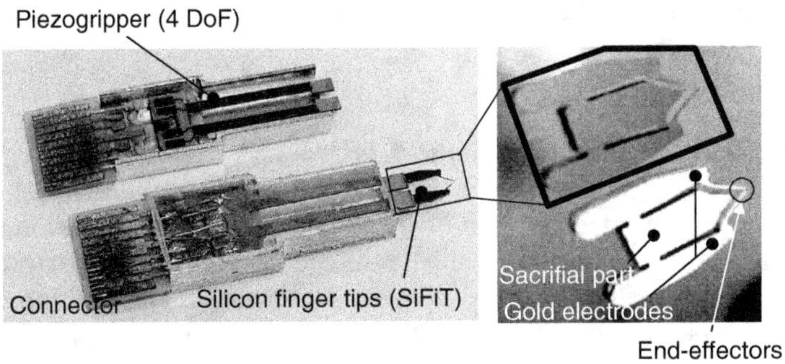

Figure 5.5. Piezomicrogripper and silicon finger tips (SiFiT).

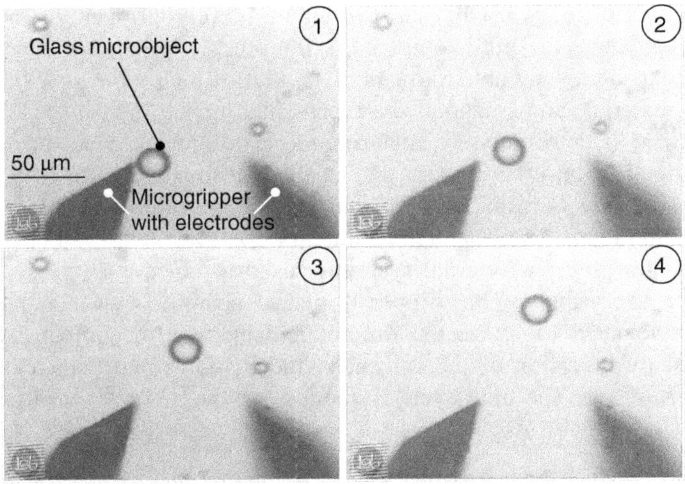

Figure 5.6. Experimental DEP release.

in Agnus et al. [2]. The silicon end-effectors and microgripper are presented in Figure 5.5. Thickness of the end-effectors is 12 μm and its shape is presented in Figure 5.6. Gold electrodes are sputtered on the silicon end-effectors to applied alternating electric field.

An example of glass microsphere release is presented in Figure 5.6. The electric voltage used was a sinusoidal signal ±20 V peak-to-peak. The release and the trajectory of the microobject are visible in Figure 5.6. Experimentations show a high reliability on glass microobject releases. The control of the release is easy to perform via the tension of electrodes. This result demonstrates the interest in using dielectrophoresis release in submerged micromanipulations. However, at present, the final position of the released microobject is not controlled. Further work must be done to purchase the modeling of the microobject behavior after release to control its final position. The shape, number, and architecture of electrodes must be studied and tested to optimize and control this release principle.

5.3 SUBMERGED FREEZE GRIPPER

The study of a new medium (liquid medium) in micromanipalution opens the way to original handling methods. The submerged freeze gripper presented in this section is one of these new strategies. The usual ice microgrippers that act in the air are first presented. The principle of the submerged gripper is explained and some experimentations are presented.

5.3.1 Ice Grippers in the Air

The properties of ice is usually used in air by the cryogrippers or ice grippers. The gripping forces are high compared to the other gripping principle. They are flexible because they are able to manipulate a lot of microobjects, whatever their shapes or material. Some examples are presented next.

The CSEM (Centre Suisse d'Electronique et Microtechnique) in Switzerland has developed some cryogrippers that enable the manipulation of microobject whose typical size is from 0.1 to 5 mm [16]. For example, the MicroGRIP is presented in Figure 5.7. The temperature is controled with a Peltier module. Before the grasping, an external system is used to place a droplet (about 0.1–0.3 μL) on the gripper. The gripper is placed near the object and the Peltier module is actuated to freeze the droplet and grasped the object. The release is obtained by inversion of the current in the Peltier effect, which induces the thawing of the ice. The time cycle is about 2 s, which is relevant for small and

Figure 5.7. The MicroGRIP, CSEM, Switzerland [16].

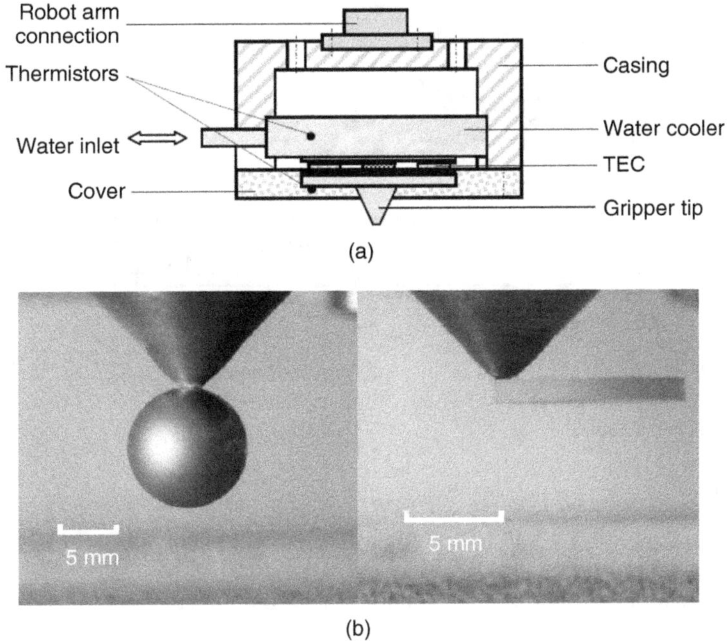

Figure 5.8. Ice gripper of the Laboratory for Precision Manufacturing and Assembly, The Netherlands [27]: (a) scheme of the device and (b) example of microhandling task.

medium volumes of production. Recently, the Laboratory for Precision Manufacturing and Assembly, in the Netherlands, has developed another ice gripper [27, 28]. It uses also a Peltier module to control the temperature of the end-effector whose surface is about 1–2 mm^2. The design of the system includes the feeding of water. The maximal grasping force is 4.5 N on an object whose size is about 5 mm. Figure 5.8 shows a principle scheme and two examples of microhandling tasks. One example shows the ability to manipulate different types of objects with a unique ice gripper.

These examples show the relevance of ice gripping on millimetric objects. However, these systems are not able to manipulate micrometric objects. In fact, in the microscale after thawing a capillary bridge stays between the object and the grippers. This droplet induces capillary force (see Section 1.2.2), which disturbed the release. In this case, release strategies should be used to break the capillary bridge [6, 46, 47]. The major limitation in the miniaturization of this kind of gripper is the impact of capillary force during the release. One way to solve this limitation consists in micromanipulating in submerged medium. During the thawing, the ice is mixed with the surrounding water and capillary force does not appear. Moreover, the use of a submerged medium solves the problem of the feeding of the water. Submerged ice gripper is thus a principle that enables the manipulation of micrometric objects without disturbances during the release.

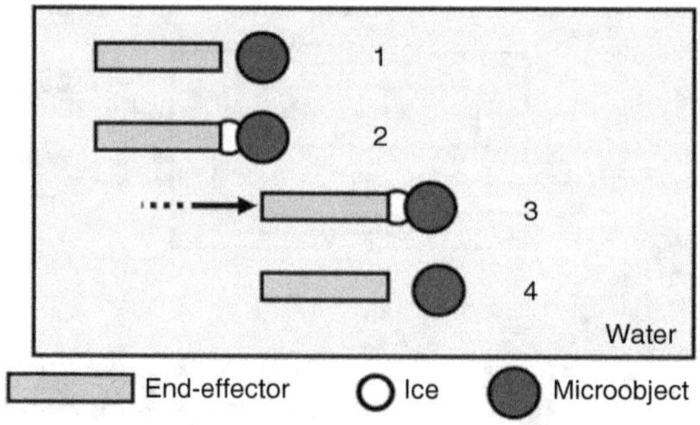

Figure 5.9. Handling strategy: (1) the microgripper approaches; (2) an ice droplet is generated and catches the object; (3) the object is manipulated; and (4) the ice thaws and the object is positionned.

5.3.1.1 Submerged Freeze Gripper

The handling strategy of the submerged freeze gripper is shown in Figure 5.9. First, the gripper comes close to the object without touching it. Second, an ice droplet is generated holding just a small part of the object. The object can be then picked and positioned. Finally, the ice droplet thaws, mixing with the water, and the object is released without any influence of capillary force [32]. As described below, the submerged freeze gripper uses the water environment to create an ice droplet. The cooling energy for freezing water is provided by two Peltier thermoelectric components.

A Peltier module is a transducer that converts an electrical current to a generation or absorption of heat. The direction of the heat flow depends on the direction of the current, and the difference in temperatures induced by the heat transfer defines two faces: a cold one and a hot one. The hot face must be associated to a heat sink in order to dissipate the heat flux.

As illustrated in Figure 5.10, the submerged freeze system consists of two Peltier module stages and a forced convection system. The first stage contains a Peltier micromodule named MicroPelt (μP). The end-effector is directly attached to its cold side. In this way, the MicroPelt can cool it and consequently generate the ice droplet on its acting part. The freezing process increases the temperature of the MicroPelt's hot face. Convection heat flow in water is thus so important that the whole system (liquid, gripper, and Peltier micromodule) could warm up. To actively decrease the temperature at the MicroPelt's heat sink, a second Peltier element (called MiniPeltier) is connected. The temperature of its hot face must be constant to optimize its performance: It is maintained at the ambient temperature by forced convection using a liquid cooling system [30, 32]. As MicroPelt's maximal cooling capacity is not sufficient to freeze the end-effector from ambient temperature, the liquid cooling system cannot be used directly on its hot face. The end-effector and the MicroPelt are completely submerged and electrically insulated. The MiniPeltier and the cooling liquid system stay in air to dissipate heat outside water.

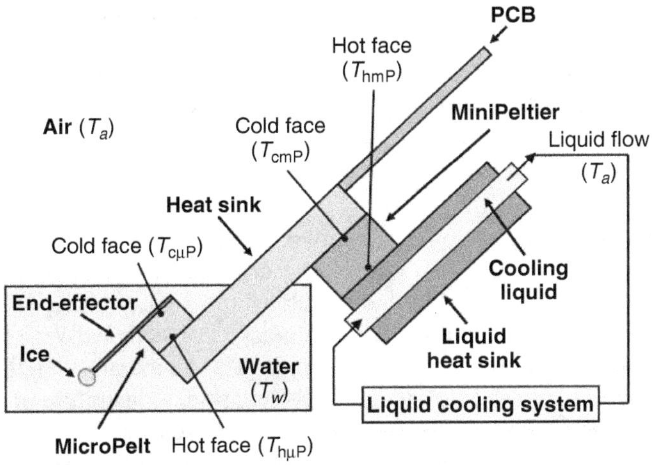

Figure 5.10. Submerged freeze system principle.

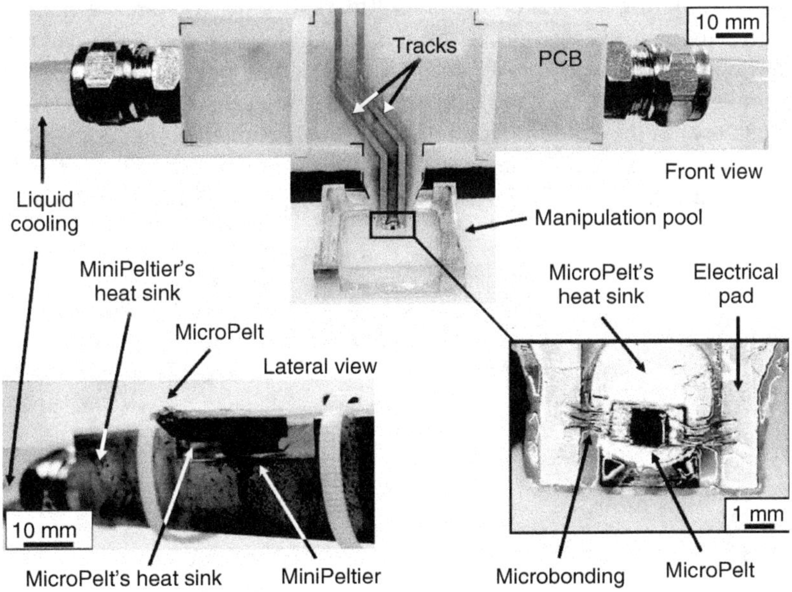

Figure 5.11. Experimental freeze gripper.

5.3.1.2 Physical and Technical Characteristics

The first prototype of the submerged freeze gripper (without the end-effector) is shown in Figure 5.11. The MicroPelt (Infineon Technologies AG) has as dimensions $720 \times 720 \times 428\ \mu m^3$. Its hot face is fastened to a copper heat sink (MicroPelt's heat sink). The MiniPeltier (Melcor FC0.6-18-05), with dimensions of $6.2 \times 6.2 \times 2.4\ mm^3$, is fixed on its cold face to the MicroPelt's heat sink;

and on its hot face to the copper liquid heat sink of the cooling liquid system. A specific PCB has been fabricated to establish the electrical connections of both Peltier modules. Because of the very small dimensions of the MicroPelt, microbonding technology was used for its connections.

5.3.1.3 Thermal Behavior

The relationship between the current in the Peltier module and the temperature on the surface of the end-effector is a complex link that depends on several parameters (e.g., initial temperatures, convection constants). Figure 5.12 presents an example of the evolution of the end-effector temperature during several cycles. This measurement has been done without grasped objects and with a microthermocouple placed behind the gripping surface. The temperature of the water and the air are, respectively, 6.5°C and 21.7°C. The initial temperature of the gripping surface is between both previous temperatures and reaches 14.5°C

First at $t = 0$, the MiniPeltier is actuated with a current of 0.9 A. This value has been determined in static tests in order to obtain a sufficient precooling of the gripping surface. The current of the MiniPeltier is not modified along the experiments. During the precooling phase [mark (a)] the MicroPelt is not actuated and the temperature is decreasing by the MiniPeltier. Second, the MicroPelt is actuated with a current of 0.5 A in order to decrease locally the temperature of the gripping surface. In fact, the temperature is decreasing rapidly at t_b. However, the water stays in the liquid phase below 0°C (surfusion phenomenon [1, 35]). At t_c, the ice is forming and the temperature is increasing because of the latent heat. During phase (c), the ice is growing on the gripping surface. Third, the current

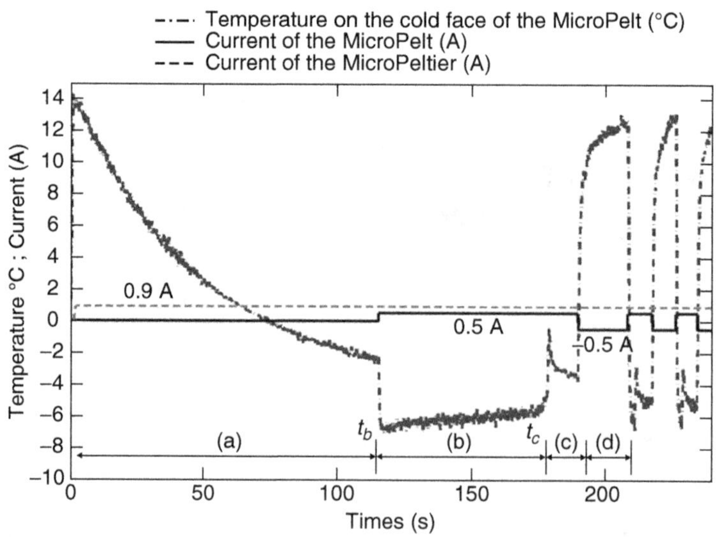

Figure 5.12. Temperature variation in the submerged freeze gripper: (a) precooling, (b) local cooling of water, (c) crystallization of water, and (d) thawing of the water.

of the MicroPelt is inverted to −0.5 A in order to thaw the ice. Temperature is increasing rapidly because of the addition of the Joule effect and Petlier effect in the MicroPelt. After that, cycles of ice generation and thawing are created by switching the current of the MicroPelt from 0.5 to −0.5 A. The thermal modeling of this device is proposed Lopez Walle et al. [30, 31]. This experiment shows the ability of the gripping device to control the generation of ice by the current of the Peltier modules.

5.3.1.4 Experimentations

The experimentations using the prototype described above were performed in distilled water at 2°C. They show the good working of the handling system. Figure 5.13 describes the telemanipulation of a silicon object whose dimensions are 600 × 600 × 100 μm^3. A precooling phase is necessary to decrease the temperature of the MicroPelt's heat sink. During this phase, only the current in the MiniPeltier (i_{mP}) is applied and set constant at 0.9 A [Fig. 5.13(a)]. When the temperature is about 0.5°C (this temperature is sufficiently close to 0°C but it prevents the heat sink to freeze), the MicroPelt is approached to the microobject and its current ($i_{\mu\mathrm{P}}$) is turned on at 0.5 A. The cooling energy generates the ice droplet (4 μL), which involves a part of the object in 3 s [Fig. 5.13(b)]. The freeze gripper can thus displace it toward a new position [Fig. 5.13(c)]. To release it, the MicroPelt's current is inverted at −0.3 A. The ice droplet thaws in 7 s and melts with the aqueous medium, liberating the microobject without adhesion perturbations [Fig. 5.13(d)]. The micromanipulation has been performed in 30 s. As previously mentioned, the cycle time for pick and place, obtained for optimal

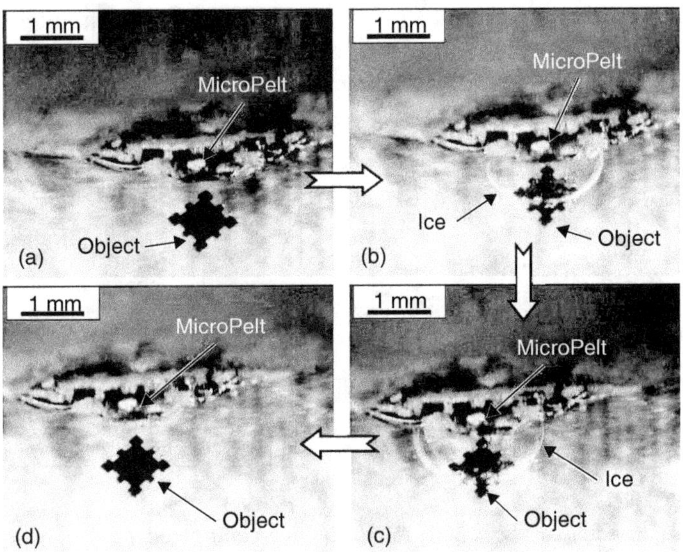

Figure 5.13. Micromanipulation of a 600 × 600 × 100 μm^3 silicon object with the submerged freeze gripper.

working conditions of the Peltier modules, is $3 + 7 = 10$ s. The rest of the time, that is, 20 s of transportation time in this case, depends principally on operator's ability or microgripper speed in the case of full automation. Contrary to the cryogenic grippers in air, capillary force does not perturb the release because the object and the MicroPelt are submerged. The Peltier currents choice is based on the thermal simulation presented [29, 32]. The same experiment was successfully repeated several times. The submerged freeze principle seems thus a promising approach to manipulate microobjects.

5.4 CHEMICAL CONTROL OF THE RELEASE IN SUBMERGED HANDLING

In submerged contact handling consists in chemically controlling the surface forces between the object and the gripper [12, 13]. The major objective is to control the adhesion force or to create a repulsive force to guarantee a reliable release. Now, the surface properties of a material can be controlled by surface functionalization in a liquid using pH. The charge density on functionalized surfaces is effectively linked to the pH. The general microhandling principle is presented in Figure 5.14. The grasping can be done at pH_1 where the surface charge on the gripper and the object induces an attractive force. In order to release the object, the pH is modified to a second value pH_2 where the charge of the object is changing. The electrostatic force becomes repulsive and the object is released.

The microhandling method proposed is based on two chemical functions: anime and silica (see in Fig. 5.15). On the one hand, the anime group is in the state NH_2 in basic pH and in $NH_3{}^+$ in acidic pH. On the other hand, the surface charge of the silica in water is naturally negative except for very acidic pH where the surface is weakly positive [15].

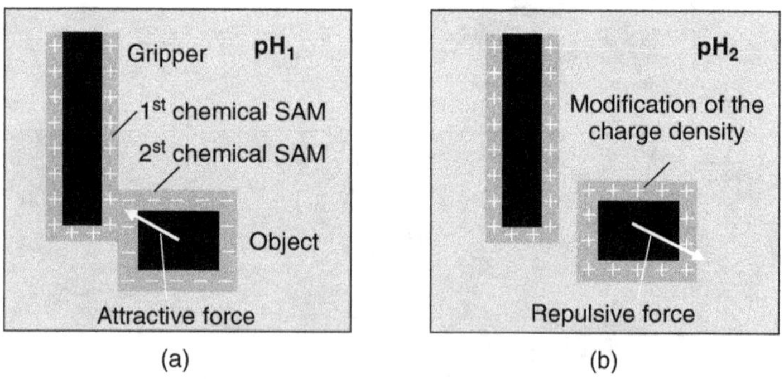

Figure 5.14. Principle of the robotic microhandling controlled by chemical self-assembly monolayer (SAM): (a) handling in pH_1 where charges in SAM induces electrostatic attractive force and (b) release in pH_2 where charges in SAM induces electrostatic repulsive force.

	Anime function	Hydroxyl groups on silica
Acidic pH	$-NH_3^+$	Si(OH)
Basic pH	$-NH_2$	SiO^-

Protonation

Figure 5.15. Modification of the electrical charges on chemical elements in function of the pH.

5.4.1 Chemical Functionalization

5.4.1.1 General Principles

The surface functionalization of both object and gripper can be obtained by different methods. The two more usual methods are the physisorption of polyelectrolyte (polyelectrolyte with positive or negative charges) [11, 39] or the grafting of molecules on the surface (covalent bond between the substrate and the molecules) [14, 33]. The second method has been investigated because it generates covalent bond between substrate and molecules. These molecules must contain silanol, thiol, azide, allyl, or vinyl groups [14, 33] in an extremity. These molecules have to be used in organic solvents such as toluene, acetone, methanol, ethanol, and the like. The silanol creates a bond Si–O–Si with the silica substrate [14] while allyl or vinyl generates Si–O–C (or Si–C) bond [34] and the azide groups produce Si–N bond [44].

5.4.1.2 Materials and Chemicals

Two chemical functionalizations have been tested (see Fig. 5.16):

- The silane, 3 (ethoxydimethylsilyl) propyl amine (APTES)
- The silane, (3 aminopropyl) triethoxysilane (APDMES)

Both chemical compounds (APTES, APDMES) used to surface functionalization are amine functions NH_2, which can protonated or ionized to NH_3^+ according to pH. In acidic pH, the anime is totally ionized, then the ionization decreases and

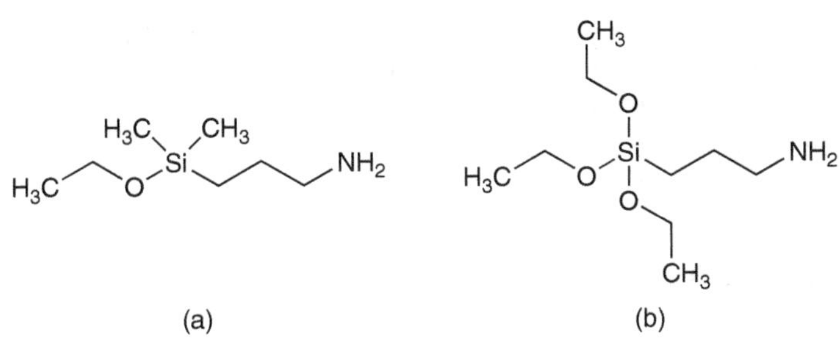

Figure 5.16. Molecules used for the silica functionalization: (a) APDMES and (b) APTES.

is null in basic pH (between pH 9 and 12). The silanes (APTES and APDMES), ethanol, sodium chloride (NaCl), sodium hydroxyde (NaOH), and chlorydric acid (HCl) were purchased from Sigma Aldrich. The deposits were made on silicon surfaces. The Milli-Q water was obtained with the Direct-Q 3 of Millipore. The pH of the solution was measured with a pH meter (Satorius, PT-10) and an electrode (Sartorius, PY-P22) and adjusted with the addition of sodium hydroxyde and chlorydric acid just before measurement.

5.4.1.3 Surface Functionalizations

Before being functionalized, the wafers were cleaned by immersion in a piranha solution (2 parts H_2SO_4, 1 part H_2O_2) during 25 min at 70°C. Then, the wafer was rinsed in Milli-Q water and in ethanol before silanization (functionalization by silane: APDMES, APTES). Solutions were freshly prepared by direct dissolution in milli-Q quality water and in ethanol. The final silane concentration was of 1%. The surfaces were functionalized by immersion in solutions during one night at room temperature. In the silane solution, the silanes were gratfed on the substrate (covalent bond). The excess of ungrafted silanes was removed by ultrasonication during 2 min in ethanol.

5.4.1.4 Functionalization Mechanisms: Grafted Silanes

The mechanism of self-assembled monolayer formation during silanization process is depicted in Figure 5.17, which takes place in four steps [45]. The first step is physisorption, in which the silane molecules get physisorbed at the hydrated silicon surface. In the second step, the silane head-groups arriving close to the substrate hydrolyse, in the presence of the adsorbed water layer on the surface, into highly polar trihydroxysilane $Si(OH)_3$ or hydroxysilane $Si(OH)(Me)_2$ for, respectively, triethoxysilane $Si(OEt)_3$ (APTES) and ethoxysilane Si(OEt) (APDMES). These polar $Si(OH)_3$ or Si(OH) groups form covalent bonds with the hydroxyl groups on the SiO_2 surface (third step), subsequent to which the condensation reaction (release of water molecules) goes on between silanol functions of neighbor molecules. Self-assembly is driven by lipophilic interactions between the linear alkane moieties. During the initial period, only a few molecules will adsorb (by steps 1–3) on the surface, and the monolayer will definitively be in a disordered (or liquid) state. However, at longer times, the surface coverage eventually reaches the point where a well-ordered and compact (or crystalline) monolayer is obtained (step 4), for APTES only, by the condensation reaction between the APTES molecules.

5.4.2 Experimental Force Measurements

5.4.2.1 Atomic Force Microscope

Force measurements were performed in order to characterize the functionalizations. Force–distance curves were performed using a stand-alone SMENA

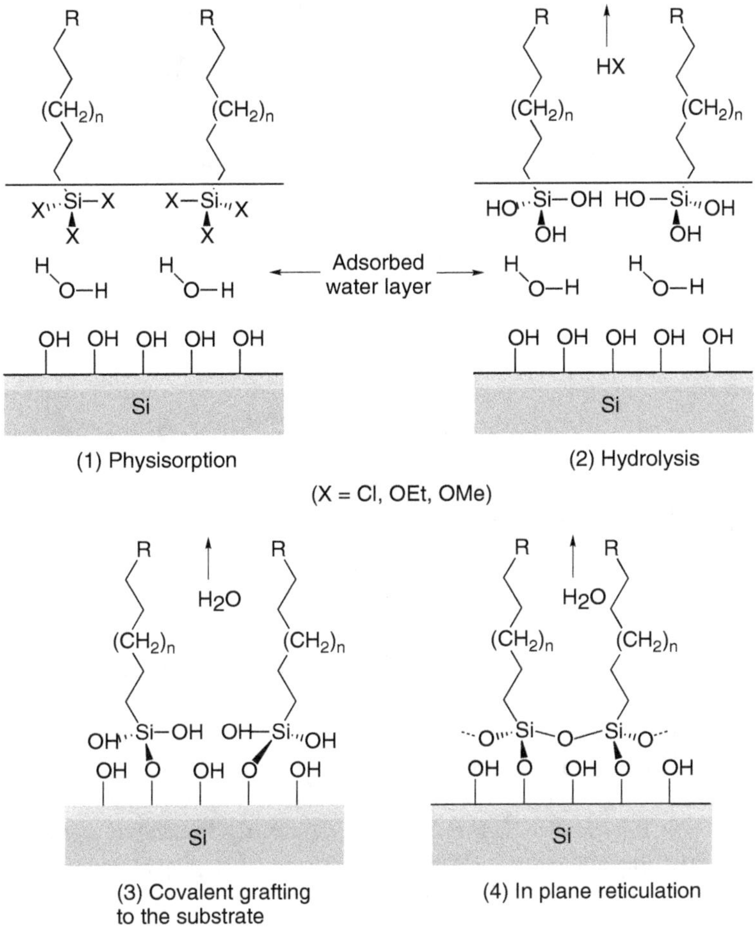

Figure 5.17. Steps involved in the mechanism of SAM formation on a hydrated silicon surface [45].

scanning probe microscope (NT-MDT). The measure of the force performed on this atomic force microscope (AFM) is based on the measure of the deformation of the AFM cantilever with a laser deflection. The silicon rectangular AFM cantilever, whose stiffness is 0.3 N/m, was fixed and the substrate moved vertically. As the applicative objective of this work is to improve reliability of microobject manipulation, interactions have been studied between a *micrometric* sphere and a plane. Measurements were in fact performed with a cantilever where a borosilicate sphere[1]. ($r_2 = 5$ μm radius) was glued. All measurements were done at the driving speed of 200 nm/s to avoid the influence of the hydrodynamic drag

[1]Cantilevers and beads provided by Novascan Technologies, Ames, Iowa, under the reference PT.BORO.SI.10.

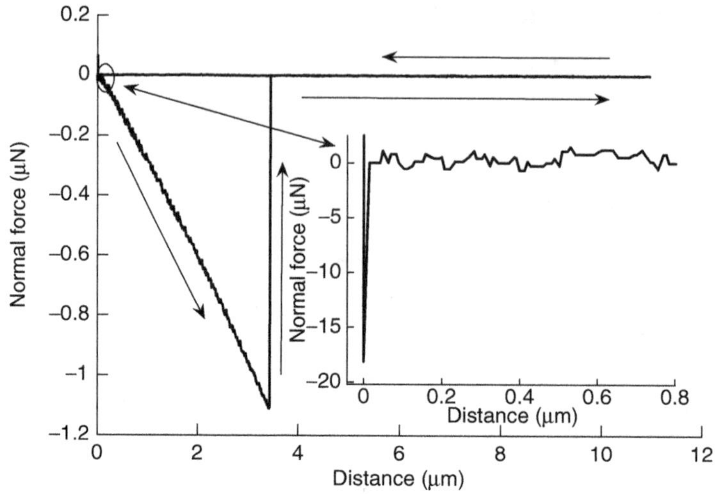

Figure 5.18. Force–distance curves on functionalized APTES in dry medium obtained with a tip whose spring constant is 0.3 N/m.

forces [43]. For each sample, nine measures were done in different points. The repeatability of all the measures of pull-off and pull-in forces was better than 10%.

5.4.2.2 Typical Distance–Force Curves

The first type of behavior is presented in Figure 5.18. In this case, an attractive force (pull-in force) is measured when the sphere is coming close to the substrate (near −20 nN; Fig. 5.18). In Figure 5.18, we clearly measured a pull-off force that represents the adhesion between the borosilicate sphere on the tip and the functionalized substrate. In this example, the pull-off force is reaching −1.1 μN. This behavior represents an attraction between surfaces.

The second type of behavior is presented in Figure 5.19. In this case, there are repulsion between surfaces. We observe a repulsion (positive pull-in force near 0.75 μN) and no pull-off force between both surfaces. Experiments have been done in wet medium with the functionalized surface and:

- A cantilever grafted with APTES or
- A nonfunctionalized cantilever

The pH of the solution varied by the addition of sodium hydroxide or chlorydric acid. The surface was stay in the solution 2 min before the measurement in order to equilibrated the system. Force measurements in liquid have been also compared with measurement done in air.

5.4.2.3 Influence of pH on Functionalized Surface

First, the measurements were done with a cantilever with a nonfuntionalized sphere. The results of the pull-in and pull-off forces is presented in Figure 5.20.

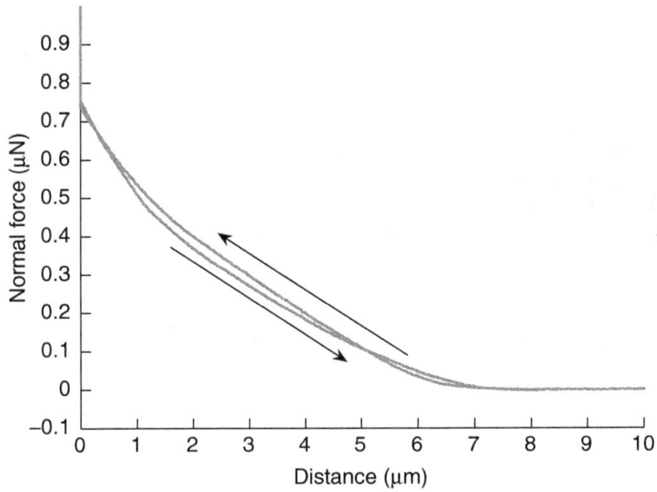

Figure 5.19. Force–distance curve for the APDMES functionalized substrate in wet medium obtained with a tip whose spring constant is 0.3 N/m.

We noted that the pH influences significantly the forces between the cantilever and the surface. At natural pH (pH nat), an attractive pull-in is measured [near −60 nN, Fig. 5.20(a)] with an important pull-off [−350 nN; Fig. 5.20(b)]. When the pH increases the pull-in force is inverted and becomes repulsive respectively 280 nN and 770 nN at pH 9 and 12. Moreover, the adhesion forces disappear. The average values of the different measurements (pull-in and pull-off forces), at different pH, are summarized in Table 5.1. We observe that the phenomena described above for APTES is the same for APDMES. In fact at natural pH (near 5.5), the interaction is attractive with an important adhesion force and at basic pH, above 9, the interaction is repulsive. At pH 2, pull-in force was not detectable, perhaps because the charge density on the silica cantilever was too low. We show that the forces measured with APDMES grafted are lower than APTES. We can explain this by the fact that the quantity of molecules grafted on the substrate is more important for APTES than APDMES. As the charges on the surface of the silica cantilever is negative or null (see Fig. 5.15), the surface density σ of APTES and APDMES verifies:

$$\begin{aligned} &\text{For natural pH or pH 2}, \sigma \geq 0 \\ &\text{For pH 9 or pH 12}, \sigma \leq 0 \end{aligned} \tag{5.16}$$

In fact, in acidic pH, the positive charges induced by the functionalization are greater the negative charges induced by the hydroxyl groups. In basic pH, the negative charges are predominant. The inversion of the interaction forces during the variation of the pH of the solution represents a great interest in micromanipulation. Indeed, the control of the pH is able to switch from an attractive behavior (grasping) to a repulsive behavior (release).

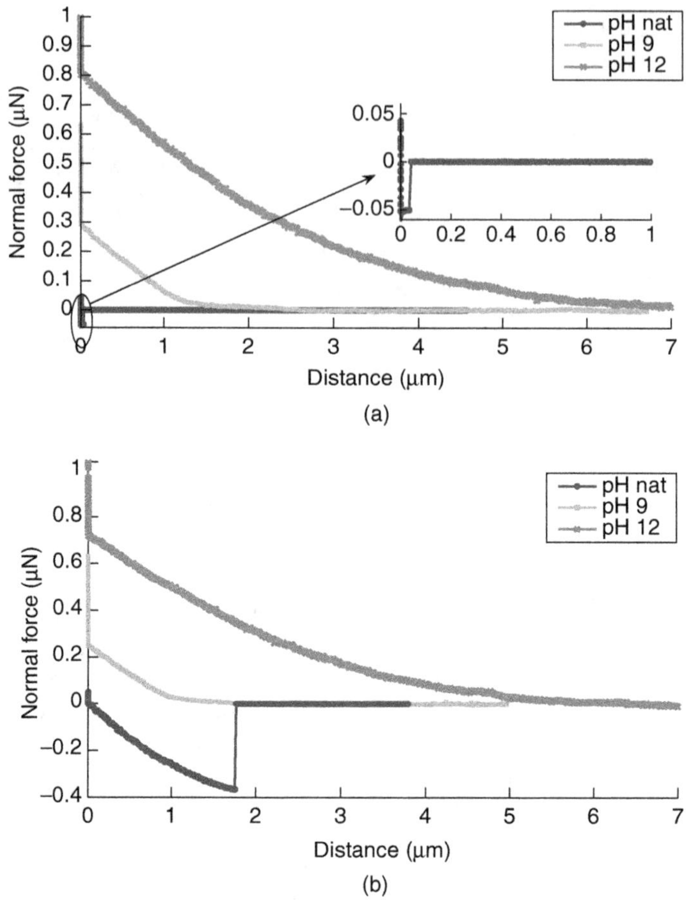

Figure 5.20. Force–distance curve for the APTES functionalized substrate in wet medium at different pH obtained with a tip whose spring constant is 0.3 N/m: (a) approach measurement and (b) retract measurement.

TABLE 5.1. Influence of pH on Pull-in and Pull-off Forces Obtained with a Tip Whose Spring Constant is 0.3 N/m for APTES and APDMES Grafted on Surface

	APTES		APDMES	
Medium	Pull-in (nN)	Pull-off (nN)	Pull-in (nN)	Pull-off (nN)
pH 2	0	−176	0	−93
pH nat	−59.5	−387	−29.8	−353
pH 9	282	0	377	0
pH 12	768	0	1100	0
Air	−13.2	−1150	−4.97	−769

5.4.2.4 Influence of pH on Functionalized Surface and Cantilever

Second, the cantilever was functionalized with the APTES silane and without the sonification step. Similar experiments were previously done in aqueous solution with a pH that varied between 2 and 12. The force–distance curves obtained with a APDMES grafted on the substrate is presented in Figure 5.21.

Contrary to the case in air, in liquid the forces measured were always repulsive between the functionalized cantilever with APTES and the APDMES grafted on the surface. Any pull-off forces were detected. There was in fact no adhesion between both functionalized objects. A cantilever deformation was observed on an important distance (typically several microns) when the sphere is approaching from the surface. This large interaction distance is typical from electrostatic interactions. The average values of the force measurements at different pH are summarized in Table 5.2. We note that the pH of the medium changes the value of the repulsive force between the cantilever and the surface, but the behavior stays always repulsive. For acidic and natural pH, the repulsion can be explained by the positive charges of the aminosilane grafted on the surface. For basic pH, repulsion is induced by the negative charges of the silicon substrate down to the functionalization. Indeed, from pH 9, the positive charge of the aminosilane is not sufficient to totally screen the negative charge of the silicon. However at pH 9, the screening of the charge induced by some NH_3^+ explains why the repulsion is lower with a functionalized cantilever (pH 9 in Table 5.2) than with a nonfunctionalized cantilever (pH 9 in Table 5.1). Moreover, at pH 12, the behavior of the functionalized surface and the nonfunctionalized surface are quite similar. In fact, the aminosilane has no more positive charge and the repulsion is only induced by the negative charge on silicon and borosilicate. In micromanipulation, the repulsive charge between two objects, whatever the pH of the solution, is an interesting behavior in order to make easier the separation of two objects. Indeed, the release of microobjects will be easier if both microobject and gripper are functionalized with aminosilane, which induces a repulsive force.

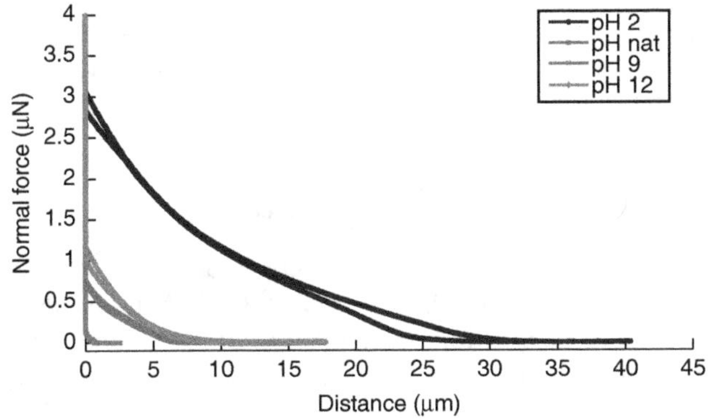

Figure 5.21. Force–distance curve for the APDMES functionalized substrate in wet medium at different pH obtained with a tip-functionalized APTES whose spring constant is 0.3 N/m.

TABLE 5.2. Influence of pH on Pull-in and Pull-off Forces (nN) Obtained with Tip-Functionalized APTES whose Spring Constant Is 0.3 N/m for APTES and APDMES Grafted on Surface

	APTES		APDMES	
Medium	Pull-in (nN)	Pull-off (nN)	Pull-in (nN)	Pull-off (nN)
pH 2	3190	0	3080	0
pH nat	655	0	735	0
pH 9	150	0	114	0
pH 12	983	0	989	0
Air	0	−91	0	−136

5.4.3 Modeling of Surface Charges

In order to be able to extrapolate this result to other geometries, the electrical surface density induced by the functionalization has been studied. Based on the force measurements, the surface charges on the substrate could be estimated. Let us assume that the surface is large enough to be considered as an infinite plane compared to the sphere whose radius is $r_2 = 5$ μm. The electric field $\overrightarrow{E_1}$ induced by the surface charge density σ_1 of the substrate is uniform:

$$\overrightarrow{E_1} = \frac{\sigma_1}{2\epsilon_3\epsilon_0}\overrightarrow{n_1} \tag{5.17}$$

where ϵ_0 is the electric permittivity of the vacuum and ϵ_3 the relative permittivity of the medium (for water, $\epsilon_3 = 80$), and $\overrightarrow{n_1}$ is the unit vector perpendicular to the substrate. The repulsive electrostatic force applied by the gripper on the object whose charge is q_2 is thus

$$\overrightarrow{F}_{\text{pull-in}} = q_2\overrightarrow{E}_1 = 2\pi r_2^2\frac{\sigma_1\sigma_2}{\epsilon_3\epsilon_0}\overrightarrow{n_1} \tag{5.18}$$

where σ_2 is the charge density on the sphere whose radius is r_2.

If both objects have the same surface density σ_1, this later can be deduced from the force measurement:

$$|\sigma_1| = \left(F_{\text{pull-in }1-1}\frac{\epsilon_3\epsilon_0}{2\pi r_2^2}\right)^{1/2} \tag{5.19}$$

The sign of σ_1 should be determined by considering the chemical functions equation (5.16). Equation (5.19) has been used to determine the charge density of APTES (see Table 5.3). Moreover, in case of an interaction between two different functionalized surfaces, the charge density σ_2 of the second surface is done by

$$\sigma_2 = \frac{F_{\text{pull-in }1-2}}{\sigma_1}\frac{\epsilon_3\epsilon_0}{2\pi r_2^2} \tag{5.20}$$

TABLE 5.3. Electrical Surface Density of Functionalized Surface in Function of pH

pH	APTES σ_1 (μC/cm^2)	APDMES σ_2 (μC/cm^2)
pH 2	+0.38	+0.36
pH nat	+0.17	+0.19
pH 9	−0.08	−0.06
pH 12	−0.21	−0.21

Equation 5.20 has been used to determine the electrical surface density of APDMES (see Table 5.3). The sign of the charge density was determined in Section 5.4.2.4. Buron et al. have also found a positive charge density at pH natural (5.5) [7]. The value of the charge density was weak (less than 1 μC/cm^2), so grafted amino group density was about 0.2 sites/nm^2 and 1% the of silanol group was grafted by silane. This value can be explained by the important influence of the grafted condition and more particularly of the water in the solution and in the atmosphere [5].

5.4.4 Application of Functionalized Surfaces in Micromanipulation

The behavior described in Table 5.1 shows a transition between attraction in natural pH and repulsion in pH 9. This switching behavior can be used to control the grasping and the release of a microobject manipulated with a microgripper. Figure 5.22 shows the first experiments made with AFM with a tipless cantilever (PointProbe Technology), functionalized with APTES. Using attractive force (natural pH), a glass sphere whose diameter is around 50 μm is grasped with the cantilever [Fig. 5.22(a)]. The increase of pH inverts the behavior and at pH 9, the sphere has been released [Fig. 5.22(b)].

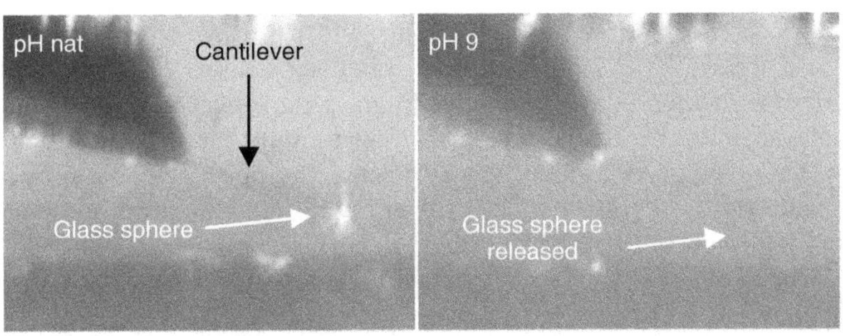

(a) (b)

Figure 5.22. Grasping and release of the sphere with functionalized cantilever: (a) grasping of the sphere with cantilever at natural pH and (b) release of the sphere at pH 9.

We have shown that the pH can be used to control the release of a nonfunctionalized object during micromanipulations. Furthermore, the use of functionalized grippers and objects enables to simply cancel adhesion on microobjects. As adhesion is the current highest disturbance in micromanipulation, functionalization is a promising way to improve microobject manipulation in the future.

5.5 RELEASE ON ADHESIVE SUBSTRATE AND MICROASSEMBLY

This section is dedicated to the presentation of a lasting handling strategy and its application in microassembly. It is a transition between Part II on microhandling strategies and Part III on microassembly systems. The release on adhesive substrate is not strictly dedicated in liquid medium and is usable in all media.

5.5.1 Handling and Assembly Strategy

The general principle consists in assembling microparts in two steps. The first one is the positioning of the first object on the substrate and its blocking during assembly. The objective of the second step is to grasp the second object and perform the assembly. Both steps require specific strategies adapted to the microworld.

5.5.1.1 First Microobject Micromanipulation Principle

The proposed method enables a reliable and reversible positioning of the microobject on a substrate. The principle is a hybrid strategy between adhesion manipulation (see Section 4.2.3) and gripping (see Section 4.4.1) and is based on a hierarchy of forces. On the one hand, to guarantee object's release, the adhesion force between object and substrate must be higher than the adhesion force between object and gripper along the normal vector $\overrightarrow{n}$ of the substrate [see in Fig. 5.23(a)]:

$$F^{\text{adhesion}}_{\text{object}-\text{substrate}} \gg F^{\text{adhesion}}_{\text{object}-\text{gripper}} \tag{5.21}$$

To reduce the impact of external perturbations, $F^{\text{adhesion}}_{\text{object}-\text{gripper}}$ must be as low as possible and $F^{\text{adhesion}}_{\text{object}-\text{substrate}}$ must be as high as possible. The major drawback of this release method is the difficulties to grasp the object on the substrate [22]. A reliable grasping cannot be obtained by using only the adhesion force of the gripper. This method is a good way to release the object but not for grasping. On the other hand, to grasp the object, a gripping force higher than the adhesion force between substrate and object along the direction $\overrightarrow{n}$ is required [see in Fig. 5.23(b)]:

$$F^{\text{gripping}}_{\text{object}-\text{gripper}} \gg F^{\text{adhesion}}_{\text{object}-\text{substrate}} \tag{5.22}$$

One of the best technological solutions is to use a gripper with two fingers where the gripping force could be easily higher than the adhesion between the object and

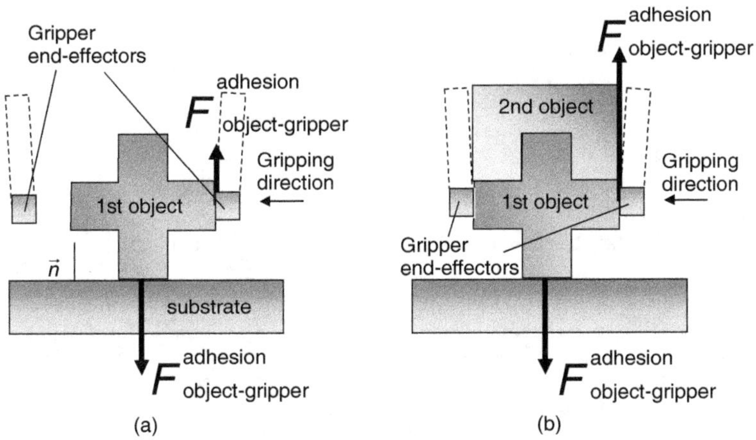

Figure 5.23. Principle of the release and grasping of the first object: (a) releasing the first object on the substrate and (b) grasping the first object from the substrate.

the substrate. This hydrid method uses advantages of both adhesion manipulation and gripping. It induces a reliable release and grasping of the microobject. To guarantee, the conditions (5.21) and (5.22), the gripper must have a high ratio between its gripping force and the adhesion force object–gripper. Technological solutions are proposed in Section 5.5.3.1.

5.5.1.2 Microassembly of Both Objects

The compliance is one of the major stakes in assembly and especially in the microworld [see Fig. 5.24(a)]. Without compliance, both objects have to have a very high accuracy typically up to hundreds of nanometers for a microscopic object. By using compliance, both objects could be positioned with a lower accuracy, typically around 1 μm. Compliance can be obtained by mechanical elastic structures on the object (see Section 7.2.1), on the gripper (see Section 7.2.1), or on the substrate [8, 37]. The three solutions have advantages and drawbacks. In a general way, the component (object, gripper, or substrate), where the compliance is placed, requires usually complex microfabrication capabilities. When the elastic structure is on the object, the object should be specific and the microassembly principle cannot be extended to other objects. Elastic structures could be consequently also placed on the gripper or on the substrate. The gripper requires degrees of freedom, microfabricated end-effectors, and is a more complex component than the substrate. To split up the design constraints into both components, it seems to be relevant to place the compliance on the substrate.

The release of the second one requires a specific strategy. Two cases can be considered:

- Both objects have to be locked during assembly. In this case, both objects can be considered as the same object, and the release strategy of the first object [see Fig. 5.23(a)] can be used.

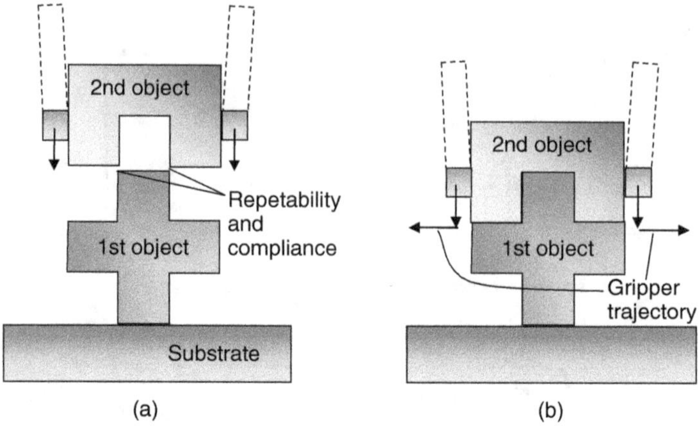

Figure 5.24. Principle of the positioning, assembly, and release of the second object: (a) positioning of the second object before assembly and (b) release of the second object.

- Both objects do not have to be locked during assembly. It could be the case, in the construction of a larger product, where, for example, a third object is used to lock the whole assembly. In this case, the previous strategy cannot be used. In this case it is possible to induce a dedicated trajectory of gripper to be able to release the second object without adhesion perturbation. An example of trajectory is proposed in Figure 5.24(b).

The implementation and experimentation of these strategies are presented in the following sections.

5.5.2 Robotic Microassembly Device

The robotic microassembly device is composed of a robotic structure, optical microscope, and a piezogripper. These elements and the microobjects used are presented in this section.

5.5.2.1 Robotic Structure

Performing serial microassembly tasks requires adapted robotic structures, able to position microobjects with sufficient accuracy and repeatability, typically up to 1 μm for microparts whose typical size is about 10 μm. These performances are mainly reachable by closed-loop robotic microstages. Nevertheless, in case of complex robotic structure with a gripping device, robotic joint sensors are not sufficient to determine microobject positioning. Then, using a videomicroscope with a dedicated vision computer is an important way to perform closed-loop control on the entire robotic structure, including the microgripper [41]. Moreover, it enables teleoperated control of the robot by a human operator.

The robotic structure (Fig. 5.25) is composed of three linear micropositioning stages (Physik Instrumente—M111.1DG). The robot is divided into two independent mobile structures. The first part is the robot's arm, a vertical stage carrying

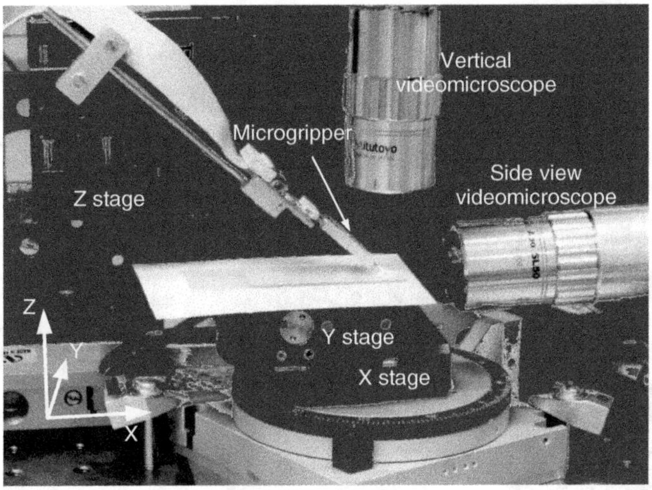

Figure 5.25. Robotic device.

the microgripper. The second part, composed by two stages on the X and Y axes, is used to move microobjects under a vertical videomicroscope. A lateral view is also added to enable manual teleoperations. The robotic structure is built to make videomicroscopes motionless: The vertical view is focalized on the substrate and the lateral view is focalized on the gripper's end-effectors (see Fig. 5.25).

5.5.2.2 Piezoelectric Microgripper

The MMOC piezomicrogripper [3] used in this robotic structure was developed in the FEMTO-ST Institute in France (see Fig. 5.5) and is commercialized by Percipio Robotics, France. It has 2 independent degrees-of-freedom on each finger, which can perform open–close motion of 320 μm and up–down motion of 200 μm. The resolution of the actuator is close to 1.6 μm/V; then submicrometric accurate motions are controllable. Several kinds of fingertips can be glued on this piezoelectric actuator. Up–down motion of the gripper's actuator is in fact used to align them before manipulation. The fingertips [2] used for microassembly have been designed to handle microscopic objects. They are build in single crystal SOI wafer by a well-known microfabrication process: DRIE [23, 24]. These end-effectors have a long and thin beam (12 μm) designed to handle objects from 5 μm to a few hundred micrometers [25].

5.5.2.3 Design of Microobjects

Testing microassembly needs microobjects that could be mechanically fastened to the others. Thus, microobjects have been designed with mechanical fastener structures discussed in Chapter 6. To supply a challenging benchmark, objects' shape are squares of 40-μm sides with a thickness of 5 μm. SOI wafers of 5-μm device layer thickness and DRIE process have been used to build these microparts. Many shapes, fastener designs, and sizes were tested (Fig. 5.26). Two

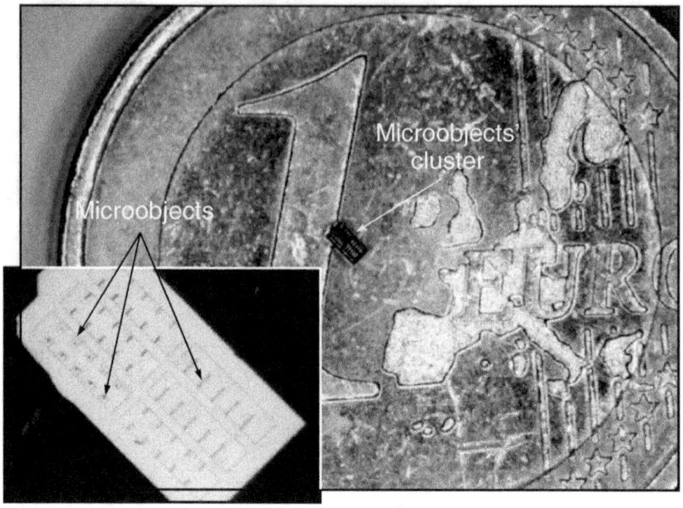

Figure 5.26. Microobjects designed for assembly.

examples of parts can be quoted: The first one is 40 μm square puzzle parts, with four notches of 5 μm. The second one is a mechanical plug device between two 40-μm squares. The male part has a key that is able to lock the female part after assembly as proposed by Dechev et al. [10] in Chapter 6.

5.5.3 First Object Positioning

5.5.3.1 Adapting Adhesive Effects

As presented in Figure 5.23, two ways have been chosen to guarantee the first object's manipulation: Increase adhesion forces between the substrate and the object and decrease adhesion force between the object and the gripper. A relevant adhesive substrate is a transparent gel film well known in microelectronics: Gel-Pack. This material is in fact transparent and softly adhesive; it consequently enables accurate pick-and-place tasks. Moreover, the low mechanical stiffness of this polymer induces natural compliance of the substrate required for microassembly. In fact, efforts have been made on end-effectors shaping. First, the surface in contact with the microobject has been reduced by using end-effectors with a small thickness. Second, the fabrication process, called DRIE, has been used to give the gripping surface a specific texture. Etching anisotropy of this process is made by a short succession of isotropic etching/protection cycles. These cycles create a phenomenon called *scalloping* illustrated in Figure 5.27. In this way, contact shape between object and end-effectors is a succession of microscopic contact points. As proved by Arai et al. [41], the roughness induced by DRIE is able to highly reduce pull-off force. Force measurements will be performed in the near future to validate the surface force reduction and the adhesion of the Gel-Pack.

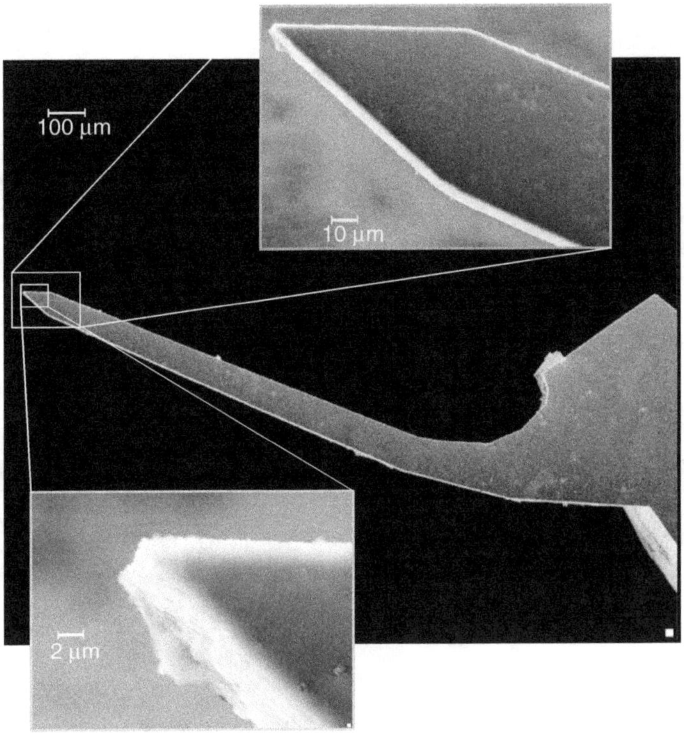

Figure 5.27. End-effectors' shape in SEM view. Scalloping is visible in lower picture.

5.5.3.2 Pick and Place

A microobject is placed on the substrate. First, the gripper is moved above and fingers are opened enough to grip the object. Then the object is held by the end-effectors and the gripper is used to separate the object from the substrate. The substrate is moved to a new position (target position). Finally, release is performed by moving down the gripper to create a contact between object and adhesive substrate, then opening gripper induces the release of the object. All the micromanipulation sequence is shown in Figure 5.28. Without adhesive substrate (e.g., on silicon or glass), it is very difficult to release object because during the gripper opening, the microobject still stick on one on both end-effectors.

5.5.4 Experimental Microassembly

Robotic agility of the presented microassembly station has been tested in tele-operate mode to assemble benchmark microobjects. Two kinds of mechanical assembly have been tried to make a three-dimensional microproduct. The first one is made by an insertion of two identical puzzle parts. The second one is a reversible assembly of two different parts.

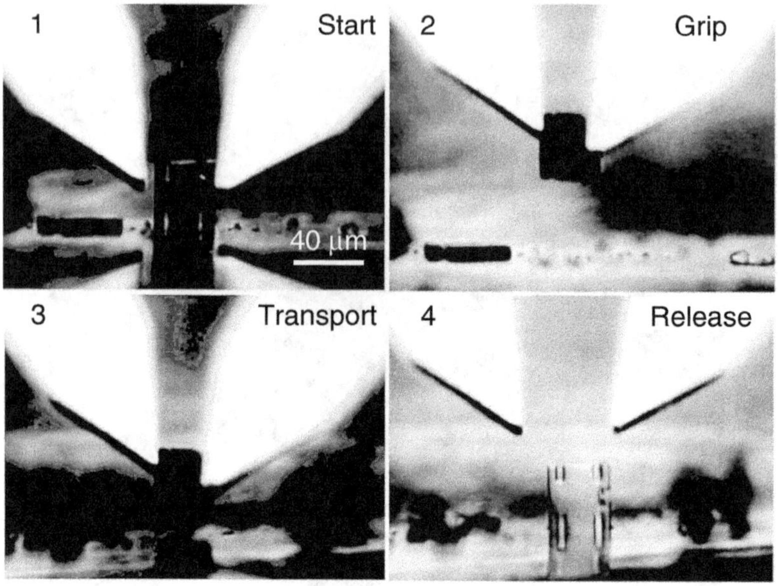

Figure 5.28. Pick-and-place of 40-μm microobjects.

5.5.5 Insertion

Each puzzle piece has four notches, close to 5 μm wide and 10 μm long. The part's thickness is 5 μm, assembly of two pieces requires to insert perpendicularly (Fig. 5.29).

The first part is gripped and placed vertically on the substrate. The second part is taken perpendicularly to the first one (step 1). Then the second part is gripped and is accurately positioned above the first part (step 2). Assembly clearance is very small and evaluated to 200 nm by scanning electron microscope (SEM) measurement and accuracy can be made up by substrate compliance. Indeed, compliance of adhesive substrate enables small rotative motion of the first part; thus insertion is easily performed without any fine orientation of the gripper (step 3). When insertion is complete, the microgripper is opened and the trajectory presented as in Figure 5.24 and used to release the assembled part (step 4).

5.5.5.1 Reversible Assembly

The second assembly benchmark requires more steps and more accuracy. Both mechanical parts are different but have the same square shape of 40 μm sides. The first part has a small key joint with a T shape on one side. The second part has a T-shaped imprint in the center of the square (Fig. 5.30). To perform the assembly, the key must be inserted in the imprint and then a lateral motion of the second part locks the assembly. This benchmark is inspired from Dechev et al. [10]. This benchmark has been tested with the robotic structure (Fig. 5.31). Parts' orientation is very important, especially for the relative orientation between

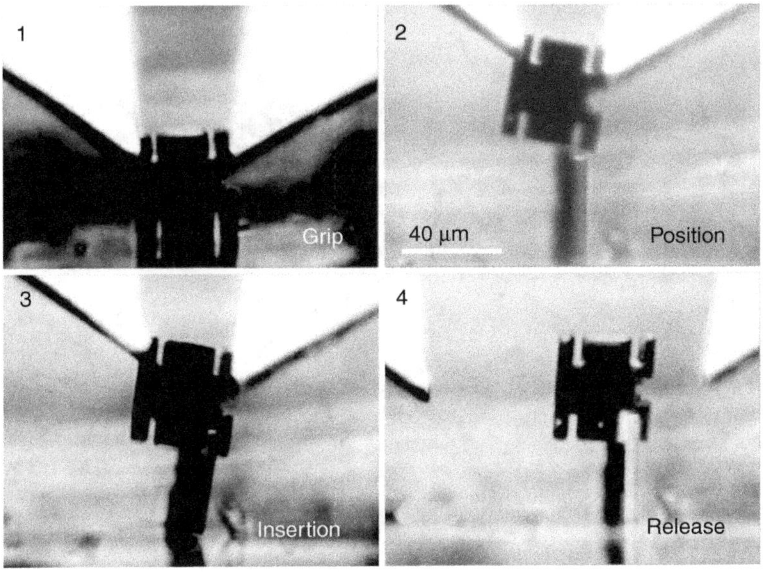

Figure 5.29. Insertion assembly.

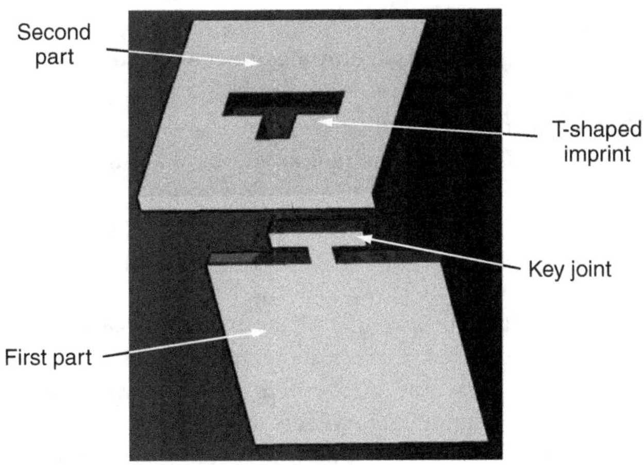

Figure 5.30. Lock joint design.

both microobjects. The first part is set vertically on the substrate. The gripper is used to grasp and align the second part above the key (step 1). When the key is in the imprint (visible on the vertical view), a vertical motion puts the key in the hole (step 2). Finally a lateral motion locks the key and the assembly is performed (step 3). After the locking motion, the 3D microassembly built can be extracted from the substrate and moved to another place (step 4). Moreover, the major interest of this kind of assembly is the possibility to disassemble it. Indeed

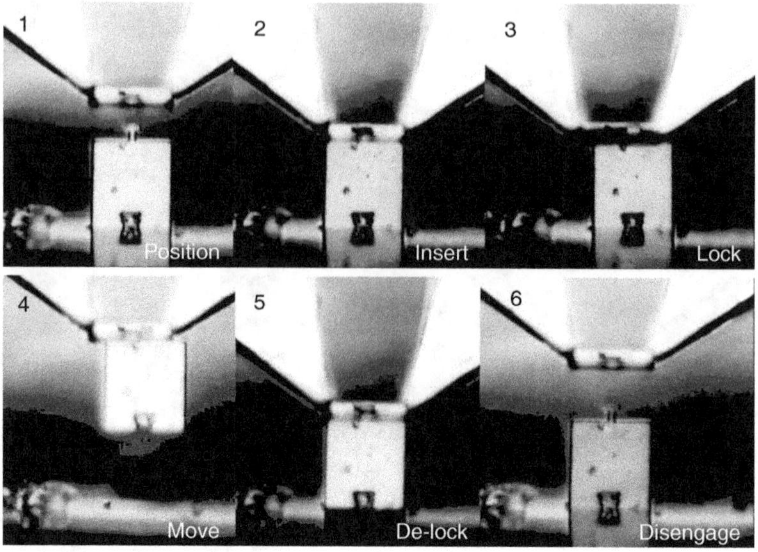

Figure 5.31. Reversible assembly.

motions can be repeated in an opposite way: a lateral motion to unlock the key (step 5) and a vertical motion to disengage the key from the imprint (step 6). Several cycles of assembly–disassembly have been tested.

5.5.5.2 Analysis of the Reliability

In order to show the reliability of the handling method, numerous pick-and-place operations have been performed in teleoperation and in an automatic cycle. The tests have been done on a silicon microobject whose dimensions are $5 \times 10 \times 20\,\mu m^3$. The objective of the pick-and-place operation is to grasp the object placed on the substrate, move it along 100 μm, and release it on the substrate. To evaluate the reliability, the success rate of the pick-and-place operations and the time cycle have been measured. First, tests have been done in teleoperation. The operator sees the lateral view and the vertical view on two screens. He controls the trajectories and the gripper movements with a joystick without force feedback: 60 operations have been done. The time cycle stays always between 3 and 4 s. Second, tests have been done in an automatic cycle without force and position feedback. The pick-and-place trajectory was repeated 60 times and the time cycle was 1.8 s. In both tests, the reliability reached 99%. As only some articles in the litterature quote the reliability of micromanipulation methods, it is quite difficult to compare this value with other works. However, tests of the reliabilty of microhandling strategies have been presented in Dafflon et al. [9] and Gauthier et al. [20]. Both tests have been done on polystyrene spheres whose diameter is 50 μm. The success rate was between 51 and 67% on around 100 tests in Gauthier et al. [20] and was between 74 and 95% on 60 tests in Dafflon et al. [9]. Consequently, this handling method enables a higher reliability on smaller objects.

5.5.5.3 Discussion

Based on several experiments performed with the robotic device, some key points have been highlighted: First, an adhesive substrate enables the reliable release of microobjects, but gripping strength has to be large enough to unstick the handled object. A smart substrate where adhesion could be controlled during assembly could be interesting to reduce the gripping force. Moreover, one of the limitations of the handling method is the difficulties to assemble more than three parts in the current configuration. Second, microobject orientation is very important to perform an assembly. Increasing DoF, especially for rotation motions, will improve the robotic agility of the device. Third, the microscopic side view is highly useful for teleoperated assembly. This view could be also used for visual servoing in addition to the vertical view.

5.6 CONCLUSION

Development of new robotic microhandling methods is a key point to fabricate hybrid microsystems as well as micromechatronic products. At present, the release task is the most critical and unreliable phase because of the impact of the surface forces and adhesion forces. Theoretical and experimental comparative analysis between the water medium and the air indicate that both types of medium show the potential interest of the liquid in micromanipulation applications. In fact, surface and adhesion forces decrease significantly in water, while the hydrodynamic force increases. Both phenomena are able to reduce, respectively, the electrostatic and adhesion perturbations and the loss of microobjects. Manipulation of artifical objects in water is consequently a promise to obtain reliable handling.

Some submerged microhandling strategies are currently available. Negative dielectrophoresis can be used to control the release of an object grasped with a two-fingered gripper. Submerged freeze gripping enables grasping with high blocking force and the release of microobjects without adhesion perturbations. At least chemical properties of the medium (e.g., pH) can be used to directly control the surface behavior of functionalized objects and gripper. This chemical control is able to switch interactions between gripper and object from attraction to repulsion.

As a transition between Part II dedicated to microhandling and Part III on microassembly, a last handling method, which is based on adhesive release and its application in microassembly, has been presented.

Acknowledgment

The author wishes to thank the FEMTO-ST Institute, graduate student Elie Gibeau; research engineers Joël Agnus, David Hériban, and Patrick Rougeot; Ph.D. students Beatriz Lopez-Walle and Mohamed Kharboutly; Dr Jérome Dejeu and Prof. Nicolas Chaillet for their contributions to the work presented in this chapter.

This work have been supported by the French National Agency (ANR) under NANOROL contract ANR-07-ROBO-0003: Nanoanalyse for micromanipulate, and PRONOMIA Contract ANR No. 05-BLAN-0325-01, and the European Union under HYDROMEL contract NMP2-CT-2006-026622: Hybrid ultra precision manufacturing process based on positional- and self-assembly for complex microproducts.

REFERENCES

1. S. F. Ackley, K. Itagaki, R. P. Murrmann, W. F. Limpert, H. H. Jellinek, Ice adhesion studies: Properties of defects in the interfacial region, *Highway Res. Board Special Rep.,* (115):87–96, 1970.
2. J. Agnus, D. Hériban, M. Gauthier, and V. Pétrini, Silicon End-Effectors for Microgripping Tasks, *Precision Eng.,* 33(4):542–548, 2009.
3. J. Agnus, P. Nectoux, and N. Chaillet, Overview of Microgrippers and Design of a Micromanipulation Station Based on a MMOC Microgripper, in Proc. of IEEE CIRA, Finland, June 2005.
4. F. Arai, D. Andou, Y. Nonoda, T. Fukuda, H. Iwata, and K. Itoigawa, Integrated Microendeffector for Micromanipulation, *IEEE/ASME Trans. Mechat.,* 3:17–23, 1998.
5. D. K. Aswal, S. Lenfant, D. Guerin, J. V. Yakhmi, and D. Vuillaume, Self Assembled Monolayers on Silicon for Molecular Electronics, *Anal. Chim. Acta*, 84–108, 2006.
6. F. Becker, P. Gascoyne, Y. Huang, and X-B Wang, Method and Apparatus for Manipulation Using Spiral Electrodes, in Patent 5,858,192, Etats-Unis, 12 janvier 1999.
7. C. C. Buron, C. Filiâtre, F. Membrey, C. Bainier, D. Charraut, and A. Foissy, Effect of Substrate on the Adsorption of Polyelectrolyte Multilayers: Study by Optical Fixed-Angle Reflectometry and AFM, *Colloids Surface A*, 305:105–111, 2007.
8. C. Clévy, A. Hubert, and N. Chaillet, Flexible Micro-assembly System Equiped with an Automated Tool Changer, *J. Micro-Nano Mechat.,* 4(1–2), 2008.
9. M. Dafflon, B. Lorent, and R. Clavel, A Micromanipulation Setup for Comparative Tests of Microgrippers, in International Symposium on Robotics (ISR), 2006.
10. N. Dechev, W. L. Cleghorn, and J. K. Mills, Microassembly of 3d Microstructures Using a Compliant, Passive Microgripper, *J. Microelectromech. Syst.,* 13(2), April 2004.
11. J. Dejeu, B. Lakard, P. Fievet, and S. Lakard, Characterization of Charge Properties of an Ultrafiltration Membrane Modified by Surface Grafting of Poly(allylamine) Hydrochloride, *J. Colloid Interface Sci.,* 33(1):335–340, 2009.
12. J. Dejeu, P. Rougeot, M. Gauthier, and W. Boireau, Reduction of Micro-object's Using Chemical Functionalisation, *Micronanoletters*, 2009.
13. J. Dejeu, M. Gauthier, P. Rougeot, and W. Boireau, Adhesion Forces Controlled by Chemical Self-assembly and pH, Application to Robotic Microhandling, *ACS Appl. Mat. Interfaces*, 1(9):1966–1973, 4(2), 2009.
14. G. Demirel, M. O. Caglayan, B. Garipcan, and E. Piskin, A Novel DNA Biosensor Based on Ellipsometry, *Surf. Sci.,* 602:952–959, 2008.

15. P. M. Dove and C. M. Craven, Surface Charge Density on Silica in Alkali and Alkaline Earth Chloride Electrolyte Solutions, *Geochim. Cosmochim. Acta*, 69(21):4963–4970, 2005.

16. S. Droz et al., New Generation of Grippers for the Manipulation of Miniaturized Components, in Proc. of Mechatronics, Besançon, France, October 2001, pp. 572–575.

17. M. Frenea, S. P. Faure, B. Le Pioufle, Ph. Coquet, and H. Fujita, Positioning Living Cells on a High-Density Electrode Array by Negative Dielectrophoresis, *Mat. Sci. Eng.,* 23:597–603, 2003.

18. P. Gascoyne and J. Vikoukal, Dielectrophoresis-based Sample Handling in General-Purpose Programmable Diagnostic Instruments, *IEEE Proc.,* (1):22–42, 2004.

19. M. Gauthier, E. Gibeau, and D. Hériban, Submerged Robotic Micromanipulation and Dielectrophoretic Micro-object Release, in Proc. of the IEEE ICARCV 2006 Conference, Singapore, 2006.

20. M. Gauthier, B. Lopez-Walle, and C. Clévy, Comparison between Micro-objects Manipulations in Dry and Liquid Mediums, in Proc. of CIRA'05, June 2005.

21. M. Gauthier, S. Régnier, P. Rougeot, and N. Chaillet, Forces Analysis for Micromanipulations in Dry and Liquid Media, *J. Micromech.,* 3(3–4):389–413, Sept. 2006.

22. D. S. Haliyo, Y. Rollot, and S. Régnier, Manipulation of Micro-objects Using Adhesion Forces and Dynamical Effects, in Proc. of the Int. Conference on Robotics and Automation, Washington DC, May 2002, pp. 1949–1954.

23. D. Hériban, V. Pétrini, J. Agnus, and M. Gauthier, Mechanical De-tethering Technique for Silicon MEMS Etched with DRIE Process, *J. Micromech. Microeng.,* 2009.

24. D. Hériban, J. Agnus, and M. Gauthier, Micromanipulation of Silicate Micro-sized Particles for Biological Applications, *J. Micromech. Microeng.,* 19(5):05011, 2009.

25. D. Heriban, J. Agnus, J-R. Coudevylle, M. Gauthier, and N. Chaillet, Design of Silicon Finger Tips for a MOC (Microrobot on Chip) Microgripper, In Proc. of the Int. Workshop on Topica Meeting on Microfactories (TMMF05), Tsukuba, Japan, October 2005.

26. A. Kochan, European Project Develops "Ice" Gripper for Micro-sized Components, *Assembly Automation*, 17(2):114–115, 1997.

27. D. Lang, I. Kurniawan, M. Tichem, and B. Karpuschewski, First Investigations on Force Mechanisms in Liquid Solidification Micro-gripping, in Proc. of ISATP, July 19–21 2005, pp. 92–97.

28. D. Lang, M. Tichem, and S. Blom, The Investigation of Intermediates for Phase Changing Micro-gripping, in Proc. of IEEE IWMF, France, October 2006.

29. B. Lopez-Walle, M. Gauthier, and N. Chaillet, Submerged Freeze Gripper to Manipulate Micro-objects, in Proc. of the IEEE Int. Conf. on Intelligent Robots and Systems—IROS'06, Beijing, China, Oct. 2006, pp. 784–789.

30. B. Lopez-Walle, M. Gauthier, and N. Chaillet, Dynamic Modelling of a Submerged Freeze Microgripper Using a Thermal Network, in Proceedings of the 2007 IEEE/ASME International Conference on Advanced Intelligent Mechatronics, Sept. 2007.

31. B. Lopez-Walle, M. Gauthier, and N. Chaillet, A Submerged Freeze Microgripper for Micromanipulations, in Proc. of the 2007 IEEE International Conference on Robotics and Automation, Roma, Italy, April 2007, pp. 10–14.

32. B. Lopez-Walle, M. Gauthier, and N. Chaillet. Principle of a Submerged Freeze Gripper for Microassembly, *Robotics, IEEE Trans.,* 24(4):897–902, Aug. 2008.
33. J. H. Maas, M. A. Cohen Stuart, A. B. Sieval, H. Zuilhof, and E. J. R. Sudhölterb, Preparation of Polystyrene Brushes by Reaction of Terminal Vinyl Groups on Silicon and Silica Surfaces, *Thin Solid Films*, 426:135–139, 2003.
34. O. Moreau, C. Portella, F. Massicot, J. M. Herry, and A. M. Riquet, Adhesion on Polyethylene Glycol and Quaternary Ammonium Salt-grafted Silicon Surfaces: Influence of Physicochemical Properties, *Surface Coatings Tech.,* 2001:5994–6004, 2007.
35. A. Okuda, T. Nagasawa, S. Okawa, and A. Saito, Research on Solidification of Water on Surface, in Proc. of International Conference on the Properties of Water and Steam (ICPWS), Japan, August 2004, pp. 690–695.
36. H. A. Pohl, *Dielectrophoresis*, Cambridge University Press, Cambridge, 1978.
37. D. O. Popa and H. E. Stephanou, Micro and Meso Scale Robotic Assembly, in WTEC Workshop: Review of U.S. Research in Robotics, WTEC, 2004.
38. A. Rosenthal and J. Voldman, Dielectrophoretic Traps for Single-Particle Patterning, *Biophys. J.,* 88, March 2005.
39. J. B. Schlenoff and G. Decher, *Sequential Assembly of Nanocomposite Materials*, Wiley-VCH, Hoboken, NJ, 2003.
40. A. Subramanian, B. Vikramaditya, B. J. Nelson, D. J. Bell, and L. Dong, Dielectrophoretic Micro/nanoassembly with Microtweezers and Nanoelectrodes, in Proc. of the 12th Int. Conf. on Advanced Robotics, Seattle, July 2005.
41. B. Tamadazte, N. Piat, and S. Dembélé, Robotic Micromanipulation and Microassembly Using Monoview and Multiscale Visual Serving, *IEEE/ASME Trans. on Mechatron.,* pp 99:1083–4435, 2010.
42. N. Tsukada, K. Kudoh, A. Yamamoto, T. Higuchi, M. Kobayashi, K. Sato, K. Oishi, and K. Iida, Development of Oocyte Rotation System for Biological Cell Manipulation, in Proc. of the 32nd International Symposium on Robotics—ISR2001, Seoul, Korea, 19–21 April 2001.
43. O. I. Vinogradova and G. E. Yakubov, Dynamic Effects on Force Measurements. 2. Lubrication and the Atomic Force Microscope, *Langmuir*, 19:1227–1234, 2003.
44. J. Wang, D-J. Guo, B. Xia, J. Chao, and S-J. Xiao, Preparation of Organic Monolayers with Azide on Porous Silicon via sine Bonds, *Colloids Surface A*, 305:66–75, 2007.
45. S. R. Wasserman, Y.-T. Tao, and G. M. Whitesides, Structure and Reactivity of Alkylsiloxane Monolayers Formed by Reaction of Alkyltrichlorosilanes on Silicon Substrates, *Langmuir*, 5:1074–1087, 1989.
46. T. Watanabe, N. Fujino, and Z. Jiang, Micromanipulation Using Squeeze Effect, in Proc. of IEEE IROS, vol. 4, Japan, 2004, pp. 3357–3362.
47. J. Wu, J. Lu, and J. Chu, Handling Micro Objects by Adhesion Forces Using Gilt Probe, in Proc. of IEEE IWMF, China, 2004.

PART III

ROBOTIC AND MICROASSEMBLY

CHAPTER 6

ROBOTIC MICROASSEMBLY OF 3D MEMS STRUCTURES

NIKOLAI DECHEV

6.1 INTRODUCTION

Microassembly is a method of manipulating microparts from their original fabrication location to a "final" assembly location. It allows for the construction of complex microsystems, which cannot be fabricated using micromachining processes alone. In the scope of our work, microassembly is useful for the creation of (i) out-of-plane microstructures with smooth microparts at various large angles to the substrate and (ii) microstructures that require microparts from two or more different chips or sources. Microassembly approaches can generally be categorized into three main groups, which are parallel microassembly, self-assembly, and serial (sequential) microassembly. Parallel microassembly is used to simultaneously assemble microdevices at multiple assembly sites and includes "flip–chip" [21] and "batch transfer" [16] methods. Self-assembly systems work by enabling the constituent microparts to assemble themselves spontaneously, by subjecting them to the influence of an external driving force such as heat, magnetism, centrifugal force, liquid surface tension, or other. Some examples include solder surface tension self-assembly [12], plastic deformation magnetic assembly (PDMA) [24], thermo-plastic polyimide joint assembly [9], centrifugal force assembly [14], and ultrasonic vibration/electrostatic field sorting/aligning assembly [3]. Serial microassembly is a sequential process in which assembly tasks are performed one after the other. To complete one assembly, a series of subtasks are required, such as grasping microparts with a grasping tool, manipulating them, and joining them to other microparts. Our work consists of a serial microassembly

Robotic Microassembly, edited by Michaël Gauthier and Stéphane Régnier

method using a robotic micromanipulator. Other serial microassembly systems that make use of robotic micromanipulators are those equipped with microgrippers [11, 22, 23] or those equipped with microtweezers [15, 20].

6.2 METHODOLOGY OF THE MICROASSEMBLY SYSTEM

6.2.1 Purpose of the Microassembly System

The creation of useful microelectromechanical system (MEMS) devices (microsystems) is the end goal of any microassembly process. However, the use of microassembly generally places limits on the microparts that can be assembled. When developing a microsystem that will employ this microassembly process, the design of the microparts must take into account the following three factors: (i) the function/role of the micropart within the microsystem, (ii) the joining method to fasten the microparts together to create a functioning whole microsystem, and (iii) ensuring that the microparts can be compatibly handled by the microassembly process. Note that these three factors are ranked in relative order of importance. In other words, the function of the micropart is more important than the joining method, which is in turn more important than the compatibility with the microassembly process. Generally, there would be no point to assemble microparts with good microassembly compatibility, if this compatibility reduces the function of the microsystem or inhibits the joining process. In a sense, a balance must be reached between these three factors.

6.2.2 System Objectives

The goal of this work is to develop a general microassembly process that can be used to construct a wide variety of three-dimensional (3D) microsystems. In working toward this goal, four major objectives have been identified. The first major objective is to develop the microassembly process so that it minimizes the impact on the function/role of the microparts, and hence the finished microsystem. The second objective is to maximize the number of possible assembly configurations by allowing the microparts to be moved and oriented to any possible position in space. As part of this objective, the system should accommodate microparts of different shapes, microparts fabricated from different materials, and combine microparts from multiple chips to assemble a single microstructure. The third objective is to develop a joining system that can mechanically and electrically join together microparts at any orientation or position in 3D space. The fourth objective is to develop the microassembly process so that it is capable of rapid automatic assembly.

6.2.3 Microassembly versus Micromanipulation

It is important to make a clear distinction between microassembly and micromanipulation since both terms involve the "handling" of microscaled objects. In making this distinction, we can develop the basic concept for the design of all microparts and microgrippers of this microassembly method. Micromanipulation is generally the act of translating and rotating microobjects from one location and orientation to another. It is important to note that micromanipulation does not imply that the microobjects are to be assembled. Many researchers use micromanipulation equipment to manipulate biological cells or other microobjects for the purposes of examining them or altering them. As such, the end effectors (e.g., probes, pipettes, or grasping tools such as tweezers or grippers) used for micromanipulation are usually designed to handle a variety of microobjects. This is because the microobjects may have unknown properties or variable shapes. In addition, the microobject may behave in an unknown manner during manipulation. Microassembly is the act of building, constructing, or collecting two or more microparts into a microstructure, in a permanent manner. Microassembly is deterministic in the sense that the final result is specified by design. Since the geometry of the desired microstructure must be known in advance, it follows that the constituent microparts of a microassembly will have a known shape, size, material property, and final position. Therefore, we propose that the microparts and the end-effector (microgripper) that will handle them can both be designed in advance such that their designs are mutually complementary to each other. As such, we propose that the end-effectors for microassembly can be more specialized than those for general micromanipulation.

6.2.4 Microassembly Concept

The microassembly performed by this work is serial, or "sequential," as one micropart is handled at a time. Hence the following assumptions are made about the microassembly method. (1) Microparts are of known standard sizes and geometries. (2) Microparts can have their design altered slightly to accommodate the microassembly system. (3) Conventional robotic assembly systems use components that are prearranged in trays or on feed tapes. Therefore, it would follow that the microparts are arranged in ordered, specific locations on the substrate, to speed the grasping process. (4) Microassembly implies that microparts would be joined in some way to other microparts, and not released freely into the workspace. Based upon assumptions 1–4, it is proposed that each micropart used by this system be designed in such a way that it incorporates three "modular design features." These modular features are based on geometrical shapes and can be adapted to a variety of microparts. They are: (a) an interface feature that allows the micropart to be grasped by the microgripper; (b) a tether feature that secures the micropart to the substrate and breaks away after the microgripper has grasped the micropart; and (c) a joint feature that is used to join the micropart to other microparts.

6.2.5 Interface Between Microassembly Subsystems

The serial microassembly system of this work was developed on the basis of four interconnected subsystems, as illustrated by the blocks in Figure 6.1. The

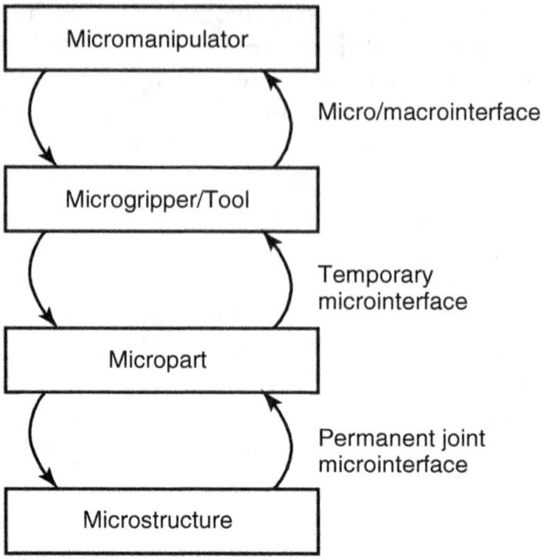

Figure 6.1. Subsystems and interfaces for serial microassembly system.

arrows in the diagram represent the interface between these blocks, which plays a critical role in determining the functional requirements of these subsystems. Notice that the blocks influence each other's functional requirements. In other words, the micromanipulator design is tied to the design of the microgripper/tool, which is in turn tied to the design of the microparts, which are in turn tied to the design of the microstructure, and vice versa. Although these subsystems are tied together, the system has been developed as a general process, so that it can be used to assemble a wide variety of microstructures.

6.3 ROBOTIC MICROMANIPULATOR

A micromanipulator is required for handling all the microparts used in this work due to their small size. The microparts range from 60 to 400 μm in length or width and 5 μm in thickness. They must generally be positioned to within ±2 μm at their assembly sites for assembly to succeed. The robotic micromanipulator (RM) used in this work is motorized, designed with a dual control mode strategy. In one mode, it is designed as a telerobotic system that is controlled by a human operator. In the other mode, it is designed for automatic task execution [1, 19]. This dual approach is pragmatic for this system. Due to the experimental nature of grasping, manipulating, and joining microparts while developing this microassembly system, human operator control is critical for experimental work. However, once the parameters for a particular assembly process are defined, the system can be programmed to execute that procedure automatically. Automation of microassembly tasks is not trivial and is the subject of much research [10, 17, 18]. Note that for topics in this chapter all tasks are performed by telerobotic control, using a human operator and a video microscope system for visual feedback. The RM used for this microassembly process employs six independent axes of motion (i.e., 6 degrees of freedom, or 6 DoF), as illustrated in Figure 6.2(a). The six axes of the RM are split into two groups with the x, y, z, and α axes mounted on a granite base (kinematic ground), and the β and γ axes mounted on a granite post (also kinematic ground). The usable workspace is 360° about α, 180° about β, 110° about γ, and 25 mm in translation for each of the x, y, and z axes. During telerobotic operation mode, a human operator uses a joystick and a keyboard to enter motion commands. The operator relies on a video microscope for visual feedback, and a GUI (graphical user interface) [8], which displays encoder positions of the six axes. Figure 6.2(b)

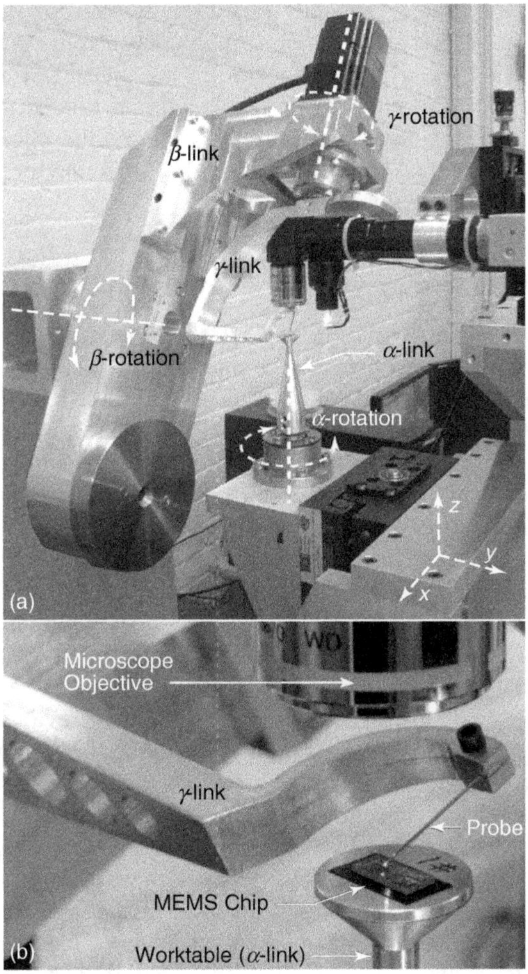

Figure 6.2. 6 DoF robotic manipulator for microassembly operations.

shows a closeup view of the worktable holding the MEMS chip and the probe pin held by the γ link (γ axis) of the RM. The MEMS chip is a loose die that is attached to a 20-mm-diameter disk using double-sticky tape. This disk is in turn attached onto the worktable (α axis) of the RM. A detailed description of the development, design, and operation of the RM is provided in Dechev et al. [8].

6.4 OVERVIEW OF MICROASSEMBLY SYSTEM

A general overview of the microassembly system is now presented to introduce the basic concepts and the terminology used with this system. Prior to microassembly operations, a microgripper must first be attached to the probe pin held by the RM. After preparing the RM, the assembly process to construct a microstructure involves the sequential addition of microparts. Each micropart added to the microstructure must go through five steps, which are: (1) the micropart is grasped by a microgripper, (2) the micropart is removed from the chip substrate, (3) the micropart is translated and rotated through space, (4) the micropart is joined to the target joint site, and (5) the micropart is released. After the micropart is released, the sequence of (1–5) may be repeated as often as necessary to complete a microstructure.

6.4.1 Bonding a Microgripper to the Probe Pin of the RM

The tip of the probe pin represents the interface between the macroworld and the microworld. This is an important distinction. All objects up to and including the probe pin are macroscopic and can be handled by the human hand, whereas all objects past the tip, such as the microgripper and microparts are all microscopic. Figure 6.3 illustrates the procedure to bond a microgripper to the tip of the probe pin. Note that the elements shown in Figure 6.3 are not to scale. The bonding process consists of seven tasks, which are: (1) Locate the microgripper to be bonded in the x and y coordinates with respect to the probe tip using a "visual calibration," z using a "touch calibration," and record the location, as shown in Figure 6.3(a). (2) Lower the worktable that carries the MEMS chip away from the probe to provide clearance to apply an adhesive to the probe tip. (3) Apply an ultraviolet (UV)-curable adhesive to the probe tip. (4) Reposition the worktable under the probe tip to the previously recorded location at the microgripper bond pad. (5) Allow the adhesive to flow from the probe tip onto the microgripper bond pad for a few seconds. (6) Radiate the probe tip with UV light for 30 s to cure the adhesive. (7) Move the MEMS chip down and away from the probe tip, thereby detaching the microgripper from the MEMS chip, as shown in Figure 6.3(b). The microgripper is now ready to be used for grasping tasks. Figure 6.4(b) shows an SEM (scanning electron microscope) image of a microgripper successfully bonded to the tungsten probe tip. There are usually a dozen or so microgrippers fabricated alongside the microparts on the same chip. Of these, there may be a few different designs corresponding to

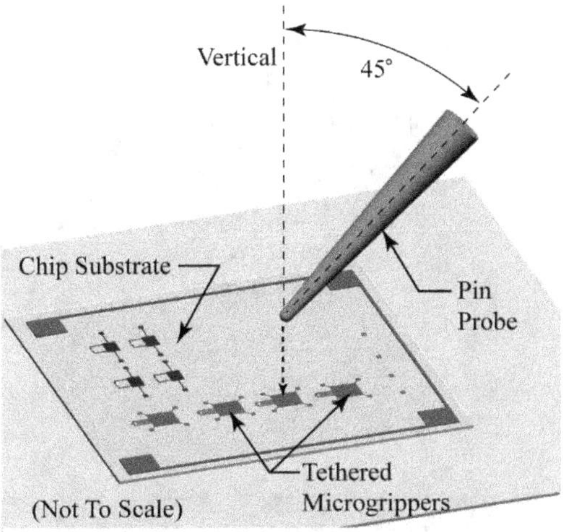

(a)

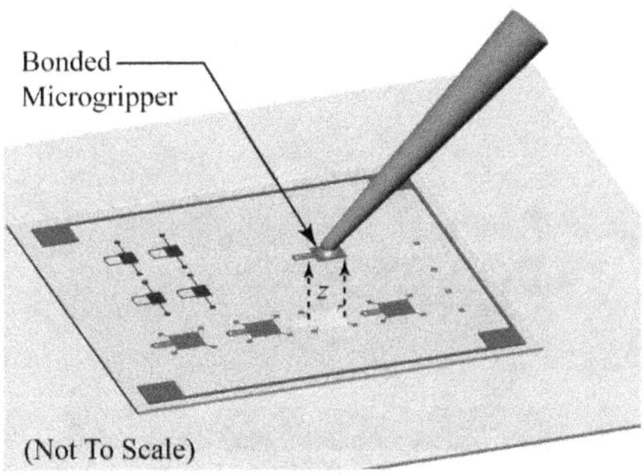

(b)

Figure 6.3. Illustration of bonding microgripper to the probe tip: (a) pin probe in bonding orientation and (b) microgripper bonded to probe tip.

different grasping or assembly needs [6] for different styles of microparts. In this sense, the microgrippers of this work can be considered as "disposable tools" that are bonded to the RM, used as needed, and then discarded. The same microgripper may be used for hundreds of operations or for only one operation, before it is changed with an alternate design for another type of procedure. For this reason, it is important for the bonding system to be reversible, fast, and reliable. As such, the UV adhesive system was adopted since the bond is rapid (45 s) and the adhesive can be dissolved rapidly when changes are required. All of the microgrippers used in this system are attached to the chip substrate by tethers, as shown in Figure 6.3(a) and 6.4(a). The tethers are attached to anchor pads,

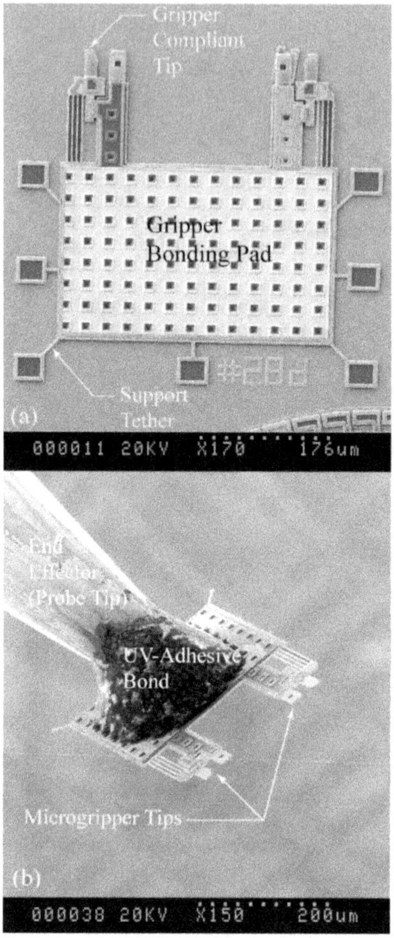

Figure 6.4. SEM image of microgripper with modular interface feature: (a) microgripper on the chip substrate held by tethers and (b) microgripper bonded with UV adhesive to the end-effector (probe pin) of the micromanipulator.

which are permanently attached to the substrate. The tethers are designed to be strong enough to hold the microgripper onto the substrate during transportation of the chips. They also immobilize the microgripper during the adhesive bonding operation. However, the teathers are weaker than the adhesive bond and hence are easily broken off when a bonded gripper is pulled off the substrate by the probe pin.

6.4.1.1 Grasping a Micropart with a Microgripper

After a microgripper is bonded to the probe pin, it is ready for grasping microparts. To grasp a micropart, the microgripper is translated along the x, y,

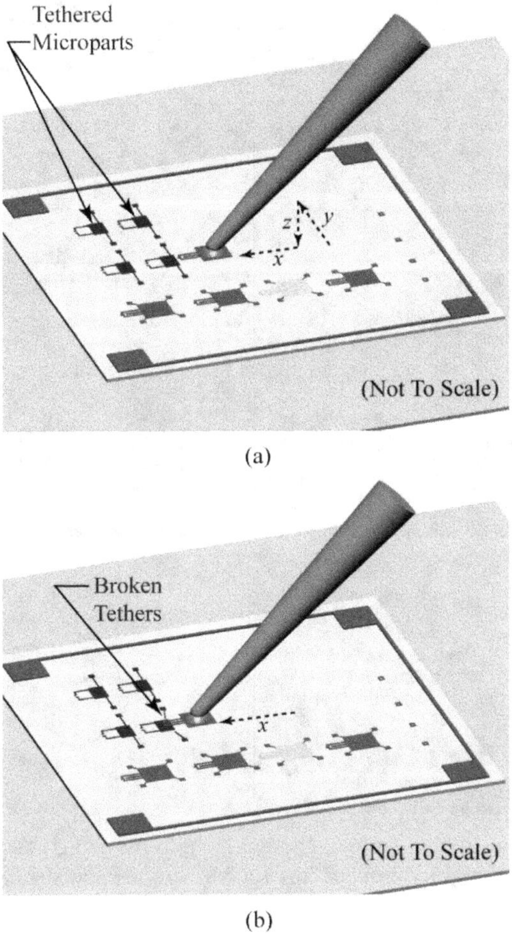

Figure 6.5. Illustration of grasping a micropart with a microgripper: (a) microgripper alignment with micropart and (b) microgripper grasping micropart.

and z axes to align the tips with the interface feature of a micropart, as shown in Figure 6.5(a). After alignment, the microgripper is translated in the x direction, causing it to grasp the micropart. Details of the design and operation of the microgripper are provided in Dechev et al. [5]. Figure 6.6 shows a number of different video images of microparts prior to being grasped by a microgripper.

6.4.1.2 Removing a Grasped Micropart from the Chip

After a successful grasp, the microgripper must remove the micropart from the substrate. Each micropart is attached to the substrate by tethers, as shown in Figure 6.5(a). In order to remove a micropart, the tethers that hold it to the substrate must be broken. The tethers are designed to break/rupture when the microgripper is translated further in the x direction beyond the grasping position,

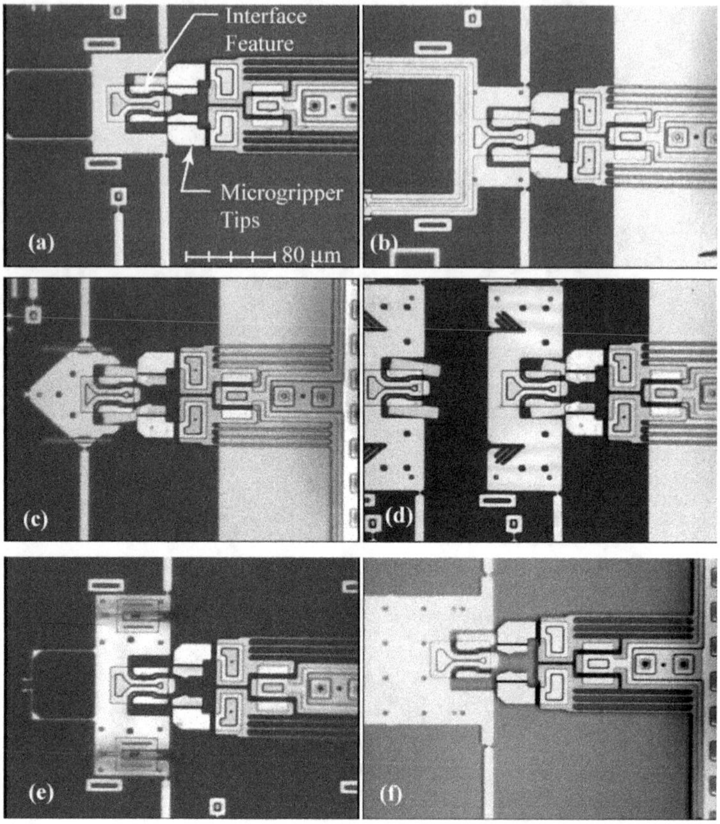

Figure 6.6. Video images of various microparts, prior to being grasped by a microgripper. All microparts and microgripper tips shown here use the standard interface feature geometry design.

as shown in Figure 6.5(b). Details of the tether design and operation are provided in Dechev et al. [4].

6.4.1.3 Manipulating the Micropart

After grasping a micropart and breaking the tethers that hold it, the micropart is securely held by the microgripper and is free from the substrate. The micropart is lifted from the substrate in the z direction, as shown in Figure 6.7(a). To join the micropart perpendicularly to another micropart on the substrate, the probe pin must be rotated 90° counterclockwise about the β axis, as shown in Figure 6.7(b). Note that in this new orientation, the microgripper and the micropart are now perpendicular to the substrate. Also, note that the micropart (held by the microgripper) is the lowest point on the probe pin. In this orientation, the microgripper is able join the micropart to the base structure as shown in Figure 6.8(a). Note that this described manipulation is quite simple, consisting of a single β-axis rotation and an x, y, and z translation. However, the manipulation

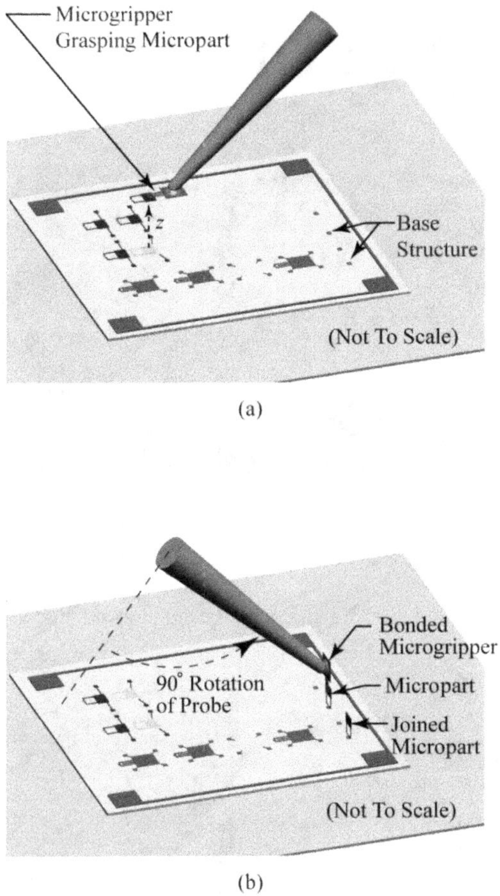

Figure 6.7. Illustration of manipulation of a micropart and microgripper: (a) microgripper grasping micropart above chip and (b) probe and microgripper in joining orientation.

may involve five or even all six axes, depending on the nature of the microjoint that must be made.

6.4.1.4 Joining the Micropart to Another Micropart

After orienting the probe pin to the joining orientation, as shown in Figure 6.7(b), the video microscope will view the micropart and microgripper while they are perpendicular to the focal plane. In this viewing configuration, the micropart will appear highly "out of focus" due to the limited focal depth of the microscope. Therefore, the microscope is focused on the "joint feature" of the perpendicular micropart. With the joint feature in focus, the micropart is aligned in the x and y axes with the "target joint feature" located on the base structure, as shown

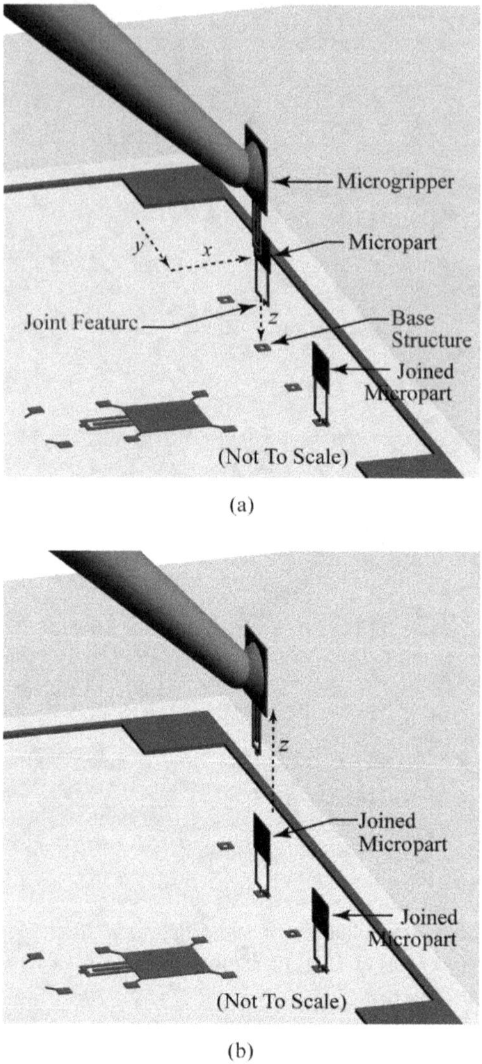

Figure 6.8. Illustration of joining a micropart into another (base structure): (a) micropart joint feature alignment with base structure and (b) micropart joined to base structure.

in Figure 6.8(a). When the alignment is deemed good, the microgripper is commanded down along the z axis. The insertion of the micropart joint feature into the target joint feature can be observed using the video microscope.

6.4.1.5 Releasing the Micropart from the Microgripper

After a successful joint is formed, the probe pin is commanded away in the z axis, which causes the microgripper to release the micropart, as shown in Figure 6.8(b). The micropart remains joined perpendicularly to the base structure. The probe pin is reoriented back to the grasping position, and the assembly operation steps 1–5 can be repeated as often as required to complete the assembly.

6.5 MODULAR DESIGN FEATURES FOR COMPATIBILITY WITH THE MICROASSEMBLY SYSTEM

Given the four assumptions outlined in Section 6.24, a requirement is imposed that all microparts used with this microassembly system, regardless of their shape or function, must be designed to incorporate three "modular design features." These modular features ensure that the microparts are compatible with the microassembly process. First, and most important, all microparts have an interface feature, which allows them to be grasped by the microgripper. The microgripper tip geometry is specifically designed to interface with the interface feature geometry on the micropart. This work makes use of two different interface feature designs known as the "standard" as shown in Figures 6.6 and 6.9, and also the "modular" interface feature [6] as shown on the microgrippers of Figure 6.4. Second, all microparts are designed with tether features [4] protruding from their sides, which allow them to be securely held and accurately located on the surface of a silicon chip. Yet, as described in the Section 6.4.1.2, these features are designed to "break away" after a micropart is grasped by a microgripper. A tether feature is shown on the micropart of Figure 6.9(c). Third, all microparts have a built-in joint feature [7] used for joining them to other microparts during assembly. These joint features will be described later in this chapter since they tend to be application specific.

6.6 GRASPING INTERFACE (INTERFACE FEATURE)

Ideally, a microgripper used for microassembly should be designed to handle a range of microparts of various shapes and sizes. As noted in Section 6.2.4, the design of microparts must be altered and standardized in some way to allow the microgripper to grasp them. In this work a single, standard microgripper [5] was initially developed. It can handle various microparts, which are equipped with a corresponding standard interface feature. Figure 6.9(a) shows an SEM image of the grasping tips of the original, standard microgripper. This microgripper has a "passive" design, which requires no "active" actuation of the microgripper tips. Most other microassembly research groups use active microgrippers. In other words, the controller actively sends a "close command" to the end-effector to grasp a microobject or an "open command" to release a microobject. We propose a passive microgripper that can be used to accomplish microassembly operations. A passive design refers to the fact that there is no active control of the gripper tips to open or close. Instead, the microgripper "self-opens" during microcomponent grasping and "self-releases" microcomponents after they are joined to other microstructures. Details of the passive microgripper operation are available in

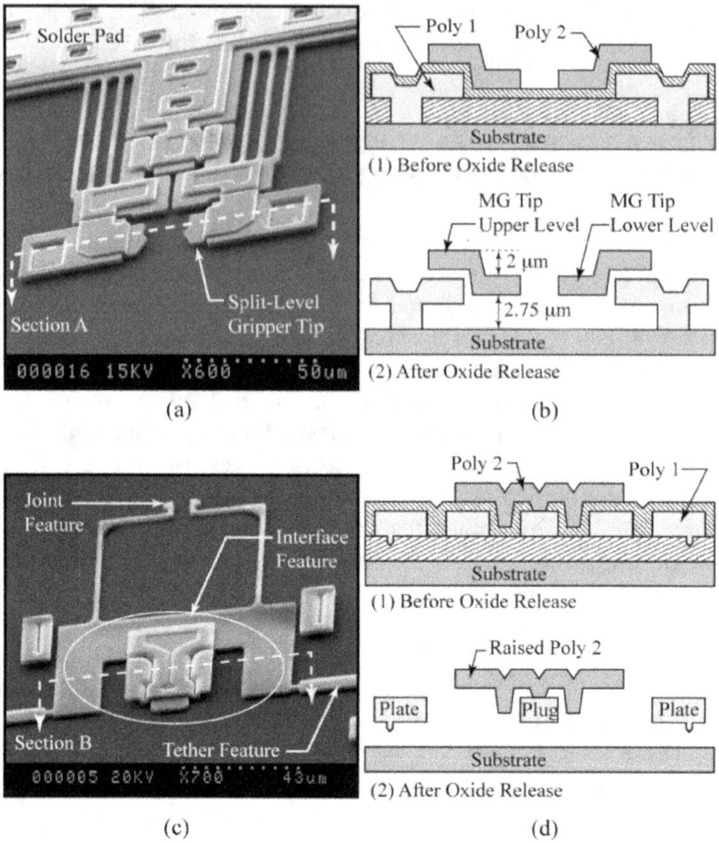

Figure 6.9. SEM image of microgripper tips and micropart interface feature with cross sections of both: (a) SEM of standard microgripper, (b) cross section A of microgripper, (c) SEM of standard interface F, and (d) cross section B of interface F.

Dechev et al. [5]. Figure 6.9(b) illustrates the cross-sectional view of the microgripper tips along Section A. Note that the microgripper tips have a split-level design consisting of an upper level and a lower level. This split-level design assists in providing a secure grasp of the microparts. Figure 6.9(c) shows an SEM image of the original, standard interface feature located in the center/back of microparts. Figure 6.9(d) illustrates the cross-sectional view of the interface feature along Section B. Note that the interface feature is comprised of two layers of polysilicon, denoted Poly 1 and Poly 2. The Poly 2 layer has been intentionally fabricated to create the raised Poly 2 structure shown in 6.9(d-2). This allows the split-level microgripper tips to interlock with the interface feature to create the secure grasp. This interface feature design has been applied to a wide variety of microparts. Figure 6.6 shows some video microscope images of various microparts used in this research, which are about to be grasped by the standard microgripper tips. Note that the microparts have different shapes, sizes, and different joint features, yet all are equipped with an identical, standard interface feature, located on the center/back of the microparts.

The passive microgrippers used in this work have a number of advantages. They do not require complex and bulky actuator designs and do not require electrical power. Since they do not require power, they have a single bonding pad, which makes them easy to bond to the RM probe pin, and makes them smaller in size than active microgripper designs. As passive designs, they employ a grasping and releasing strategy [5] that overcomes the problems of stiction, which can occur when attempting to release microobjects with active microgrippers. As such, passive microgrippers are ideally suited for this microassembly work. However, passive microgrippers also have a number of limitations. Passive microgrippers can only grasp objects that are adequately restrained (i.e., by tethers or other restraint) and can only release objects after they are joined to other objects. Further, they can only grasp objects that have specific interface feature geometries. These are reasonable preconditions for microassembly purposes, however, these requirements are not reasonable for general micromanipulation of nonstandard, unrestrained, or irregular microobjects, which can be better handled by active microgrippers.

6.7 PMKIL MICROASSEMBLY PROCESS

The process of assembling an actual microsystem is now described. This will serve to demonstrate the practical aspects of using this microassembly approach, including the benefits and limitations. The approach described uses passive microgrippers, in combination with "key and interlock" joint features, and is hence referred to as the PMKIL microassembly process. One of the more recent and novel microsystems constructed using this microassembly process is a 3D microelectrostatic motor/mirror [2], with applications for microoptical switching. A working prototype of this 3D micromirror assembly is shown in Figure 6.10. In creating this microsystem, a number of new techniques were developed and build upon the general microassembly method. Therefore, the microassembly process will be described in the context of constructing this 3D micromirror and will describe some of the novel aspects. The process to construct the micromirror involves the sequential addition of microparts, to first assemble the microparts onto the electrostatic motor rotor and, subsequently, to assemble more microparts into those previously assembled. To do this, each micropart must go through the five tasks described in Section 6.4. As a summary, these tasks are: (1) the micropart is grasped with the microgripper, (2) the micropart is removed from the chip substrate, (3) the micropart is rotated and translated through space, (4) the micropart is joined to other micropart(s), and (5) the micropart is released by the microgripper. Figure 6.11 shows SEM images of the constituent microparts used to construct the 3D micromirror, as they lay in their fabricated positions on the chip substrate. Figure 6.11(a) shows the mirror microparts and Figure 6.11(b) shows the mirror support posts. These

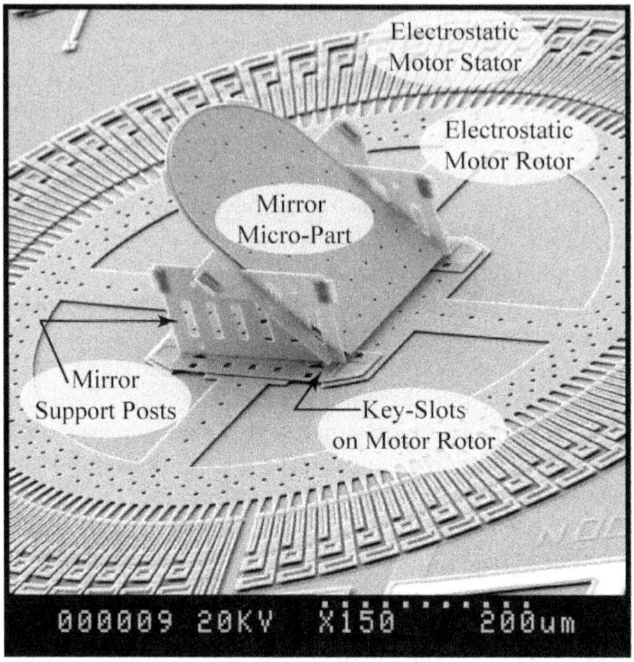

Figure 6.10. SEM image of assembled 3D micromirror.

microparts and the microgrippers are fabricated using the PolyMUMPs surface micromachining process [13]. Note that all these microparts are attached to the chip substrate via tethers. These tethers will securely hold the microparts onto the chip during transportation and during grasping but are also designed to break away subsequent to the grasping operation. Also note the built-in interface feature on each micropart, which is specifically designed to mate with the modular microgripper tips shown in Figures 6.4(a) and (b). These modular microgripper tips and modular interface features have been redesigned in comparison to the standard interface feature shown in Figure 6.9. This wider modular configuration is better suited for grasping and holding relatively wide microparts that comprise the 3D micromirrors. The five assembly tasks are now described in relation to assembling the 3D micromirror. It is assumed that a microgripper suitable for grasping a micropart has been bonded to the probe pin of the RM, using the procedure described in Section 6.4.1.

6.7.1 Grasping a MicroPart

The first step prior to grasping is to align the microgripper tips with the interface feature on the micropart. Figure 6.12 shows a sequence of video microscope images captured during a grasping operation. Figure 6.12(a) shows the microgripper tips positioned approximately 30 μm above the interface feature of the

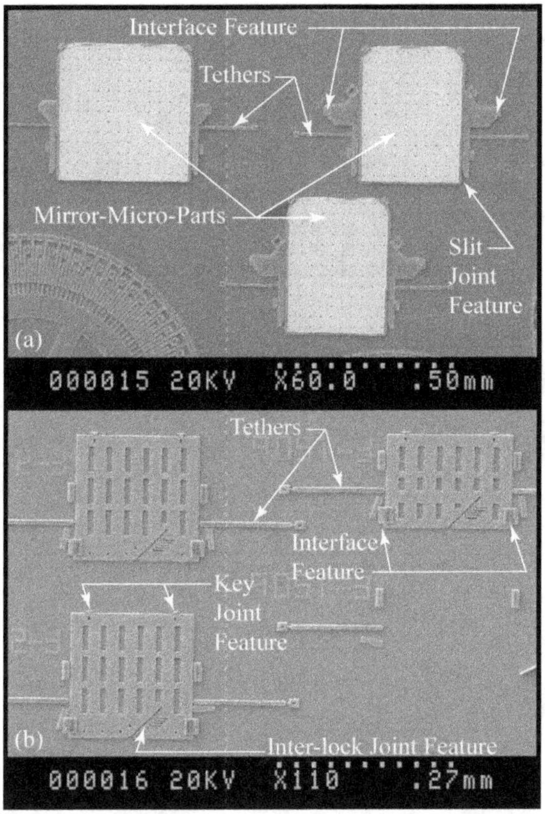

Figure 6.11. SEM image of microparts secured to substrate via tether features: (a) micromirror parts and (b) mirror support posts. Note the interface features with which the microgripper tips mate.

micropart to be grasped. The depth of focus of the microscope system is only 1.5 μm, and the microscope remains in focus with the microgripper tips at all times. Therefore, all other objects either closer to or further from the focal plane will appear out of focus. The field of view of these images is 427 μm horizontally by 320 μm vertically. Figure 6.12(b) shows the initial insertion of the microgripper tips into the interface feature. As the tips are inserted into the interface feature in the x direction, they passively open outward (y direction). Figure 6.12(c) shows the completed grasp. Note that although the micropart is now grasped, it still remains tethered to the chip substrate.

6.7.2 Removing the Micropart from Chip

To remove the micropart from the chip, force is applied in the x direction, as shown in Figure 6.12(d), to break the tethers. The deflection of the tethers can be observed in the image. The tethers have a narrow "notch" at each end to create a

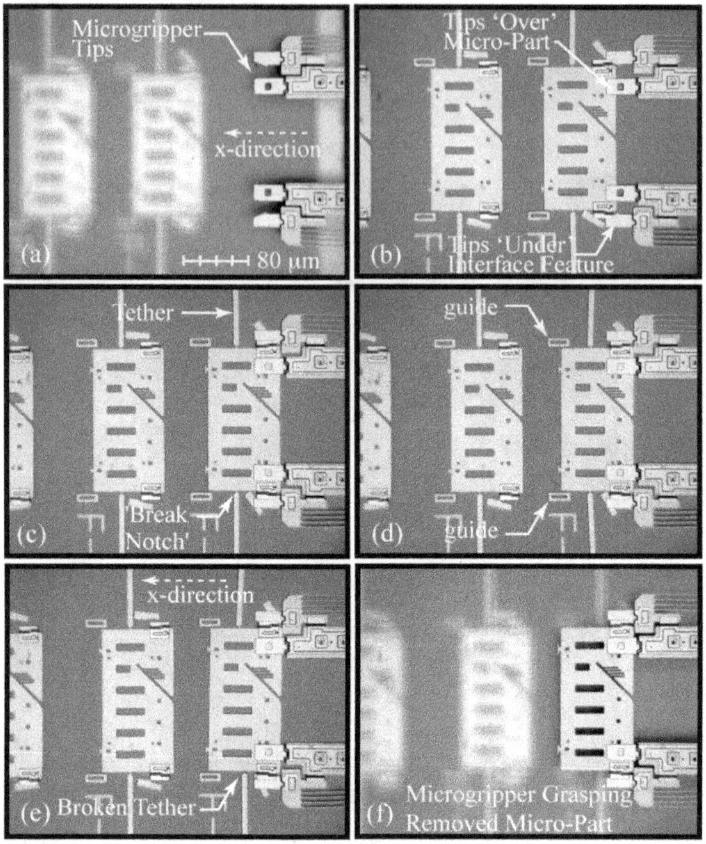

Figure 6.12. Sequence of video images showing the grasp of a micropart equipped with a standard interface feature, using a passive microgripper.

stress concentration point. The tethers are designed to break at these notches when 100 μN or more is applied at the interface feature in the x direction. Figure 6.12(e) shows the micropart after the first tether is broken. Continued motion in the x direction will result in the break away of the second tether. Figure 6.12(f) shows the released micropart held by the microgripper approximately 30 μm above the chip substrate.

6.7.3 Translating and Rotating the Micropart

The microgripper exerts a "holding" force upon the micropart it grasps to keep it from shifting during manipulation. When the microgripper is in the rest position, as shown in Figure 6.4(b), it has a space of 298 μm between the compliant (flexible) tips. The space between the "grasp edges" of the interface feature on a micropart is 302 μm wide, therefore, the microgripper tips are elastically deflected by 2 μm each during a grasp. When designing the microgripper and the microparts, different interference values can be selected, allowing for suitable holding forces for a particular application. It can be seen from Figure 6.12 that the microparts are grasped in such a way that they are planar with the microgripper and with the chip substrate. In order to build the 3D micromirror, the microparts must be reoriented and translated in various ways to ensure that they are lined up with the joint feature with which they must be jointed. For the support posts, they must be perpendicular to the motor rotor on the substrate (i.e., 90° to the substrate), as shown in Figure 6.10. For the mirror micropart, it must be perpendicular with respect to the interlock-joint axis on the support post, which is 45° to the substrate.

6.7.4 Joining Microparts to Other Microparts

This work has developed a number of different mechanical joining strategies. These joint methods involve an interference fit between joint features on mating microparts, causing the joints to elastically deflect and push against each other. In combination with the effects of stiction, the resulting joints become very secure. For the 3D micromirror assembly, two different microjoint systems are used. One is the "key-lock" joint system and the other is the "interlock" joint system. The design details of these joint systems are provided in Basha et al. [2] and Dechev et al. [7]. The 3D micromirror shown in Figure 6.10 consists of three microparts, which are two support posts and one mirror micropart. Figure 6.13 illustrates the assembly sequence to construct the 3D micromirror. The first step involves creating a double key-lock joint between the motor rotor and one support post. The next step is joining a second support post to the motor rotor. The third step is to create a double interlock joint between the mirror micropart and the two vertical support posts. Figure 6.14 shows a sequence of video images of the actual assembly to construct the 3D micromirror. The images show the final operation, which is to insert the mirror micropart into the two support posts. Prior to this operation, the two support posts were key-lock joined into the motor rotor (this operation is not shown). The preassembled support posts can be seen in Figure 6.14(a) where the microscope is focused on their top edges. Figure 6.14(a) also shows the microgripper grasping the mirror micropart and holding it in an orientation parallel to the chip substrate, at about 90 μm above the substrate. This orientation is not suitable for the joint operation, and the mirror micropart must be rotated so that its plane is at 45° to the substrate. Figure 6.14(b) shows the mirror micropart reoriented at 45° to the substrate. Due to the limited depth of focus of the microscopy system, the mirror micropart is out of focus and appears as a dark blur since the coaxial light from the microscope is not reflected back into the microscope. As a result, it can become difficult for the human operator controlling the micromanipulator to clearly see the mirror micropart at this stage of the assembly task.

In order to properly insert the mirror micropart into the support posts, its position with respect to those support posts must be accurately localized. This is done with a touch-based and visual target-based calibration procedure that

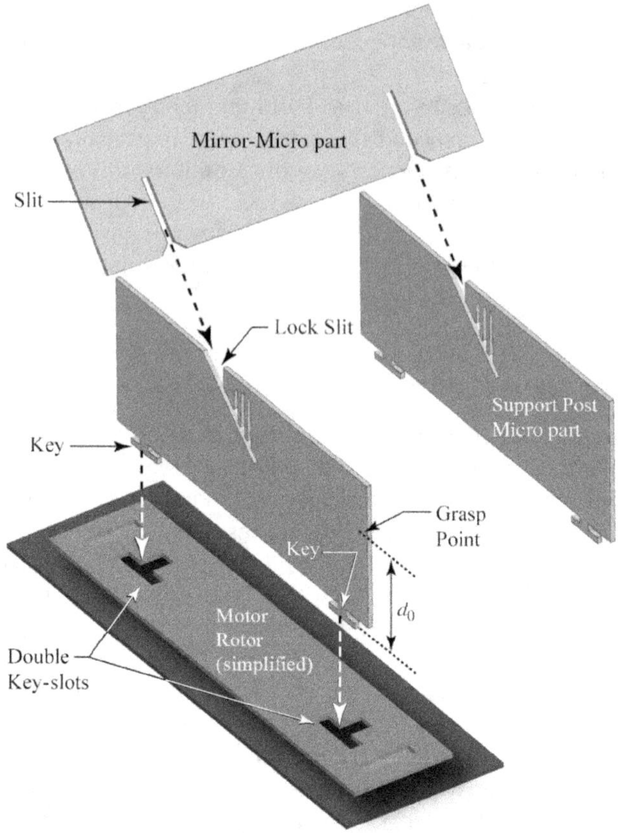

Figure 6.13. Illustration of key-lock joint and interlock joint operation.

relies on the digital encoders and high repeatability of the micromanipulator. After the localization procedure, the mirror micropart and the support posts are brought into alignment, as shown in Figure 6.14(c), based entirely on the numerical localization data. The insertion trajectory vector is programmed into the RM (in this case a simple vector at 45° to the substrate), and the microgripper (holding the mirror micropart) is commanded to move along that vector. This joining procedure relies heavily on the initial calibration, and on the RM to maintain the insertion vector. Interestingly, it relies very little on the operator skill. The operator is indeed in the "control loop," but only to the extent as to permit the RM to either "advance" along the insertion trajectory or to "retract" along it. The operator visually watches for anything unusual during the joint attempt, which may indicate something is wrong. If that is the case, the joint attempt is aborted, the calibration procedure is performed again, and the microparts are realigned for another joint attempt. Figure 6.14(d) shows the mirror micropart successfully inserted into the support posts, to form a double interlock joint.

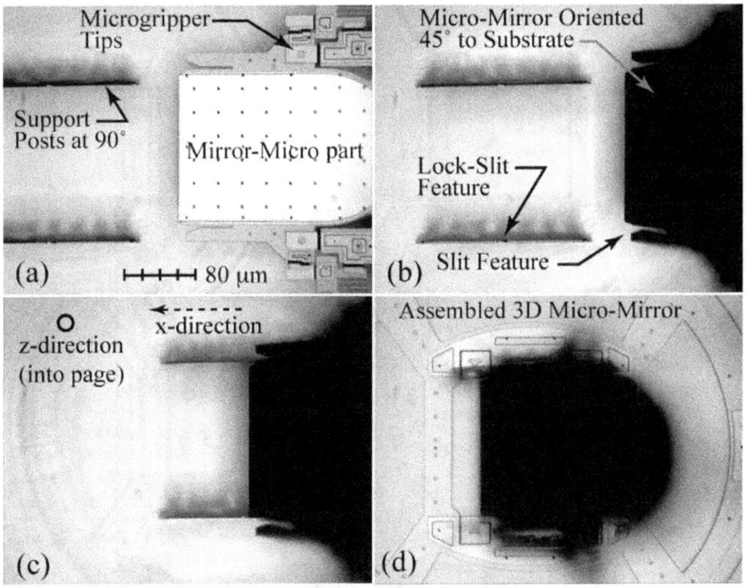

Figure 6.14. Sequence of video microscope images showing the process of joining the mirror micropart into two support post microparts.

6.7.5 Releasing the Assembled Micropart

Releasing a micropart from the grasp of the microgripper is straightforward, as long as that micropart has been joined to another object. After an interlock joint has been achieved, as shown in Figure 6.14(d), the microgripper is retracted along a vector opposite to the initial insertion vector. This retraction causes the interlock joint to lock in, thereby securing the microparts together [2]. As a result, when the microgripper is retracted further from the mirror micropart, the microgripper tips self-open and release the interface feature of the mirror micropart. At this stage, the 3D micromirror is complete.

6.8 EXPERIMENTAL RESULTS AND DISCUSSION

The microassembly process was applied to construct a working prototype of a 3D micromirror mounted onto an electrostatic micromotor. In this regard, the experiment was a success in that four different mirror/motor devices were assembled [2]. Two of these designs are shown in Figure 6.15, in a configuration suitable for optical switching. The more fundamental goal of the experimental microassembly work was the development of new assembly techniques that could achieve the major objectives of this research and expand the capabilities of the PMKIL microassembly system. Prior to this work, the PMKIL microassembly system typically handled microparts from 60 × 60 μm to 200 × 300 μm in size. Since

Figure 6.15. SEM image of two different 3D micromirrors, assembled side-by-side in a configuration suitable for optical switching.

some of the microparts used for the 3D micromirror construction were larger than this, the microassembly system required the development of a few new assembly methods and a redesign of the interface features. This allows it to handle the bigger microparts and to account for some of the unique aspects of the 3D micromirror construction. However, the majority of the assembly steps, such as bonding the microgrippers, grasping the microparts, or breaking the tethers, remained the same as those used to assemble other devices in the past. The microassembly proceeded smoothly for most steps; however, a number of unique challenges were encountered when dealing with the 3D micromirror. These include (in order of importance): (a) limited depth of focus when joining microparts at oblique angles, (b) difficulty in creating key-lock joints with large microparts, (c) limited visual field of view when handling and joining large microparts, and (d) joining microparts onto a rotational base (motor rotor) that can rotate slightly during assembly. These four challenges are specifically discussed, since they will require future work to resolve/minimize them. As can be observed in Figure 6.14, any micropart whose plane is at an angle of more than 2° to the substrate appears out of focus on the video images. As the angle of its plane approaches 90° to the substrate, the ability to discern features on the microparts becomes increasingly difficult with the video microscopy system. For this reason, when attempting to create an interlock joint, identification of the joint features, such as the slit and lock slit [compare Figs. 6.13 and 6.14(d)], was difficult. This in turn caused a difficulty in correctly aligning two microparts prior to an interlock joint insertion attempt. In order to resolve this problem, a more effective video imaging system is required. A proposed system for future work involves the use of multiple microscopy cameras, to obtain images at different angles. Such a system would also require an interconnected software reference system so that common reference points, with x, y, and z position information, will appear on each video image. This would significantly help to achieve the correct alignment between microparts prior to a joining operation. This system would also aid in the development of automated visual identification techniques that could assist the human operator in target recognition. Due to the increased size of the surface micromachined microparts for the 3D micromirror, the location of the interface feature (where the micropart is grasped) on the micropart becomes important. In the case of the key-lock joint system, it is important to keep the microgripper tips (which grasp the micropart) as close as possible to the bottom edge of the micropart where the key features are located. This is necessary for a smooth insertion of the keys into the double key slots, in a single sliding motion, as illustrated in Figure 6.13. The reason for this can be explained by observing the grasp point indicated in Figure 6.13. The microgripper holds onto the micropart (via the interface feature) at the grasp point. The distance, d_0, measures from the grasp point to a line that passes through both key features. In order to create a key-lock joint, and then the keys are inserted into the wide region of the micropart must be translated parallel to the substrate, so that the keys can slide into the narrow region of the key slot. However, stiction and friction make it difficult for the keys to slide smoothly into the narrow region of the key slot. In order to drive the keys into the narrow region of the key slot, force is applied by the microgripper. When distance d_0 is small (<20 μm), this force is easily transferred, and the keys overcome stiction and friction and slide into position. However, when d_0 is large (>30 μm), a moment is developed on the microgripper. Since the microgripper tips have some flexibility, they start to bend, rather than to drive the keys to slide into the narrow region. When dealing with the support posts used in the 3D micromirror, d_0 is often 80–130 μm or more. This makes it impossible to fully slide the keys into the key slot. As a result, an inefficient and time-consuming two-step process had to be used during the assembly experiments to assemble the support posts into the motor rotor. To prevent this problem in the future, the microgripper tips must be redesigned to be (a) more rigid along the direction in which they apply the sliding force for key-lock joints and (b) be designed to grasp microparts such that the distance d_0 is at a minimum. The field of view of the video microscope system is 427 μm horizontal and 320 μm vertical, with an optical resolution of 0.8 μm. However, the largest microparts handled and joined are over 400 μm wide. Since they are held in the vertical direction, both edges of the microparts cannot be viewed simultaneously. This complicates the grasping and joining procedures since the camera must be moved back and forth (on its manual translation stage) to allow the operator to monitor the grasping and joining of large microparts. The solution is not as simple as using a microscope system with a larger field of view because this can only be achieved at the cost of having a lower resolution. The resolution required for the joint features of this work must be at least submicron. Therefore, a suitable system to view large microparts must be developed in future work. One of the interesting and unique challenges of assembling the 3D micromirror is that the base micropart (the motor rotor) is free to rotate. This means that during the assembly operation, if any forces are applied such that there is a net imbalance about the motor axis of rotation, the motor rotor will rotate. Since the key-lock and interlock joints require translation with a direction component along the plane of the motor rotor, they needed to be designed to ensure that the net force created during assembly would pass through the center of rotation and, thereby, not rotate the motor. This is the ideal case and would usually work in practice. However, there were a few cases where an initially small imbalance, due to an insertion that was not sufficiently aligned or became out of alignment, would cause a small rotation, leading to a greater imbalance of force, leading to more rotation, and the like. This situation could be corrected by aborting the joint attempt, realigning the micropart by rotating the RM α axis, and trying again. However, for future work, it is worth investigating ways to restrain the rotating motor rotors during the assembly operation. Methods under consideration could be the use of tethers that can be broken away after assembly, or a temporary layer of material that could be deposited to secure the rotors, and could then be rinsed away after assembly.

6.9 CONCLUSION

This work described a general microassembly system for constructing microsystems. Additionally, the practical aspects for the assembly of a novel 3D micromirror has were described. A robotic micromanipulator equipped with a microgripper was used to grasp microparts from the substrate of the chip. The microparts were then oriented at various angles to the chip and joined together. By inserting the key features of the support posts into the motor rotor, key-lock joints were created. By lining up and inserting the slits of the mirror micropart into the lock slits on the support posts, interlock joints were created. Together, these joints allowed for the construction of the 3D micromirror. Preliminary testing of the assembled mirror/motor MEMS device has shown good results [2]. It is important to note that since assembly is used, it is possible for the electrostatic motor, and the various microparts of the 3D micromirror, to be fabricated on different chips, by different fabrication methods. The RM would then be able to assemble all these components together. This creates various possibilities such as assembling bulk micromachined mirror elements with gold coatings onto surface micromachined devices. This would be useful in creating a flatter and more reflective mirror. The use of this microassembly system allows for many possibilities.

Acknowledgment

The author wishes to thank Dr. M. Basha of the Electrical Engineering Department, University of Waterloo, for his work in developing the 3D micromirror, to which this microassembly process was applied. Also, thanks to Prof. J. K. Mills and Prof. W. L. Cleghorn of the Department of Mechanical Engineering, University of Toronto, for providing access to the 6-DOF robotic micromanipulator used to perform the assembly experiments described in this chapter.

REFERENCES

1. Y. H. Anis, J. K. Mills, and W. L. Cleghorn, Automated Microassembly Task Execution Using Vision-Based Feedback Control, Proceedings of International Conference on Information Acquisition, (ICIA 2007), Seogwipo-si, Korea, July 8–11, 2007.
2. M. A. Basha, N. Dechev, S. Safavi-Naeini, and S. Chadhuri, A Scalable 1 × N Optical MEMS Switch Architecture Utilizing a Microassembled Rotating Micromirror, *IEEE J. Selected Topics Quant. Electr.*, 13(2):336–347, 2007.
3. K. F. Bohringer, K. Goldberg, M. Colm, R. Howe, and A. Pisano, Parallel Microassembly with Electrostatic Force Fields, International Conference on Robotics and Automation (ICRA98), Leuven, Belgium, May 1998.
4. N. Dechev, W. L. Cleghorn, and J. K. Mills, Tether and Joint Design for Micro-Components Used in Microassembly of 3D Microstructures, Proceedings SPIE Micromachining and Microfabrication, Photonics West 2004, San Jose, CA, Jan 25-29, 2004.
5. N. Dechev, W. L. Cleghorn, and J. K. Mills, Microassembly of 3D Microstructures Using a Compliant, Passive Microgripper, *J. Microelectromech. Syst.*, 13(2):176–189, 2004.
6. N. Dechev, W. L. Cleghorn, and J. K. Mills, Design of Grasping Interface for Microgrippers and Micro-Parts Used in the Microassembly of MEMS, Proceedings of the IEEE International Conference on Image Acquisition, Chinese University of Hong Kong, Hong Kong and Macau, China, June 27–July 3, 2005.
7. N. Dechev, J. K. Mills, and W. L. Cleghorn, Mechanical Fastener Designs for Use in the Microassembly of 3D Microstructures, Proceedings ASME International Mechanical Engineering Congress and R&D Expo 2004, Anaheim, CA, Nov. 13-19, 2004.
8. N. Dechev, L. Ren, W. Liu, W. L. Cleghorn, and J. K. Mills, Development of a 6 Degree of Freedom Robotic Micromanipulator for Use in 3D MEMS Microassembly, Proceeding of the IEEE International Conference on Robotics and Automation (ICRA 2006), Orlando, FL, May 2006, pp. 15–19.
9. T. Ebefors, J. Ulfstedt-Mattsson, E. Kälvesten, and G. Stemme, 3D Micromachined Devices Based on Polyimide Joint technology, presented at SPIE Symposium on Microelectronics and MEMS, Gold Coast, Queensland, Australia, SPIE Vol. 3892, October, 1999, pp. 118–132.
10. J. T. Feddema and R. W. Simon, CAD-Driven Microassembly and Visual Servoing, Proceedings of the IEEE International Conference on Robotics and Automation (ICRA 1998), Leuven, Belgium, May 16-20, 1998.
11. M. A. Greminger, A. S. Sezen, and B. J. Nelson, A Four Degree of Freedom MEMS Microgripper with Novel Bi-directional Thermal Actuators, Proceedings of IEEE/RSJ International Conference on Intelligent Robots and Systems, (IROS 2005), Edmonton, Canada, Aug 2-6, 2005.

12. K. F. Harsh, V. M. Bright, and Y. C. Lee, Solder Self-Assembly for Three-Dimensional Microelectromechanical Systems, *Sensors Actuators A*, 77: 237–244, 1999.

13. D. Koester, A. Cowen, R. Mahadevan, and B. Hardy, PolyMUMPs Design Handbook Revision 9.0, MEMSCAP, MEMS Business Unit (CRONOS), Research Triangle Park, NC, 2001.

14. K. W. C. Lai, A. P. Hui, and W. J. Li, Non-Contact Batch Micro-Assembly by Centrifugal Force, Proc. of IEEE International Conference on Micro Electro Mechanical Systems (MEMS 2002), Las Vegas, NV, Jan. 2002.

15. M. Last, V. Subramaniam, and K. S. J. Pister, Out-of-Plane Motion of Assembled Microstructures Using a Single-Mask SOI Process, International Conf. Transducers 2005, Seoul, Korea, June, 2005.

16. M. M. Maharbiz, R. T. Howe, and K. S. J. Pister, Batch Transfer Assembly of Micro-Components onto Surface and SOI MEMS, Transducers '99 Conference, Sendai, Japan, June 7-10, 1999.

17. D. O. Popa, W. H. Lee, R. Murthy, A. N. Das, and H. E. Stephanou, High Yield Automated MEMS Assembly, Proceedings of IEEE International Conference on Automation Science and Engineering (CASE 2007), Scottsdale, AZ, Sept. 22-25, 2007.

18. S. J. Ralis, B. Vikramaditya, and B. J. Nelson, Micropositioning of a Weakly Calibrated Microassembly System Using Coarse-to-fine Visual Servoing Strategies, *IEEE Trans. Electron. Packaging Manufact.*, 23(2):123–131, 2000.

19. L. Ren, L. Wang, J. K. Mills, and D. Sun, 2-D Automatic Micrograsping Tasks Performed by Visual Servo Control, Proceedings of IEEE International Symposium on Industrial Electronics, (ISIE 2007), Vigo, June 4–7, 2007.

20. E. Shimada, J. A. Thompson, J. Yan, R. Wood, and R. S. Fearing, Prototyping MilliRobots Using Dextrous Microassembly and Folding, Proceedings of ASME International Mechanical Engineering Congress and Expo (IMECE/DSCD), Orlando, Florida, Nov. 5-10, 2000.

21. G. A. Singh, D. Horsely, M. Cohn, A. Pisano, and R. Howe, Batch Transfer of Microstructures Using Flip-Chip Solder Bonding, *IEEE J. Microelectromech. Syst.*, 8(1):27–33, 1999.

22. K. Tsui, A. A. Geisberger, M. Ellis, and G. D. Skidmore, Micromachined End-Effector and Techniques for Directed MEMS Assembly, *J. Micromech. Microeng.*, 4: 542–549, 2004.

23. G. Yang, J. A. Gaines, and B. J. Nelson, A Supervisory Wafer-Level 3D Microassembly System for Hybrid MEMS Fabrications, *J. Intelligent Robotic Syst.*, 37: 43–68, 2003.

24. J. Zou, J. Chen, C. Liu, and J. E. Schutt-Ainé, Plastic Deformation Magnetic Assembly (PDMA) of Out-of-Plane Microstructures: Technology and Application, *IEEE J. Microelectromech. Syst.*, 10(2):302–309, 2001.

CHAPTER 7

HIGH-YIELD AUTOMATED MEMS ASSEMBLY

DAN O. POPA and HARRY E. STEPHANOU

7.1 INTRODUCTION

7.1.1 Automated Microassembly

Microassembly is an enabling technology for constructing heterogeneous (or hybrid) three-dimensional microsystems. In many instances, microparts fabricated using different materials and processes need to be assembled and packaged in order to achieve a desired functionality. Progress in lithographic fabrication methods, such as those used for silicon microelectromechanical systems (MEMS), metal LIGA, or polymer micromolding has enabled the mass production of 2.5-dimensional (2.5D) microparts. Concurrently, the past 15 years has seen considerable progress in "top-down" precision assembly, including gripping, handling, positioning, and bonding of parts with dimensions between a few and several hundred micrometers [1, 7, 8, 12, 17, 19, 20, 23]. Due to the small size of these parts, specialized microgrippers, fixtures, and positioning systems have been proposed and validated. Active microgrippers can be fabricated from a variety of materials, including metals, silicon, or piezoelectric ceramics (PZT) [7, 20]. Other examples are the use of passive microgrippers through the use of mechanical compliance [11, 12] or the use of adhesive forces [1]. In addition to serial, single-gripper methods, others have pursued parallel manipulation with gripper arrays [2, 6].

Numerous studies describe and classify the architecture and algorithms used in high-precision robotic cells for the purpose of directed microscale assembly

Robotic Microassembly, edited by Michaël Gauthier and Stéphane Régnier

[3, 10, 17, 19]. Classifications can be based on throughput (serial or parallel), deliberate intervention (deterministic or stochastic), type of end effectors (contact, noncontact), or level of human intervention (manual, teleoperated, or automated). Microrobotic assembly cell design is a challenging task because it requires appropriate precision, throughput, and yield across multiple scales of tolerance, part dimension, and workspace sizes. Traditional serial microassembly requires a high-precision micromanipulator and motion control, either by offline programming with calibration, or by online sensory feedback control. The later is traditionally accomplished with a microscope, or a force sensor integrated with the gripper, or both [17, 23]. In order to lower the precision requirements of the robotic micropositioner, it is critical to introduce engineered compliance into the MEMS components to be assembled, and in order to increase the positional repeatability of robots at all scales, it is critical to use flexure mechanisms [8].

7.1.2 Compliant Microassembly

Directed assembly at the microscale requires real-time feedback using visual and/or force information as demonstrated by many research groups [5, 6, 8, 10, 17, 19, 23]. Recently, incorporating compliant MEMS snap-fasteners into microparts has emerged as a viable way to reduce the amount of feedback necessary during assembly [5, 17, 21]. The advantages are fast assembly, disassemblability, self-guidance and alignment, and added mechanical strength of assembled structures through interaction forces generated by small deflections of mating parts.

Bohringer and Prasad were among the first to introduce the concept of snap-fasteners using MEMS as early as 1995 [18]. This approach, however, was limited to "in-plane" assemblies, which today are used in a variety of applications, for instance, MEMS inertial guidance and safety switches. An important turning point for the practical feasibility of constructing 3D assemblies using MEMS snap-fasteners occurred after metal LIGA and silicon DRIE (deep reactive ion etching) machining allowed machining of thicker 2.5D part geometries.

An example of a very well designed fastener allowing compliant assembly with silicon-on-insulator (SOI) MEMS parts is the Zyvex connector [21]. This connector is easy to assemble, and relative large friction forces generated in grippers and assembly "sockets" firmly hold parts during and after assembly. Recent work with this connector showed that it is possible to assemble MEMS parts quickly, without force and vision feedback. However, prior to our research, it was not clear why both the assembly yield and speed can be improved using this connector.

7.1.3 Focus of This Chapter

In this chapter we discuss design and yield aspects in sequential automated microassembly, though many of the concepts can be extended to other assembly methods. The size of MEMS parts discussed in this chapter have been demonstrated as being between 50 and 500 μm. It is very likely, however, that the framework discussed in this chapter also applies to some extent to 1–2 orders of magnitude smaller parts. We present results from recent research aimed at formulating assembly-related design guidelines and trade-offs for 2.5D snap-fastener design and precision robot cell configuration. We use several examples of assemblies evaluated experimentally through tolerance and strength measurements. We also describe a 3D microassembly station with three precision robots—μ^3, located at our Texas Microfactory labs, and specifically configured to assemble compliant microsnap fasteners.

Several other researchers have configured microassembly stations since the early 1990s. The kinematics of these systems is often based on available off-the-shelf hardware (e.g., by "stacking" precision stages to form a manipulator), and is limited due to trade-offs between required precision, speed, and workspace. We carefully tuned the kinematic configuration of μ^3 for high-yield assembly of 2.5D MEMS components. During microassembly, parts are passed between manipulator tools and substrate fixtures so that we never "let go" of the manipulated parts [13]. By using compliant passive or active fixtures, grippers, or microparts we can send end-effectors in close proximity to the parts of interest at fairly high speeds, and we complete the assembly without real-time force and vision information.

The chapter is organized as follows; in Section 7.2 we describe general guidelines for achieving high-yield assembly of 2.5D microparts; in Section 7.3 we discuss aspects related to the design criteria and tolerance analysis of microsnap fasteners; Section 7.4 describes the configuration and calibration of the assembly cell for high-yield assembly; Section 7.5 presents experimental results and characterization of several completed MEMS assemblies; finally Section 7.6 presents conculsions.

7.2 GENERAL GUIDELINES FOR 2.5D MICROASSEMBLY

Microscale robotic assembly systems share many common aspects with traditional robotic assembly. Among them are manipulator position, velocity, and jerk control, force control, tactile feedback, task planning, collision avoidance, grasping, cooperative manipulation, part orientation, peg-in-the-hole insertion, and the like. [17]. However, in contrast to macro- or mesoscale assembly, the design and precision of a microassembly cell must be tightly coupled with the parts to be assembled. Here we show how the coupling can be accomplished by means of tolerance and compliance analysis during the micropart design phase. Indeed, as microscale assembly suffers from well-known limitations due to limited robot precision, small field of view for machine vision, and difficulties in sensing small forces, we can alleviate many of these difficulties through engineered part compliance and the use of snap-fasteners.

In this section we present a summary of qualitative guidelines, starting with micropart "design for assemblability," a discussion regarding MEMS part fixturing, and of design guidelines for configurig precision assembly cells for parts with primarily 2.5D geometries. Examples of the type of parts whose assembly is being targeted in this chapter are shown in Figure 7.1. These parts are typicaly fabricated using high aspect ratio micromachining, in particular DRIE on SOI wafers. The mating features consist of flexures that ensure that an elastic energy minimum is reached during assembly. Such parts can be assembled into "scaffold structures" that are three dimensional. The Zyvex socket and connector proposed in 2003, and shown in Figure 7.1, is the first example of a high-aspect-ratio (HAR) snap-fastener [21].

The design of compliant microparts is a deliberate process in which we define an assembly tolerance budget (referred as σ_1). This value can be viewed as the maximum alowable misalignment that still guarantees mating nearly 100% of the time. During pick or place operations, microparts will be "located" using a measurement process with a statistical distribution σ_2, while the microgripper will be positioned with a robot at a nominal location, within a positioning variance σ_3. In order to guarantee a high assembly yield, components must be designed such that the tolerance budget σ_1 exceeds combined positional variance of the micropart and microgripper along all degrees of freedom and for all assembly operations. Alternatively stated, the robotic manipulator must be able to position the microgripper near the micropart within a precision higher than the tolerance budget.

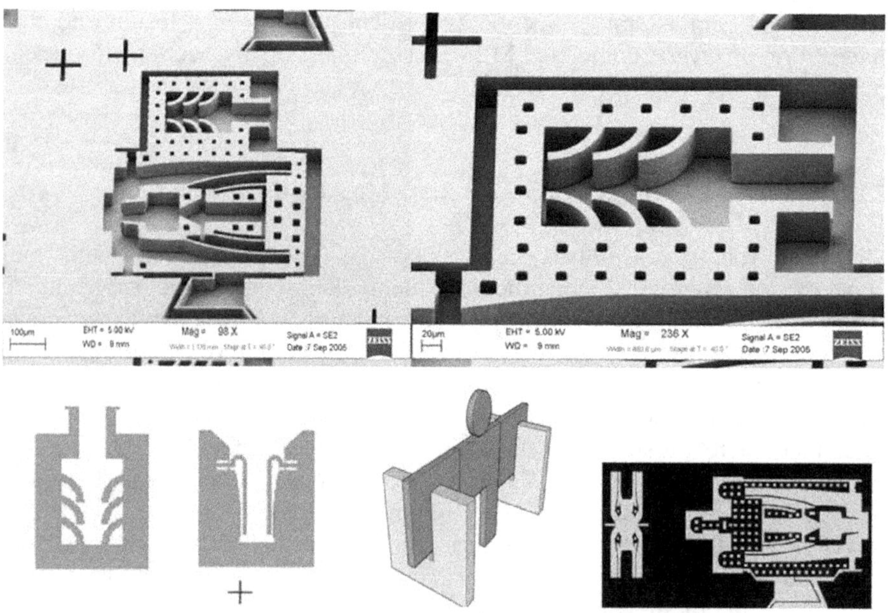

Figure 7.1. Design of compliant 2.5D MEMS parts for microassembly. Top figures: SEM pictures showing 100-μm-thick SOI flexible microparts. Bottom diagrams, left to right: female mating designs, 3D standing assemblies, and the Zyvex socket and connector.

7.2.1 Part and End-Effector Compliance

Because of the large size difference, micrograppers and microparts are a lot more flexible than the meso- and macroscale positioners, robots, or fixtures on which they are mounted. As a result, when two such parts are mated using robotic hardware, their compliance can be used to compensate for position and orientation errors, thus preventing damage. While this is a well-known technique in macroscale assembly, dating back to peg-in-hole models and the remote center of compliance (RCC) end-effector designs [22], we have pointed out in the past that its use in microassembly is even more critical [17].

Analytic models of compliant insertion can be used to represent the motion and force equations during mating [11–13]. Just as in the case of a conventional peg-in-hole insertion, the deflection and force profiles during chamfer crossing, one-point contact, and two-point contact can be predicted. Design optimization can then be used to determine the kinematics and characteristics of each flexible joint. This optimization can be based on various optimality indices, or on the mechanism Jacobian [9, 21]. In the designs presented in this chapter, we use lumped-model approximations based on cantilever models of connector flexure elements.

Since many processes for fabrication of 2.5D MEMS parts are well understood, we can produce such parts from elastic materials such as silicon and metals. Both compliant or noncompliant 2.5D microparts and microgrippers can be fabricated on common substrates, such as on SOI wafers. We will assume that all parts and end-effectors have either engineered compliance or are rigid; however, we require that for any mating chain composed of microgripper and microparts must contain at least one flexible part. For instance, in a typical operation involving assembling part A into assembly site B using microgripper C, we assume that at least one of A, B, and C includes engineered compliance. In the compliant assembly framework described here, end-effectors could be either passive ("jammers") or actuated ("tweezers"), and assembly sites could be mechanical "snap-fasteners," or active substrate connectors. Due to their 2.5D geometry, such microgrippers can be mounted onto precision robots by "flip-chip" bonding processes. Finally, assembly joints can be strengthened through snap-fastener design to minimize insertion force and maximize retention force, and through the use of bonding agents such as epoxies or solder reflow.

In Section 7.3, we describe a method of deliberate design for MEMS snap-fasteners such that a chosen 3σ misalignment with distribution variance σ_1^2 will result in successful mating operations with over 99% yield. The σ_1 value is incorporated into the micropart design using simulation, and can be adjusted experimentally after fabrication.

7.2.2 Fixtures and Micropart Transfer

Typical assembly operations in any top-down assembly cell involve repeated "pick-move-place" operations, and assembly at the microscale is no different. However, in order to automate microassembly, special guidelines should be observed in order to improve the yield and speed of the process. First, in order to cut down on the time it takes to "find" MEMS parts in the assembly cell, they should always be presented for assembly in an ordered state either on die-tethered or untethered, but nonetheless in "predictable" spots. The microparts shown in Figure 7.1 are all tethered to the substrate after fabrication, and such tethers are broken by the microgripper prior to pickup.

Microparts fabricated using lithographic techniques are by default produced in an ordered state since the wafer is already the fixture! Thus, a nice "bonus" feature of using wafer-scale micromachining is not only scalability to large number of parts with required tolerances, but also the ability to make them in a fixture for the assembly system. In addition, wafer fixtures could also be fitted with special compliant or visual markers for end-effector calibration later on. Those microparts which must be singulated (e.g., microoptical glass components) should also be presented to the assembly cell in an ordered state. In this case, vibratory energy can be used for sorting, as can typical feeding techniques used in semiconductor industry such as using tapes or gel packs.

In order to cut down on assembly time, and also improve the process yield, microparts are transferred from the substrate to the gripper and back into the substrate by means of stable grasps, interference fits and snap-fastening, and at no time should microparts be "let go" to position themselves due to uncontrolled friction, stiction, gravity, electrostatic, or van der Waals forces.

Finally, in order to pick and place MEMS parts from a fixture, we first need to "locate" them. Assuming that part location in the work cell is a Gaussian statistical process, we will quantify it using a distribution variance σ_2 around a mean position dictated by their fixture. The variance of the distribution is obtained experimentally as discussed in Section 7.4.2. Mean position data can be supplied to the assembly cell from a microscope or using a wafer layout file.

7.2.3 Precision Robotic Work Cell Design

Visual servoing through a microscope is a well-established method and has been successfully used in microassembly and micromanipulation [23]. During assembly, microparts are passed between end-effectors and substrate or other assemblies, but because large field of view visual information cannot be obtained at high resolutions, more than one robot is necessary in the assembly cell. In addition to the manipulator carying the end-effector, another robot is needed to reposition either the microscope or the micropart.

Furthermore, to assemble 2.5D microparts, the number of degrees of freedom (DoFs) shared between the robots must add to at least six independent joints. This number should exceed six if redundancy/increased dexterity or increased workspace is necessary. For our assembly cell, μ^3, we use a kinematic configuration with 4 DoFs on the gripper carrying manipulators, and 5 DoFs on the substrate carrying manipulator. Each of the μ^3 robotic chains is composed of at most 3 independent rotational DoFs, and 3 independent translational DoFs in a way as to allow the decomposition of the translational and rotational calibration for these robots. We make use of calibration whenever a new microgripper or a new substrate is added to the assembly cell. Because we are assembling 2.5D microparts, a terminating roll DoF in the robotic chain and a programmable remote center of rotation of the end-effector is required for part rotations of 90°.

Force and visual feedback from the end-effector for closed-loop control is not necessary during automation, but it is necessary during calibration. One such method utilizes weak calibration by means of vision from multiple microscopes and a "hand-to-eye" configuration [because the microscope field of vision (FoV) is limited].

While the configuration of μ^3 is not unique, the desired functionality for assembly, namely "pick," "transport," "rotate," and "place" microparts is accomplished with minimal hardware. High-speed MEMS assembly is carried out through an assembly "script" that takes into account collision and workspace constraints. The high assembly yield is guaranteed through analysis of part and robot tolerances. The work cell precision is accomplished via well-known statistical methods of kinematic calibration for end-effector frames from all the robots sharing the workspace. After calibration, we will be able to locate the MEMS end-effector with a distribution variance σ_3 around a nominal commanded position. This step is described in detail in Section 7.4.2.

7.2.4 High-Yield Assembly Condition (HYAC)

The required precision of the robotic cell is dictated by the tolerance budget of the assembly. In turn, this is dictated by the manufacturing tolerances of the parts and the compliant part interference models. As a result, the maximum allowable part misalignment during assembly depends on the desired assembly yield, as captured by the high-yield assembly condition described next.

HYAC Lemma. Assuming that the process of positioning microparts in the workspace is normally distributed, a 99% assembly yield can be guaranteed if the following condition is satisfied along every degree of freedom α:

$$\sigma_{1\alpha}^2 \geq \sigma_{2\alpha}^2 + \sigma_{3\alpha}^2 \tag{7.1}$$

The HYAC condition 7.1 basically states that the combined uncertainty of locating microparts and the end-effector in the workspace should be smaller than the compliant misalignment threshold that guarantees a successful assembly operation. Thus, if Eq. 7.1 is observed, high-speed MEMS assembly can be accomplished through an assembly "script" composed of a sequence of assembly operations at nominal locations in the workspace. Scripting requires the robots to operate in open loop, and therefore, assembly speed will be much higher than the case when robot poses are adjusted using force or vision feedback. Guaranteeing a high assembly yield is depicted in Figure 7.2.

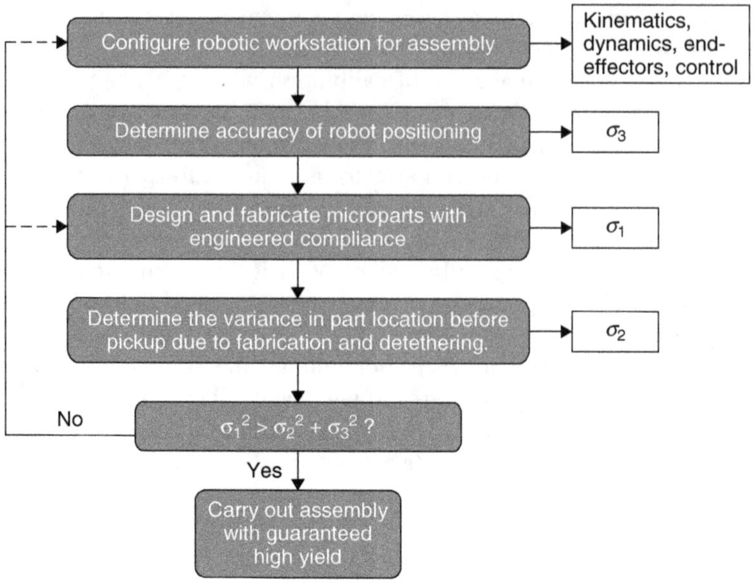

Figure 7.2. Diagram depicting the use of the high-yield assembly condition to obtain performance guarantees with a microassembly cell.

7.3 COMPLIANT PART DESIGN

7.3.1 Design Principles

Traditional concepts such as RCC (remote center of compliance), RCR (remote center of rotation), and peg-in-hole insertion models [22] can be used for both part- and gripper-compliant design. Snap-fastener insertion models can be obtained using full-fledged finite element analysis (FEA) simulation or from reduced order (lumped) models. These models relate the insertion force along the insertion direction with part misalignment. For instance, the insertion force of a 2.5D rigid part A into a compliant part B along the X direction is usually written as

$$F_{ABi}(r) = f_{ri}(\Delta_x, \Delta_y, \Delta\theta, \mu, \overline{a}) \tag{7.2}$$

where $i = x, y$ are the two components of the insertion force, Δx, Δy are misalignments between the parts in the insertion direction and perpendicular to it, $\Delta\theta$ is angular misalignment around the Z direction, μ is the coefficient of friction, $\overline{a}$ is a parametric vector describing the part geometry, and $r = r_0, \ldots, r_f$ is an insertion regime parameter describing the contact (one-point, two-point, chamfer crossing, and insertion snap). More complex models are necessary if both parts are compliant, such as in the case of the Zyvex connector in Figure 7.3.

To design an appropriate snap-fastener, the geometry design vector $\overline{a}$ is chosen such that for all misalignments in y and θ below a given design threshold $\Delta y \leq$

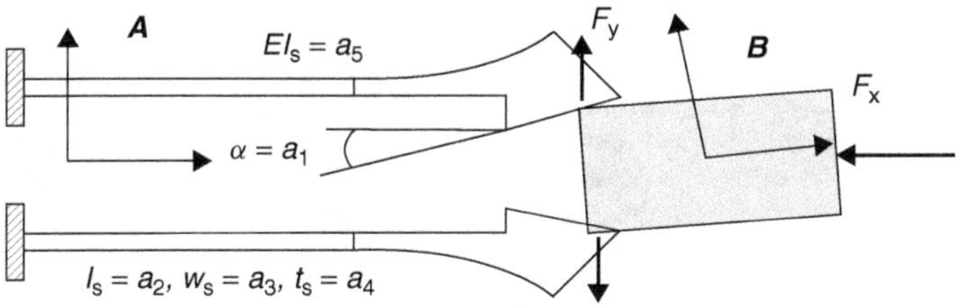

Figure 7.3. Free-body diagram depicting the insertion of rigid part *B* into compliant Zyvex snap-fastener *A*, shown in a two-point crossing state.

σ_{1y}, $\Delta\theta \leq \sigma_{1\theta}$, we have

$$\text{Min}(F_{ABx}(r_j), j < f), \ \text{Max}(F_{ABx}(r_f)) \tag{7.3}$$

and

$$F_{ABy(r_j)} \leq F_{\text{yield}}, \ j \leq f \tag{7.4}$$

where F_{yield} is the yield strength of the microstructure. In other words, the snap-fastener design criterion is based on minimizing the insertion force, and maximizing the retention force without breaking the connector.

The misalignment design thresholds σ_1 are chosen in conjunction with the part manufacturing tolerance, part B positioning tolerance prior to assembly with respect to the substrate, and the manipulator positional accuracy holding part B with respect to the substrate. Specifically, a misalignment tolerance below σ_{1y} and $\sigma_{1\theta}$ has over 99% assembly yield (or 3σ spread). For the Zyvex connector in Figure 7.1, σ_{1y}<5 μm, and $\sigma_{1\theta} < \pm 6^\circ$ [11].

7.3.2 Example of Microsnap-Fastener Design

As an example of fastener design, consider the connector in Figure 7.4, which is assembled using a lateral (e.g., parallel to the substrate) insertion operation of a vertical MEMS part into a socket. A concrete example of such assembly

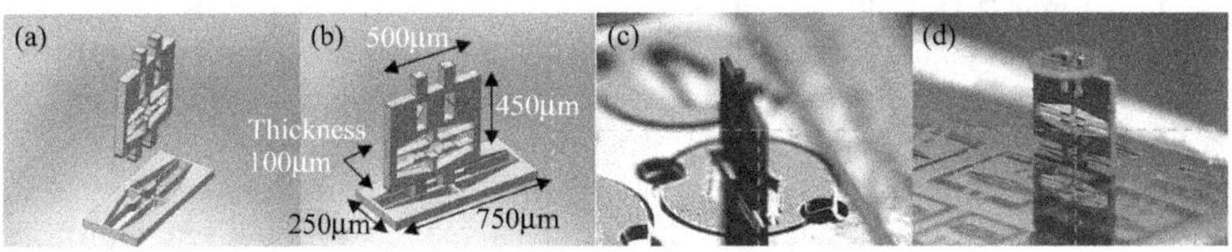

Figure 7.4. Design of compliant assemblable 2.5D microparts: (a) rigid part (top) and compliant microfastener (bottom), (b) part placed into microfastener before snap locking, (c) part snapped into a microwheel, and (d) microwheel assembly fastened into the substrate.

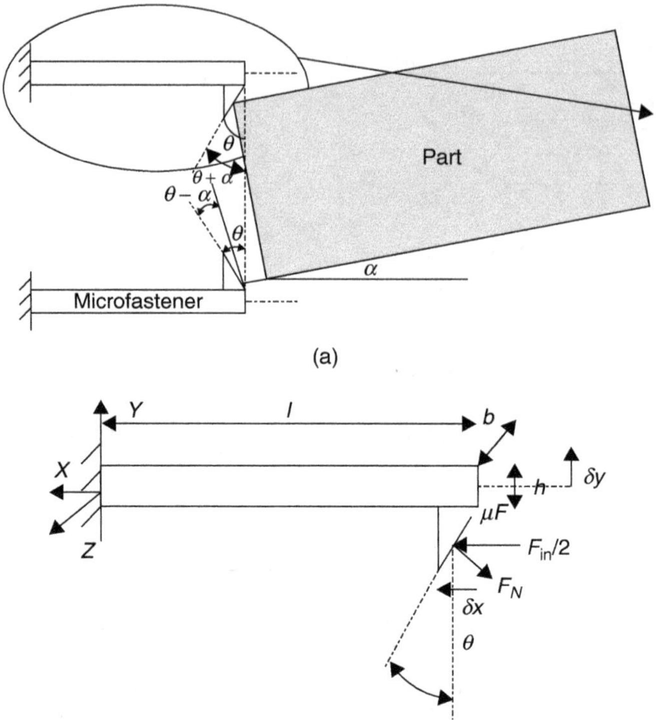

Figure 7.5. (a) Microfastener misalignment with micropart and (b) free-body diagram of insertion into the snap arm along X axis.

is the wheel/axle assembled for a MEMS-based microcar shown in Figure 7.4. The part is gripped from the substrate using a microgripper, rotated and laterally forced into the socket. Similarly, the orientation of the snap-fastener onto the part can be varied to accommodate vertical, horizontal, or any other angular assembly. The dominant forces that act upon the microparts during assembly are the insertion force (force along X axis) required to assemble the part and the retention force (force along Z axis) with which the MEMS part is retained by the joint after assembly. These forces depend on (1) socket cantilever stiffness, (2) parts interference due to design geometry, (3) coefficient of friction, and (4) positional accuracy during insertion. The aim of our design captured by Eq. 7.3 is to minimize the insertion force and maximize the retention force. Note that the insertion force acts in the plane of the wafer (X axis), while the retention force acts perpendicular to the wafer (Z axis), as shown in Figure 7.5.

The snap arm shown in Figure 7.5(a) is located on the substrate, while the can be represented as a cantilever beam with length l, width b, and height h, while dx and dy are the cantilever end deflections during assembly and θ is the guide angle.

If F_{in} is the insertion force acting on the micropart, F_N is the normal force onto the snap arm, and μ is the friction coefficient, then the cantilever deflection force F_{NY} is equal to

$$F_{NY} = -\frac{F_{\text{in}}}{4}\sin(2\theta) \tag{7.5}$$

and the resultant force along Y due to friction can be written as

$$F_{f_rY} = \mu\frac{F_{\text{in}}}{2}\cos^2(\theta) \tag{7.6}$$

where F_{f_rY} is the vertical component of the frictional force. Therefore, the net force along the Y direction is given by

$$F_{Y\text{req}} = \frac{F_{\text{in}}}{2}\left(\mu\cos^2\theta - \frac{1}{2}\sin 2\theta\right) \tag{7.7}$$

Using the cantilever bending stiffness equation, we also have that

$$F_{Y\text{req}} = \frac{3EI\,dy}{(L - dx)^3} \tag{7.8}$$

where E is the Young's modulus of silicon $E = 160$ GPa, I is the moment of inertia about the neutral axis, $b = 100$ μm is the thickness of the SOI DRIE die on which the microfastener is fabricated, h is the arm height, L the arm length, dy = deflection due to bending, θ is the "snap angle." From Eqs. 7.7 and 7.8 we can calculate the insertion force as a function of the snap arm deflection:

$$F_{\text{in}} = F_{ABx}(r_l) = \frac{6EI(dy^a + dy^b)}{(L - dx)^3(\mu\cos^2\theta - 0.5\sin 2\theta)} \tag{7.9}$$

where the a and b indices refer to the deflection experienced by the upper and lower cantilever arms. Moreover, the stress due to bending is calculated using

$$\sigma_b = \frac{M_y}{I} \tag{7.10}$$

where $y = h/2$, M is the moment due to bending force, and I is the moment of inertia about its neutral axis. The numerical values of the geometric parameters will be determined in the next section using optimization criterion 7.3.

7.3.3 Snap Arm Optimization Using Insertion Simulation

Unknown coefficients L, h, and θ make up the design vector $\overline{a}$ from Eq. 7.2. By varying these parameters, the resulting insertion and retention forces are plotted in Figure 7.6 as the micropart approaches the snap arm, engages it, and snaps

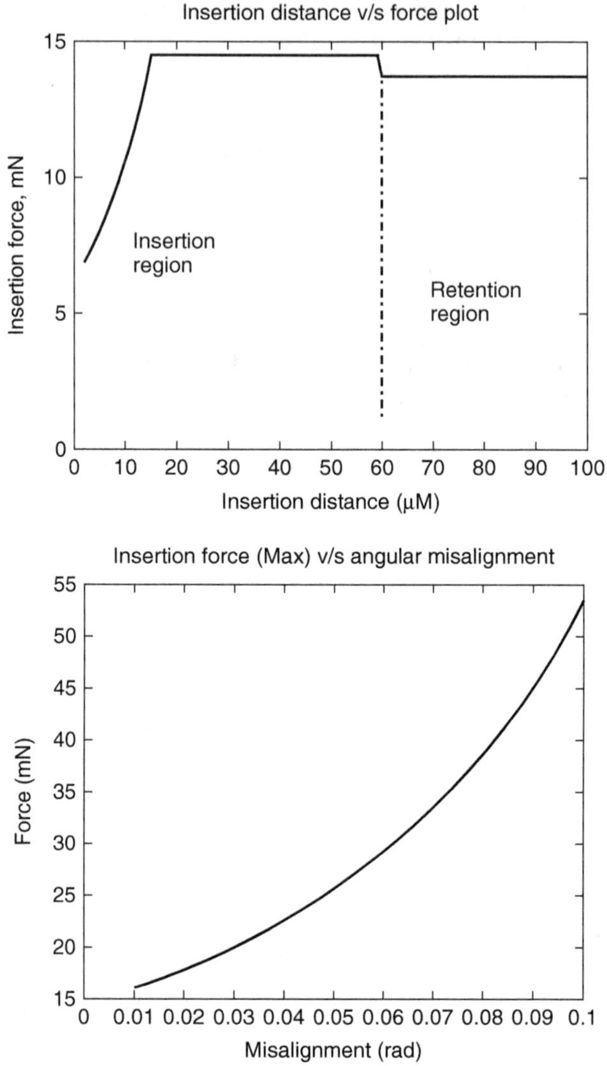

Figure 7.6. (Top) Insertion force versus insertion distance along the X axis. (Bottom) Maximum insertion force for different values of the angular missalignment angle α.

in. Since the design goal is to reach the highest retention force level while minimizing the insertion force, a parametric search reveals that good choices for the snap fastener design are $L = 600$ μm, $h = 10$ μm, and $\theta = 75^{\circ}$ for which we obtain $F_{\text{in}} = 14.5$ mN, $F_{\text{ret}} = 13$ mN assuming a friction coefficient $\mu = 0.3$.

The graph in Figure 7.6(a) is obtained for an insertion in which the part principal axis coincides with the X axis. In practice, the part may be missaligned along the Y axis (through a translation), in which case the deflections in the upper and lower cantilevers will be different. If we assume that a maximum missalignment of 10 μm needs to be tolerated, the maximum insertion force becomes $F_{\text{in}} = 24$ mN, and the maximum bending stress on the cantilever arm obtained using Eq. (7.9) is 0.17 GPa. Since this value is below the yield stress of crystalline silicon, the insertion will succeed over 99% of the time, and the variance of the missalignment along y is given by $\sigma_{1y} = 10/3 = 3.33$ μm.

Misalignment in the angle between the part and the arm's neutral axis will also cause different deflection forces being applied onto the cantilevers, and will result in a different total insertion force along the X axis for the micropart than the values shown in Figure 7.6(a). This type of misalignment arises due to the tolerance between the MEMS part width and the socket width between the two cantilever arms. For the simulated design, this tolerance is 10 μm, while the part has a length of 100 μm. Since the maximum part–socket misalignment along the Y axis is 10 μm, the maximum angular misalignment that can be observed during the part insertion is $\alpha = 5.712^\circ$ (approximately 0.1 rad). This misalignment angle adds to the θ angle on one of the arms and subtracts from angle on the opposite arm. This modifies Eq. 7.9 to

$$F_{\text{in}}^{a,b} = \frac{6EI\,dy}{(L - dx)^3[\mu \cos^2(\theta \pm \alpha) - 0.5 \sin 2(\theta \pm \alpha)]} \qquad F_{\text{in}}^{a,b} = F_{\text{in}}^{a} + F_{\text{in}}^{b} \tag{7.11}$$

where F_{in}^{a} and F_{in}^{b} are the insertion forces due to the upper and lower socket arms. Figure 7.6(b) shows a plot of the maximum insertion force for different values, indicating that the insertion force will increase significantly if the part and socket are angularly misaligned. Finally, we also obtain a value for the variance in the angular tolerance budget to be $\sigma_{1\theta} = 2.36^\circ$.

7.3.4 Experimental Validation of Insertion Force

To experimentally determine the force required for micropart insertion into the snap-fastener and the force required to knock the part off the substrate, we used a SensorOne beam element—a single-crystal silicon beam with one ion implanted resistor on each side mounted in a special miniature header. A deflection of the beam gives a resistance change fed into a Wheatstone bridge. This sensor is mounted onto a precision robot, and pushed against a snap-fastener assembly to obtain force measurements as shown in Figure 7.7(a). Figure 7.7(b) shows the insertion force vs. displacement obtained. The experimentally determined insertion force is 40 mN and the retention force is 22 mN.

Comparing the value to the simulation results, we can infer that during assembly the part was misaligned to the socket arm resulting in a sightly higher insertion force, and that the coefficient of friction must be higher than the 0.3 value used in simulation resulting in a higher retention force.

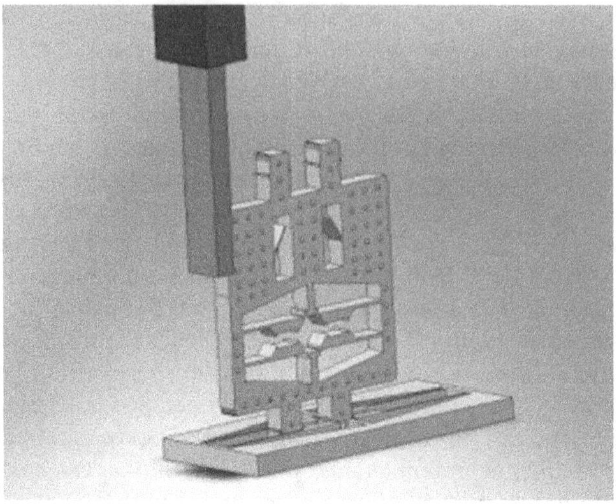

(a)

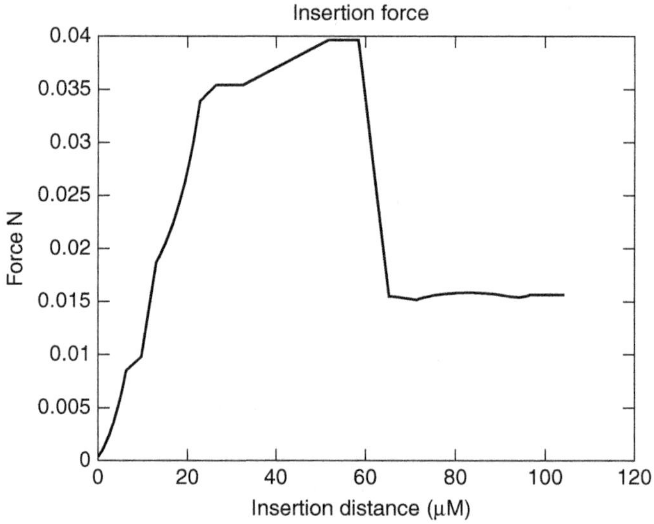

(b)

Figure 7.7. Microforce sensor with μ^3 station acting on assembly joint and the resulting force.

7.4 μ^3 MICROASSEMBLY SYSTEM

7.4.1 Kinematics of Assembly Cell

To accomplish automated compliant microassembly, we use the μ^3 multirobot cell configured using 19 DoF discrete stacked stages and arranged into 3 robotic

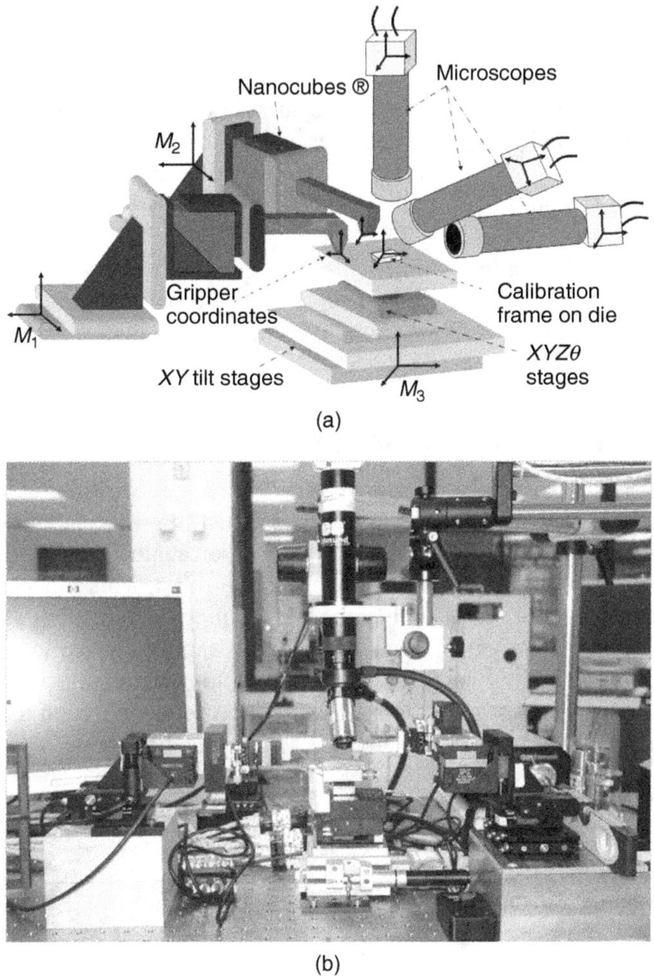

Figure 7.8. (a) Schematic diagram of μ^3 (meso–micro–nano) platforms with microgrippers and (b) photograph of the multirobot system with $M_{3,1}$ (left and right) and M_3 (center).

manipulators. Figure 7.8 shows three μ^3 manipulators (M_1, M_2, M_3) sharing a common 8-cm^3 workspace. Also depicted are three microscopes that are used for calibration and visual servoing.

Manipulators M_1 and M_2 are two robotic manipulator arms with 7 DoF each. They consist of *XYZ* coarse and fine linear stages, including the PI Nanocube for nanoscale fine motion. A rotation stage provides a terminating roll DoF (θ) axis, which is key for assemblies of 2.5D MEMS components. Mounted at the end of the manipulator chains are kinematics mounting pairs that provide for end-effector reconfigurability. The central manipulator M_3 is a high precision 5

DoF robot consisting of an XY mechanism placed on a 2-axis tilt stage. This robot carries custom-designed fixtures for microparts (the dies/substrate) and a custom-designed hotplate for interconnect solder reflow. For assembly, we mount MEMS end-effectors (jammers or active grippers) onto M_1, and end-effectors with rotational symmetry (vacuum nozzles, adhesive-based microgrippers) around the vertical axis onto M_2. Die sites with microparts are placed onto manipulator M_3. A kinematic representation of the multirobot cell can also be summarized in the frame assignment of Figure 7.9.

One important aspect in the kinematic configuration of μ^3 is that XY scanning with manipulator M_3 (center) is used to "bring" the part to the end-effector and not the other way around. As a result, the end-effector will always be in focus for the limited field of view.

7.4.2 Automation in the Assembly Cell

Calibration usually refers to a set of procedures for locating the robot end-effectors in a global coordinate frame. End-effector position data is collected through a set of measurements in the global coordinate frame and used to fit parameters of an inverse kinematic model for the robot. This model can then be used in open-loop fashion assuming that the motion of the robot is repeatable. Calibration of the μ^3 system is accomplished by expressing the local coordinate frames attached to robots M_1 and M_3 in a common frame, attached to the end-effector frame of robot M_2. In a typical calibration sequence, each manipulator is commanded to several locations, and the actual misalignments between gripper

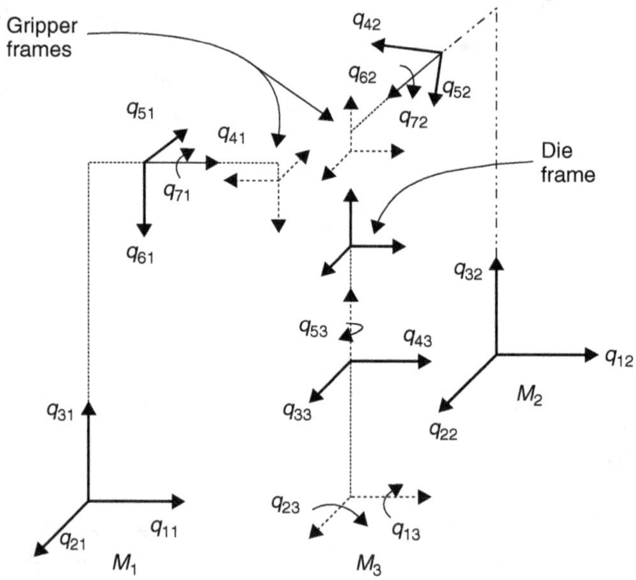

Figure 7.9. Kinematic frames for the multirobot system.

and substrate are measured using the stereo vision system. From these measurements, a mapping can be derived by doing constrained least-squares fit on the data. The number of data points should be sufficient to bring the variance of the posed estimate below the robot repeatability as shown in Popa et al. [16]. Furthermore, because each of the robots has at most 3 independent rotational DoFs, and 3 translational DoFs, we can decompose the orientation pose calibration from the translation pose calibration. The calibration steps are discussed next.

7.4.2.1 Calibration Step 1: M_1 End-Effector Calibration of the Remote Center of Rotation

Manipulator M_1 is primarily used as a pick-and-place tool for 2.5D microparts, with an additional 90° rotation for vertical part orientation for snapping into the substrate. As a result, it is important that parts that are rotated 90° be situated in close vicinity to their original position, thus requiring that the center of rotation for M_1 be located close to the microgripper tip.

In μ^3, we use machine vision with a microscope to program the remote center of rotation (RCR) of microgrippers. Due to variability in gripper designs and fabrication, it is currently necessary to repeat this step every time a new MEMS gripper is mounted on M_1. A summary of techniques for mounting MEMS grippers and program the RCR can be found in Das et al. [4] and Mayyas et al. [13].

7.4.2.2 Calibration Step 2: M_3 Angular Posed Alignment to M_1 End-Effector Through Vision

Now that the gripper is mounted on M_1, and the die containing 2.5D MEMS parts is mounted on M_3, it is necessary to calibrate the relative posed misalignment between the M_1 and M_3 end-effector coordinate frames. We accomplish this via the tilt DoFs of manipulator M_3, through the use of machine vision from microscopes. Let $q_{ij}, i \leq 7, j \leq 3$ be the joint coordinate i of manipulator j, and $\varpi_{31} \in R^3$ be the unit vector orientation of M_3's end-effector expressed in the end-effector frame of M_1, as depicted in Figure 7.9. Then ϖ_{31} is obtained via vision from stereo microscopes S_1, S_2, and S_3. We set the calibration pose of M_3 rotational joints q_{13}, q_{23}, and q_{53} to attain this pose by servoing, for example, by setting:

$$\left[\dot{q}_{13}\ \dot{q}_{23}\ \dot{q}_{53}\right] = hJ_{3o}^{-1}\left(\varpi_{31}(q_{13}, q_{23}, q_{53}) - [0\ 0\ 1]^{\mathrm{T}}\right) \tag{7.12}$$

in which h is a constant, and J_{3o} is the orientation part of the Jacobian of manipulator M_3 expressed in the end-effector frame of M_1. However, because M_3 contains three independent rotational DoFs, $J_{3o} = I_3$, and therefore Eq. DP(10) becomes a simple tilt correction exercise until the end-effector on M_1 points in the z direction. Because we servo on vision information based on a line detection algorithm, we estimate the angular calibration misalignment to be $\sigma_\theta = 0.75°$.

7.4.2.3 Calibration Step 3: M_3 to M_1 Translational Posed Alignment Through Teaching

Now that the alignment poses of M_1 and M_3 coincide, we must calibrate the remaining translational DoFs of manipulators M_1 and M_3. Assembly operations will now involve translations via $\theta = [q_{31}, q_{33}, q_{43}]^{\mathrm{T}}$, rotations by 90° via q_{71} (terminating roll of manipulator M_1 with RCR property), and arbitrary rotations via q_{53} (terminating roll of manipulator M_3). Note that we do not utilize M_1 DoFs q_{11}, q_{21} (X, Y translations) but only q_{31} (Z translation) in order to remain in the limited field of view of the microscopes. As a result, we produce q_{53}-*dependent* calibration maps based on translation by pointing with the end-effector to features (fiducials or microparts) located on the MEMS die. For each M_3 roll angle q_{53}, we can determine the required translational pose of M_1 and M_3 via a three-point teaching method, by writing

$$\theta = \theta_1 + (\theta_2 - \theta_1)\left(\frac{p - p_1}{p_2 - p_1}\right) + (\theta_3 - \theta)\left(\frac{q - q_1}{q_3 - q_1}\right)$$
$$\hat{\theta} = \theta_1 + (\theta_2 - \theta_1)\left(\frac{p_3 - p_1}{p_2 - p_1}\right) \tag{7.13}$$

where

- P_1, P_2, and P_3 are fiducials on the MEMS die, with die coordinates (p_1, q_1), (p_2, q_2), and (p_3, q_3), respectively. These values can be expressed in pixels from the charge-coupled device (CCD), or directly, in die layout coordinates, if fabrication tolerances can be neglected.
- P is an arbitrary point of interest with die coordinates (p, q). This will later become the target assembly site.
- θ_1, θ_2, and θ_3 are joint vectors, corresponding to $\theta = [q_{31}, q_{33}, q_{43}]^{\mathrm{T}}$ when the gripper tip is at locations P_1, P_2, and P_3 (by pointing to fiducials on die).
- θ is joint coordinate vector when the tip is pointing to an arbitrary assembly site on die P with die coordinates (p, q).

Figure 7.10 depicts a MEMS layout populated with calibration fiducials (P_1, P_2) as well as assembly sites (P_3). The parts themselves are thus used for both calibration and assembly. We have two options in estimating whether the gripper "points" to a given fiducial. In one case, the reference points are observed through a 5× objective lens that provides a 3.2 μm/pixel image resolution. As a result, a calibration error smaller than 6.4 μm (2 pixels) is expected. In the second case, we actually physically place the gripper tip inside a compliant feature on the die, and visually observe that it does not cause part shift along x, y, and z. The parts themselves are thus used for both calibration and assembly. The die in Figure 7.10 includes an array of SOI Zyvex connectors with dimensions 800 μm × 1300 μm × 100 μm. The microgripper is a passive "jammer" that can

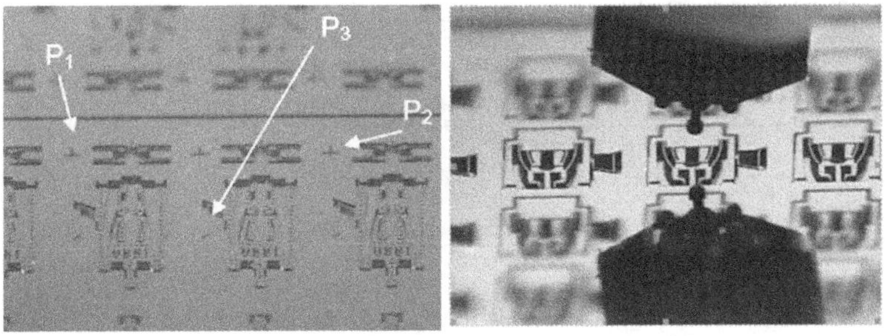

Figure 7.10. MEMS die containing calibration and assembly sites P_1, P_2, P_3 where the end-effector must be pointed to using a "jammer" gripper.

pick MEMS parts by means of compliant insertion. For the nine test points, the corresponding robot joint angles are noted by physically placing the jammer-head on these locations with visual confirmation from the stereo microscopes. Tables 7.1 and 7.2 summarize the data set of the calibration experiment including a mixed-joint data set, namely x and y axes for M_3, and z joint axis for M_1. The results show an end-effector positional accuracy $\sigma_{3,x,y,z} \leq 2.24$ μm using the three-point teaching method.

After calibration, an assembly script generated from the MEMS die layout is used to automate the assembly process. For picking and placing of a single part, the script consists of simple commands such as:

1. Point jammer to die location (P_i, Q_i), which translates into move M_3 in x,y to appropriate robot joint values derived from Eq. 7.13.
2. Pick up part by moving M_1 appropriately in z.
3. Rotate part 90° using M_1.
4. Point micropart bottom to $(P_f,\ Q_f)$. Here a small correction needs to account for the fact that we are pointing not with the gripper tip, but with a micropart that is picked up by the gripper.
5. Place part onto the substrate by moving appropriately in z.
6. Insert part into compliant socket by moving appropriately in x until snap-in.
7. Repeat to next part.

7.5 HIGH-YIELD MICROASSEMBLY

The parts to be microassembled are fabricated using DRIE on SOI wafers. At the end of the fabrication process, these parts are released from the substrate with a tether constraining them to the wafer surface to maintain part location. Prior to microassembly, the tether is broken using the same robotic end-effector that is used to pick the part. Tether breaking causes misalignment in the position of the part on the substrate. Unreleased device layer walls fabricated around the part act as hard-stop locators to limit this misalignment along directions X, Y, and θ.

TABLE 7.1. Calibration Experimental Data

Point	Actual Position (x, y, z) in μm	Derived Position (x, y, z) in μm	Position Error (x, y, z) in μm
	3552	3552.78	+0.78
1	7699.2	7698.77	+0.43
	316.8	318.388	−1.588
	4851.2	4853.11	−1.91
2	7699.2	7700.08	−0.88
	323.2	319.044	+4.156
	7449.6	7453.64	−4.04
3	7702.4	7702.21	+0.19
	320	320.111	−0.111
	2252.8	2251.33	+1.47
4	6252.8	6248.03	+4.77
	316.8	318.6	−1.8
	6150.4	6152.13	−1.73
5	6252.8	6251.23	+1.57
	320	320.2	−0.2
	3552	3550.27	+1.73
6	4697.6	4699.17	−1.57
	316.8	319.8	−3.0
	4851.2	4850.53	+0.67
7	4704	4700.24	+3.76
	323.2	320.333	+2.867
	7449.6	7451.07	−1.47
8	4697.6	4702.37	−4.77
	323.2	321.4	+1.8
	6150.4	6149.56	+0.84
9	3251.2	3251.39	−0.19
	316.8	321.489	−4.689

TABLE 7.2. Mean and Standard Deviation in Error (in μm)

	X Position	Y Position	Z Position
Mean	1.6267	2.0144	2.2457
Std Dev.	1.014	1.9067	1.5869

In order to use the HYAC lemma, we need to determine the average misalignment due to the detethering process, that is, we determine $\sigma_{xyz\theta P}$ of microparts. The subscript P refers to the variance defined when the part is on the substrate and prior to any manipulation. A typical image used for location analysis is shown in Figure 7.11. This particular micropart has moved after release to a location that contacts a hard stop. Square etch holes distributed on the part (on top end and in the gripping region) are mapped with a reference

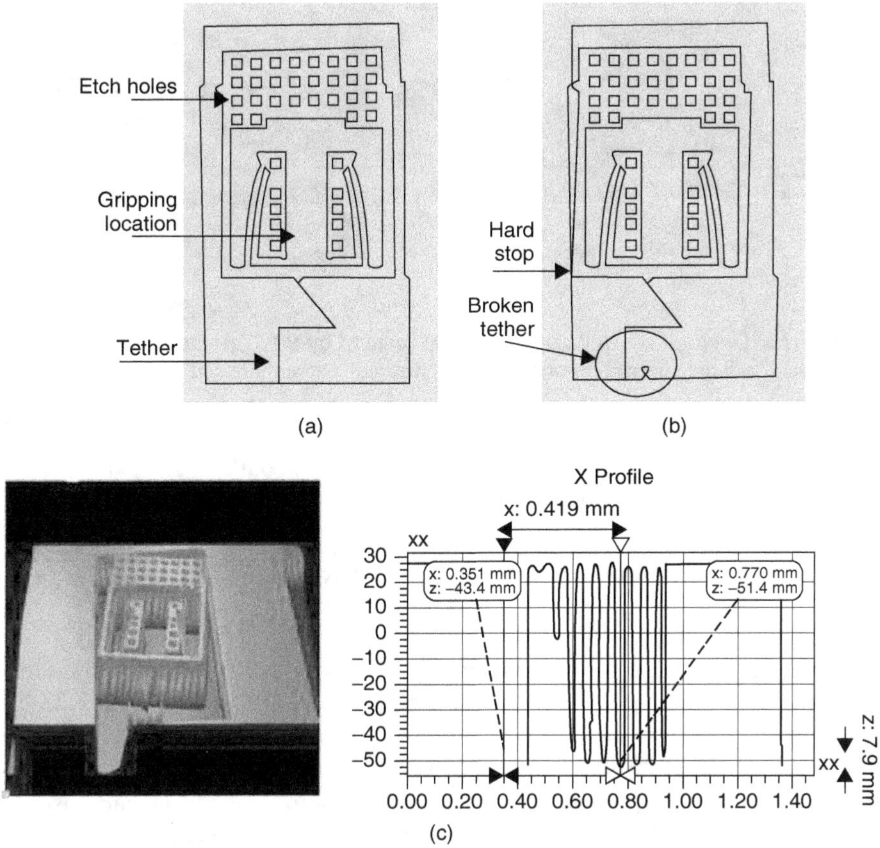

Figure 7.11. Part on substrate (a) before detethering, (b) after detethering, and (c) measurement using a surface profiler scan.

corner to determine the $XY\theta$ location of the part. Figure 7.11(c) illustrates an image of a detethered part obtained using a Veeco NT1100 surface profiler. Using this technique, we measured misalignments during 10 detethering operations and found that $\sigma_{2xP} = 1.12$ μm, $\sigma_{2yP} = 1.8$ μm, and $\sigma_{2\theta P} = 0.1^\circ$.

Figure 7.12 shows the coordinate transformation between the part-pick pose from the substrate and the part-insertion pose into the snap-fastener. As a result of this transformation, σ_{2yP} changes into σ_{2za} and σ_{2xP} changes into σ_{2xa}. For simplicity we will denote σ_{2xyza} as σ_{2xyz}.

7.5.1 High-Yield Assembly

We now use the HYAC lemma outlined in Section 7.2.4 to verify the high-yield guarantee for three basic operations: detethering, part pickup, and part assembly. Referring to Figure 7.12, the compliance-based pickup compensates

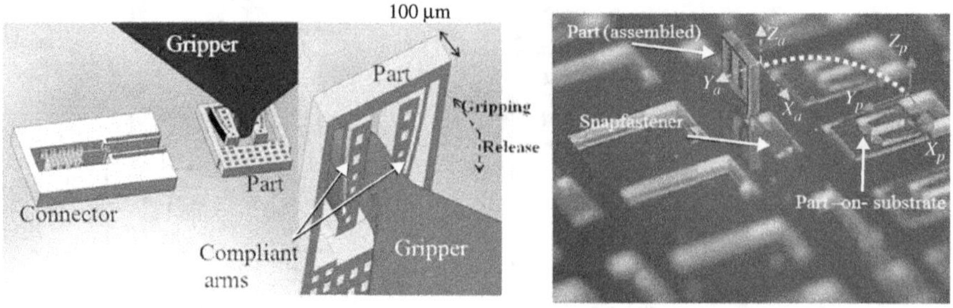

(a) (b) (c)

Figure 7.12. Diagram of (a) part pick-up, (b) rotation by 90°, (c) and coordinate system transformation between part-on-substrate and vertically assembled part.

the misalignment along the X_P axis and rotation θ about the Z_P axis. For part pickup, the HYAC condition requires that

$$\sigma_{1xyzP}^2 \geq \sigma_{2xyzP}^2 + \sigma_{3xyzP}^2 \tag{7.14}$$

The numerical values of uncertainties along X_P were $\sigma_{1xP} = 3.3$ μm, $\sigma_{2xP} = 1.12$ μm, and $\sigma_{3xP} = 2.24$ μm, and thus verify Eq. 7.14 along X. Similarly, $\sigma_{1yP} = 5$ μm, $\sigma_{2yP} = 1.8$ μm, and $\sigma_{3yP} = 2.24$ μm, and $\sigma_{1zP} = 3.33$ μm, $\sigma_{2zP} = 0$, and $\sigma_{3zP} = 2.24$ μm. Therefore eq. 7.14 is observed along Y and Z as well. After pickup, the uncertainty $\sqrt{\sigma_{2yP}^2+\sigma_{3yP}^2=3.19}$ μm is transformed into σ_{2z}, and $\sqrt{\sigma_{2zP}^2+\sigma_{3zP}^2=2.24}$ μm is transformed into σ_{2y} during part insertion into the connector. Next we look at the part–connector insertion scenario and check for agreement with HYAC. The connector design tolerance σ_1, the part misalignment tolerance σ_2, and the manipulator positional accuracy σ_3 are

$$\begin{array}{lll} \sigma_{1x,y} \leq 3.3\ \mu\text{m} & \sigma_{2x,y} \leq 2.24\ \mu\text{m} & \sigma_{3x,y} \leq 2.24\ \mu\text{m} \\ \sigma_{1\theta} \leq 2.36^\circ & \sigma_{2\theta} \leq 0.75^\circ & \sigma_{3\theta} \leq 0.75^\circ \\ \sigma_{2z} \leq 3.19\ \mu\text{m} & \sigma_{3z} \leq 2.24\ \mu\text{m} & \end{array} \tag{7.15}$$

This means that if an end-effector is positioned with repeatability σ_{3xyz} and $\sigma_{3\theta}$ relative to the substrate, the assembly yield will be over 99% since the HYAC is satisfied with values from eq. 7.15 for X, Y, Z, and θ.

7.5.2 Repeated Assemblies

The HYAC guarantee gives the assembly station operator sufficient confidence to run the μ^3 in automated mode for repeated snap-fastener insertions. Figure 7.13(a) shows an array of 8 microassemblies of parts from Figure 7.11 fastened to the substrate using consecutive insertions. Figure 7.13(b) shows an array of 12 Zyvex

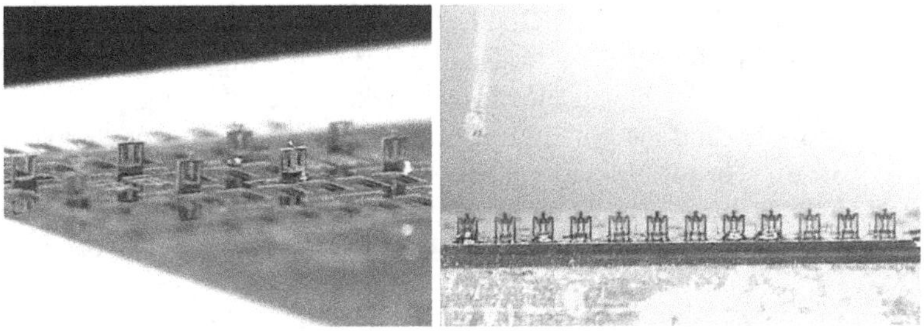

(a) (b)

Figure 7.13. (a) Array of 8 laterally assembled microparts and (b) array of 12 Zyvex jammers assembled in sequence and epoxy cured in place.

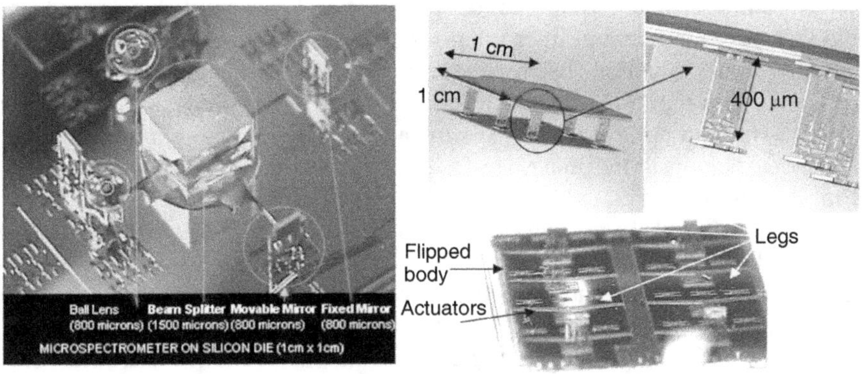

(a) (b)

Figure 7.14. (a) Assembled microspectrometer on a 1-cm × 1-cm silicon bench and (b) ARRIpede microcrawler robot with six vertically assembled legs [14].

jammers assembled in sequence using automation scripting after appropriate calibration. The total assembly time for these arrays was a few minutes and the assembly yield was 100%.

Figure 7.14(a) shows vertical parts assembled on a 1-cm × 1-cm silicon die as a microoptical bench (microspectrometer). It consists of two vertically assembled micromirrors, two vertically assembled ball lens holders, two glass microball lenses, and a glass beam splitter. Each of the ball lens assemblies consists of a rigid part (ball) inserted into a compliant holder. One of the mirrors is assembled onto a shuttle connected to an in-plane electrothermal actuator. The microspectrometer was tested using fiber-coupled laser light and a detector assembled onto the substrate. To strengthen the snap-fastener joints after assembly, epoxy was also applied using a dispensing nozzle mounted on manipulator M_2.

Finally, the ARRIpede microrobot in Figure 7.14(b) consists of an array of 1D prismatic joints on a 1-cm × 1-cm area SOI die. Just like for the microspectrometer, the prismatic joints consist of Chevron electrothermal actuators with a microsnap-fastener. The first ARRIpede prototypes consist of 4, 6, and 8 actuated legs, generating forward motion using stick-and-slip locomotion. The gait sequence is generated using an electronic backpack containing power electronics, switching logic, and a light lithium ion battery. A silicon payload weighing 4 g (made of Au–Sn alloy) was placed on the inverted robot and conveyance speeds up to 1.55 mm/s were obtained with square-wave gait motions. We are currently testing an autonomous ARRIpede for payload, speed, and controllability capabilities.

7.6 CONCLUSION AND FUTURE WORK

In this chapter we presented a systematic approach to address high-yield aspects of compliant MEMS assembly. Our methodology includes snap-fastener designs, tolerance analysis, and appropriate robot calibration to accomplish a desired target yield. We presented several microsnap-fastener designs, and their assembly yield prediction and evaluation. Microfasteners can be used as interconnects to construct more complex 3D microstructures, such as microrobots and microoptical benches. Through appropriate force simulation modeling and experimental data, it has been demonstrated that snap-fastener designs can offer good compliance during assembly, as well as good retention after assembly. The μ^3 multirobot cell was used to assemble hybrid on-die MEMS devices such as microoptical benches or microrobots. We showed experimental results confirming that high-yield, high-speed serial MEMS assembly is guaranteed if the HYAC condition is satisfied. This condition states that the combined positional uncertainties of the micropart and microgripper should be below the compliant snap-fastener tolerance budget. The condition serves as a guide to determine how much micropart compliance and precision is needed to accomplish microassembly with high yields.

Future work includes improvements to connector designs, tolerance analysis for cascaded assemblies, and their use in prototyping other miniaturized instruments and robots. We will also combine serial and parallel deterministic assembly, with stochastic sorting of microparts on the μ^3 assembly cell.

Acknowledgment

The author wishes to thank the Texas Microfactory at ARRI, graduate students Rakesh Murthy and Aditya N. Das, and Dr. Woo Ho Lee for their contributions to the design, simulation, and experimental work presented in this chapter.

REFERENCES

1. F. Arai and T. Fukuda, Adhesion-type Micron End-Effector for Micromanipulation, in Proc. of IEEE ICRA, 1997, pp. 1472–1477.
2. K. F. Bohringer, B. R. Donald, R. Mihailovich, and N. C. MacDonald, Sensorless Manipulation Using Massively Parallel Microfabricated Actuator Arrays, in Proc of IEEE ICRA, 1994, pp. 826–833.

3. M. B. Cohn, K. F. Böhringer, J. M. Noworoski, A. Singh, C. G. Keller, K. V. Goldbery, and R. T. Howe, Microassembly technologies for MEMS, *Int. Soc. Opt. Eng. Proc*, 3513:2–16, 1998.

4. A. N. Das, P. Zhang, W. H. Lee, D. O. Popa, and H. E. Stephanou, μ: Multiscale, Deterministic Micro-Nano Assembly System for Construction of On-Wafer Microrobots, in Proc. of IEEE ICRA, Rome, Italy, April 2007.

5. N. Dechev, W. L. Cleghorn, and J. K. Mills, Microassembly of 3-D Microstructures Using a Compliant, Passive Microgripper, *J. Microelectromech. Syst*., 13(2):176–189, 2004.

6. J. Fang and K. F. Böhringer, Parallel Micro Component-to-Substrate Assembly with Controlled Poses and High Surface Coverage, *J. Micromech. Microeng*., 16(4):721–730, 2006.

7. G. Greitmann and R. A. Buser, Tactile Microgripper for Automated Handling of Microparts, *Sensors Actuators A*, 53:410–415, 1996.

8. A. M. Hoover, S. Avadhanula, R. Groff, and R. S. Fearing, A Rapidly Prototyped 2-Axis Positioning Stage for Microassembly Using Large Displacement Compliant Mechanisms, in Proc of IEEE ICRA, Orlando, Florida, May 2006.

9. B. H. Kang, J. T.-Y. Wen, N. G. Dagalakis, and J. J. Gorman, Analysis and Design of Parallel Mechanisms with Flexure Joints, *Proc of IEEE ICRA*, 21(6):1179–1185, 2005.

10. K. Saitou and M. J. Jakiela, Design of a Self-Closing Compliant Mouse Trap for Micro Assembly, Proc. of ASME International Mechanical Engineering Congress and Exposition November 17–22, 1996, Atlanta, Georgia.

11. W. H. Lee, M. Dafflon, H. E. Stephanou, Y. S. Oh, J. Hochberg, and G. D. Skidmore, Tolerance Analysis of Placement Distributions in Tethered Micro-Electro-Mechanical Systems Components, in Proc of IEEE ICRA, May 2004.

12. W. H. Lee, B. H. Kang, Y. S. Oh, H. E. Stephanou, A. C. Sanderson, G. Skidmore, and M. Ellis, Micropeg Manipulation with a Compliant Microgripper, in Proc of. IEEE ICRA, Taiwan, 2003, pp. 3213–3218.

13. M. Mayyas, P. Zang, W. H. Lee, P. Shiakolas, and D. O. Popa, Design Trade-offs for Electrothermal Microgrippers, in Proc. of IEEE ICRA, Rome, Italy, April 2007.

14. R. Murthy, A. Das, and D. O. Popa, ARRIpede: A Microcrawler/Conveyor Robot Constructed via 2 $\frac{1}{2}$ D MEMS Assembly, in Proc. of IEEE IROS' 08, Nice, France, October 2008.

15. P. Zhang, M. Mayyas, W. H. Lee, D. O. Popa, and J. C. Chiao, An Active Microjoining Mechanism 3D Assembly, *J. Microme. Microeng.,* 19:035012, 2009.

16. D. O. Popa, R. Murthy, M. Mittal, J. Sin, and H. E. Stephanon, M3-Modular Multi-Scale Assembly System for MEMS Packaging, in Proc. of IEEE/RSJ IROS Beijing, China, October 2006.

17. D. O. Popa and H. E. Stephanou, Micro and Meso Scale Robotic Assembly, *J. Manufact. Process*., 6(1):52–71, 2004.

18. R. Prasad, K. F. Bohringer, and N. C. MacDonald, Design, Fabrication, and Characterization of Single Crystal Silicon Latching Snap Fastners for Micro Assembly, in Proc. of ASME International Mechanical Engineering Congress and Exposition (IMECE '95), San Francisco, CA, Nov. 1995.

19. A. Rizzi, J. Gowdy, and R. L. Hollis, Agile Assembly Architecture: An Agent Based Approach to Modular Precision Assembly Systems, in Proc of IEEE ICRA, Volume 2, 20–25 April 1997, pp. 1511–1516.
20. M. Shimada, J. A. Tompson, J. Yan, R. J. Wood, and R. S. Fearing, Prototyping Millirobots Using Dextrous Microassembly and Folding, in Proc of ASME IMECE/DSCD, vol. 69-2, 2000, pp. 933–940.
21. K. Tsui, A. A. Geisberger, M. Ellis, and G. D. Skidmore, Micromachined End-Effector and Techniques for Directed MEMS Assembly, *J. Micromec. Microeng*., 14:542–549, 2004.
22. D. Whitney, Quasi-Static Assembly of Compliantly Supported Rigid Parts, *ASME J. Dynamic Syst., Measure. Control*, 104:65–77, 1982.
23. Y. Zhou, B. J. Nelson, and B. Vikramaditya, Fusing Force and Vision Feedback for Micromanipulation, in Proc Of IEEE ICRA, Leuven, Belgium, May 1998.

CHAPTER 8

DESIGN OF A DESKTOP MICROASSEMBLY MACHINE AND ITS INDUSTRIAL APPLICATION TO MICROSOLDER BALL MANIPULATION

AKIHIRO MATSUMOTO, KUNIO YOSHIDA, and YUSUKE MAEDA

8.1 INTRODUCTION

The development of microelectronic/optical components such as microsensor components, microsemiconductor devices, microfiber optical components, microlaser diode components, or flying head components of hard disk drive (HDD) is becoming more and more active these days. Thanks to the microelectromechanical systems (MEMS) technology, these products are made in good quality and in mass quantity. Nevertheless, their assembly is time consuming and thus expensive from the viewpoint of production cost. Historically, assembly has always been a bottleneck to the progress of automation because of the complexity of the task. Microassembly is not an exception. We focused on a positioning accuracy of around 1 μm, which traditional mechanical engineering and micro/nanoelectrical engineering does not cover, but it is required accuracy for MEMS component assembly. In other words, the area of between several 0.1 μm to 1 μm has been left undeveloped. In this view, we have been developing microassembly machines that meet this demand [3–6, 9]. One of the features of our machines is that the target of positioning accuracy is achieved, yet a large working space is kept. One example is that accuracy is less than 1 μm and the working stroke is 150 mm. Moreover, this feature is obtained on normal desktop, by not using constant room temperature or heavy rigid plates. This is

Robotic Microassembly, edited by Michaël Gauthier and Stéphane Régnier

one step toward a desktop factory [4]. Thanks to this feature, the initial cost for the installation of the assembly facilities becomes very cheap, compared to traditional microassembly machines. Research on microassembly is increasing these days, and we are interested in the synthesis side rather than the analysis side of microassembly research. Research in a similar direction has been done actively, for example [1, 2, 8, 10]. In this chapter, we first explain the design considerations of the desktop assembly machines, especially how we achieved fine positioning accuracy. Next, we show one example of the industrial application of microsolder ball manipulation for the joining process of HDD head components. Last, we show our efforts toward the further improvement of better placement accuracy.

8.2 OUTLINE OF THE MACHINE DESIGN TO ACHIEVE FINE ACCURACY

8.2.1 Design Considerations

The first requirement is to realize a desktop factory [7]. We want to use a microassembly machine in a normal factory environment; in other words, without using a large and heavy granite plate or air dumper, or without a special room for constant temperature. Compact size yet high rigidity of the machine is the most important requirement. The next requirement is to achieve fine positioning accuracy (target: 1 μm) while keeping a wide working space. A 10-nm positioning is achieved by using piezoactuators, but its working space (stroke) is very small, and use of piezo is not suitable for this application. Vision technology and force control technology, which are normally used for industrial robots, can be used for this purpose. The degrees of freedom (DoF) of the assembly machine are illustrated in Figure 8.1. Basically, it has X-Y-θ table at the lower part of the machine, and X-Z axis motion in the upper part for a traveling microcomponent to be assembled. The working space of this mechanism is relatively large (from 50 to 150 mm depending on the model) compared to the size of the target parts (around 0.2 mm × 0.3 mm). The measuring resolution of the linear encoder for each axis is 50 nm, but, of course, positioning accuracy is greater than the resolution of encoders because of the stick–slip phenomenon in this area of size. Consider the positioning accuracy of three dimensional measuring machines, for example. The values range generally from 5 to 10 μm, using a constant temperature room with a rigid base. The positioning accuracy of industrial robots is worse than three-dimensional machines or numerical control (NC) machining centers because of their structure. The positioning accuracy of industrial robots ranges from roughly 10 to 100 μm depending on the structure. In this sense, the target accuracy of 1 μm by using the normal industrial machines seems to be difficult to achieve. Thanks to our experiences of designing/installing industrial robots, we adopted to use relative accuracy instead of absolute accuracy. The relative accuracy of industrial robots, for example, is better than absolute accuracy, usually 5–10 times

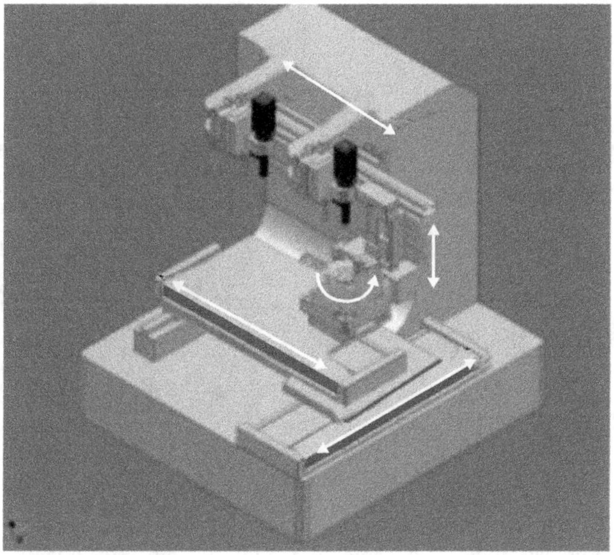

Figure 8.1. Allocation of degrees of freedom (DoF).

better. Relative accuracy is usually called repeatability. By careful machining and good encoders, the absolute accuracy of most positioning machines can usually be up to 20 μm. Of course, stress analysis by using computer aided engineering (CAE) in the mechanical design stage is indispensable. Then, the repeatability can be, say, 1–5 μm (of course, depending on the structure). But further improvement is an unknown in the world of traditional mechanical engineering. Nevertheless, if the size of the machine becomes smaller, the deformations of the machine are also small. This means that miniaturization of the machine itself is one key to the fine positioning accuracy. Additionally, we decided to use a vision system to improve positioning accuracy. If both the target position and the current position are seen in the same area in the vision system, the positional error could be compensated by the sake of the repeatability of the machine. This idea is extremely effective after many evaluation experiments. Another important point is the position of the measuring system, that is, a camera in our case. As shown in Figures 8.2–8.4, which show several models of our microassembly machines, a camera is installed in the same framework of the machine. Thus, the measuring system and positioning system vibrate in the same frequency and phase, canceling the effect of the vibration of the entire assembly machine. In addition, we adopted force control in placing the target parts onto the substrate because MEMS components are very fragile against external forces. In summary, the key to the success of the fine positioning accuracy can be realized by:

- The fine mechanical design and machining in order to miniaturize the size, yet to keep rigidity

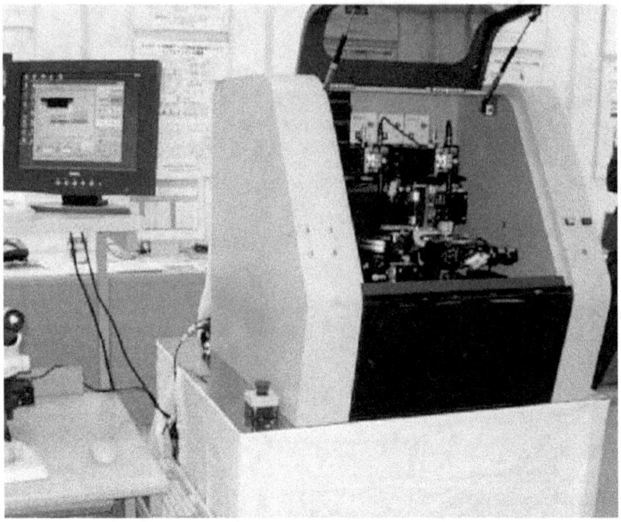

Figure 8.2. Largest model of our microassembly machines for bonding application.

Figure 8.3. Smallest model of our microassembly machines (the size of the base area is about A4 paper size).

- Use of vision measurement
- Use of force control

8.2.2 Vision Measurement Subsystem

By viewing the working space with the use of a camera, the positioning of the machine can be improved as long as both the current position and the target position are seen within the same camera area as shown in Figure 8.5. The offset

Figure 8.4. Two X-Y-θ stage model of our microassembly machines.

to be moved is calculated from the difference of the current position and the target position. Note that this process is done before the placement of the target part, and the part should be slightly over the submount that is position controlled by the X-Y-θ stage. The image processing of the vision measurement is done by special software called HexSight by Adept Technology (USA) and AJI (formerly Adept Japan), which uses a special interpolation technique to improve subpixel resolution (theoretical highest resolution is $\frac{1}{16}$ pixel). Normally the visible area is 1.6 by 1.2 mm, shown in the rectangle in Figure 8.6, and the image memory size is 640 by 480 pixels (8-bit grayscale) as shown in Figure 8.5. Then 1 pixel is equivalent to about 3 μm, but it becomes less than 0.1 μm (theoretically) for measuring accuracy by using the special interpolation technique in the HexSight. The actual measurement accuracy is experimentally evaluated as 0.1 μm [4, 9].

8.2.3 Force Control

The target microparts are handled not by grasping but by picking them up with negative air pressure. Figures 8.6 and 8.7 show the handling device that is placed

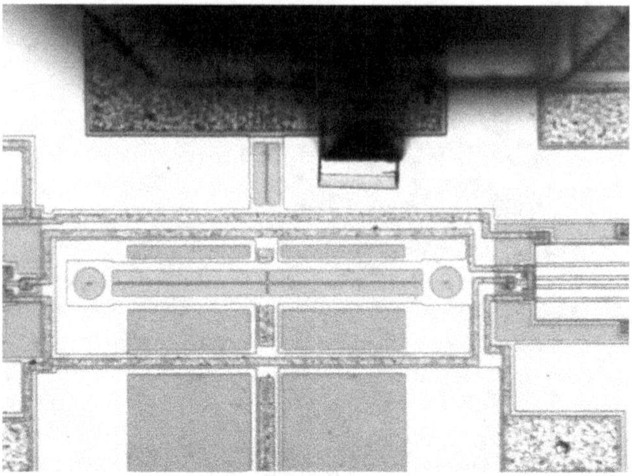

(a)

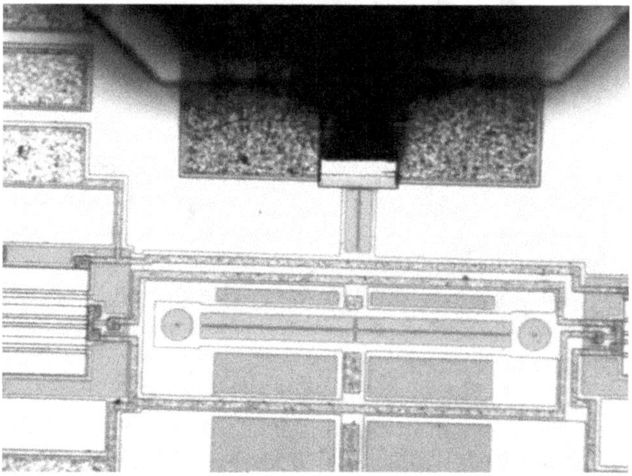

(b)

Figure 8.5. Use of image processing for vision measurement in order to compensate for positional error: (a) before and (b) after.

as the end effector of the assembly machine. This part can be replaced for other applications. The lower part of the collet is pushed or pulled by controlling the air pressure of the cylinders that are placed in the upper part of the force control unit (Fig. 8.7). The target microparts are picked by the small hole that is placed on the edge of the lowest part of the collet. After positioning improvement by vision processing, the target part is pushed onto the submount using positive air pressure control. Note that MEMS components or optical components are very fragile, so the pushing force for placement is carefully designed.

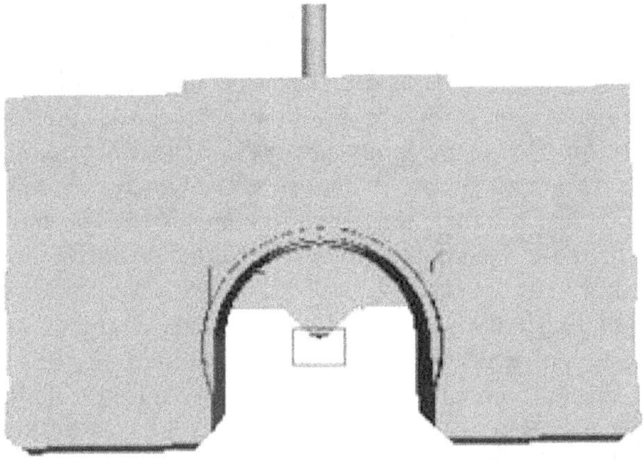

Figure 8.6. Top view of the force control unit. (Note that the camera is mounted above the control unit, and the target view area of the camera is shown in the rectangle.)

Figure 8.7. Collet (lower part) suspended by force control unit (upper part) composed by air cylinder inside. (Target part is picked up by the edge of the collet in pneumatic way.)

8.3 APPLICATION TO THE JOINING PROCESS OF ELECTRIC COMPONENTS

8.3.1 Manipulation Issue of Microsolder Balls

Reflow soldering for joining electric components may replace wire bonding in industry, mainly because MEMS components are weak to vibration with wire bonding. For this purpose, we used this microassembly machine and adopted to use microsolder balls for joining electric components. The diameter of the solder ball is 100 μm at first, then 80 μm, in order to meet the demand for microassembly. Industrial application such as HDD head component joining requires such microtechnology as shown in Figure 8.8. Figure 8.9 shows microsolder balls (Pb free) of Φ 100 μm. The shape is not always spherical. In addition, the size of the solder balls varies as shown in Figure 8.10. Of course, this is due to the difficulty in producing microsolder balls. This means that we have to take some measurements of these variation problems in order to utilize microsolder balls for microassembly.

Since manipulating only one solder ball for each circuit is not efficient, we decided to pick up multiple solder balls at the same time. For this purpose, solder ball sheets are specially prepared in joint development with Senju Metal Industry [3, 6]. Figure 8.11 shows the top view and side view of the solder ball sheet. Each solder ball is placed in the circular hole and stuck on the adhesive layer.

As shown on the left side of Figure 8.11, the position of each solder ball in the hole is random. Aligning multiple solder balls is necessary for picking them up simultaneously. Since they are picked up by "collet" using negative air pressure through the nozzle of the collet (Fig. 8.12), we move the collet by the assembly machine in order to softly push solder balls on the solder sheet for alignment. As

Figure 8.8. Use of solder balls for joining parts of HDD head. (A solder ball is to be placed on each gold pad in the indicated area by the ellipse.)

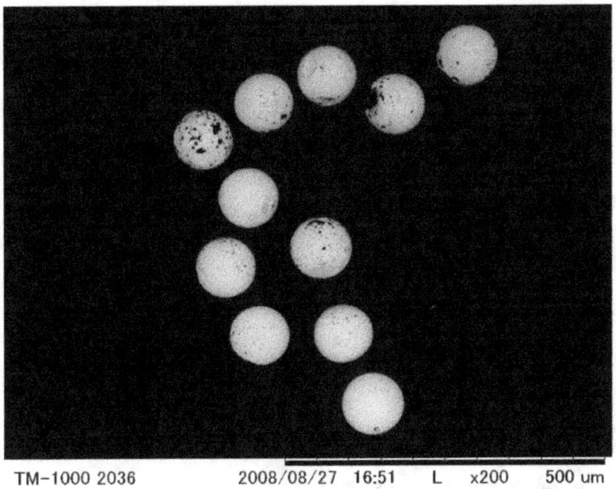

Figure 8.9. Microsolder balls of Φ 100 μm.

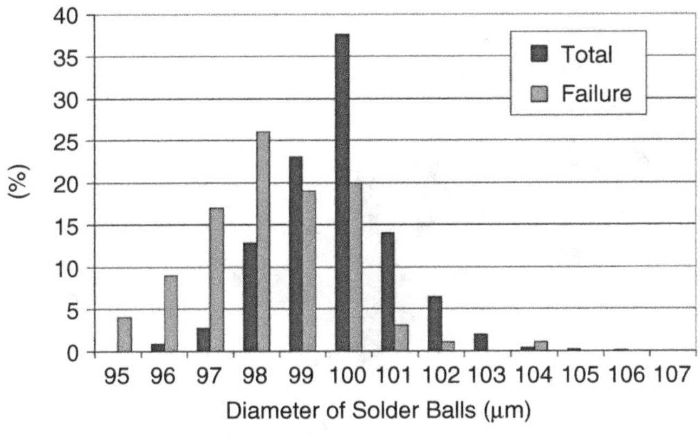

Figure 8.10. Measurement result of solder balls of Φ 100 μm.

shown in Figure 8.13, the balls roll to the end of each hole, and thus alignment is done. This motion is realized thanks to the fine positioning accuracy of the microassembly machine described in the previous chapter. Figure 8.14 is a photograph during alignment. The left of Figure 8.14 is the situation before alignment, and the right of Figure 8.14 is after alignment, which shows a good result of the alignment process. We have tested different thicknesses of resist layers of the solder sheet as shown in Figure 8.15 and confirmed that different thicknesses of resist layers and different sizes of holes should be carefully selected for the different sizes of microsolder balls. We have experimentally evaluated the performance of this idea and reported it in Kobayashi et al. [3] and Matsumoto et al. [6].

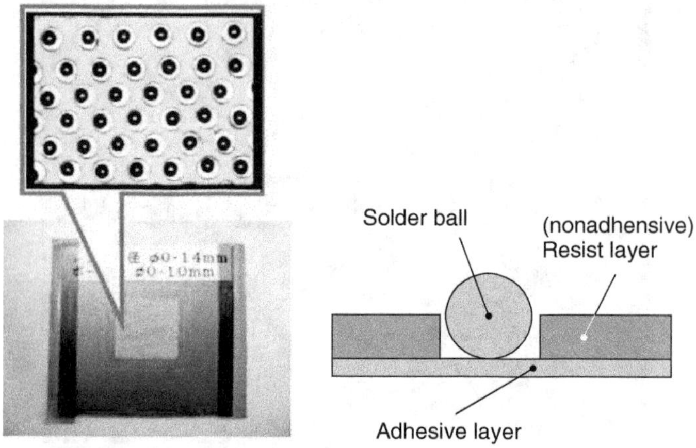

Figure 8.11. Solder ball sheet (top view and side view).

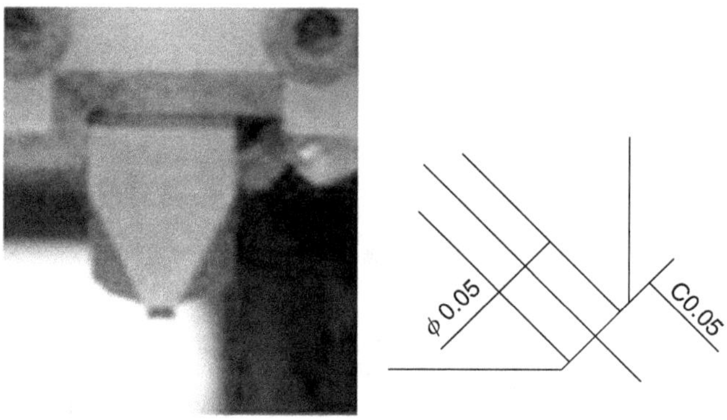

(a) (b)

Figure 8.12. Collet chuck: (a) front view and (b) side view (suction port).

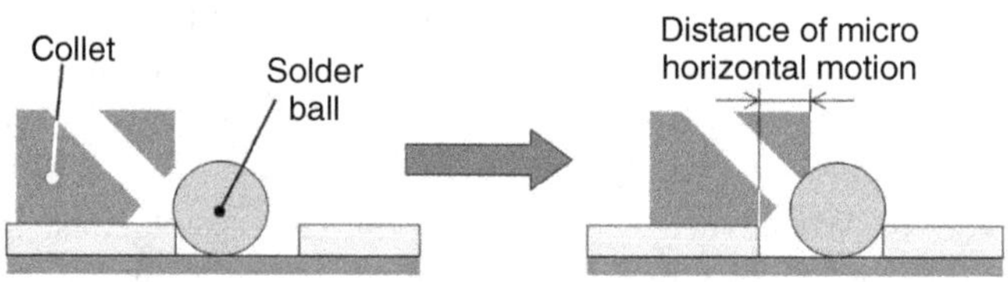

Figure 8.13. Alignment of solder balls by pushing.

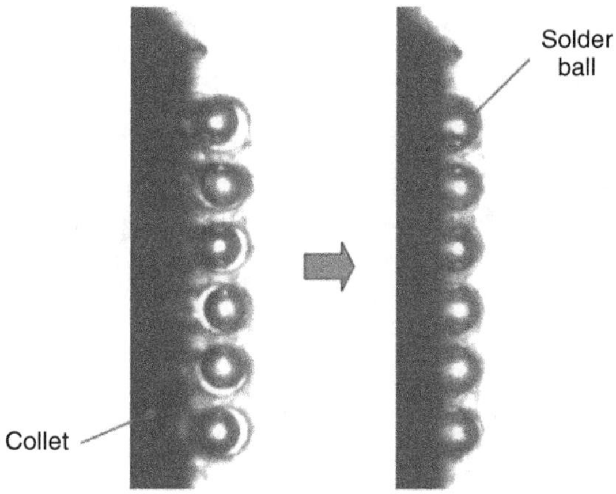

Figure 8.14. Photograph of alignment of multiple solder balls.

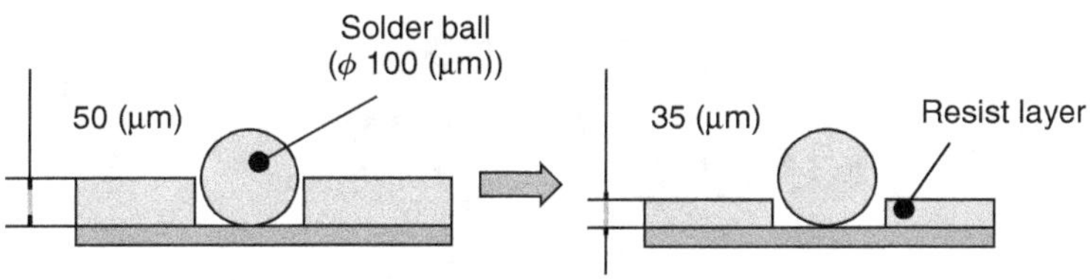

Figure 8.15. Change of thickness of resist layer.

8.3.2 Heating Issues of Reflow Soldering

Since MEMS parts are often very weak on heat, the heating of their reflow soldering must be carefully done. We used optical fiber and laser (a suitable wavelength must be selected depending on the application). In one case, we used a blue velvet laser (wavelength is 405 nm, which is suitable for heating gold). After a solder ball is placed on a gold pad the size of which is equivalent to that of the ball, then the gold pad is heated up to 230°C (the temperature of solder melt) by the laser through a fiber. By using a fiber array, which is well designed to be included in the collet, we can achieve spot heating in the selected area. A result of such heating is shown in Figure 8.16, showing the heat temperature distribution. This shows that only the solder ball is heated and the other area is cooler. An example of the final quality of reflow soldering is shown in Figure 8.17. It shows that the solder balls are well placed on the target positions. Figure 8.18 is the cross section of solder in such a case, and it shows quite uniform soldering quality. In order to use such a short wavelength of laser, we must say that the fine accuracy is needed for the mechanical design of collet with fiber arrays. In another case, we conducted experiments on the use of infrared laser (wavelength is 1110 nm). We needed more power but the projection time became short. In this case, there is a position adjustment problem

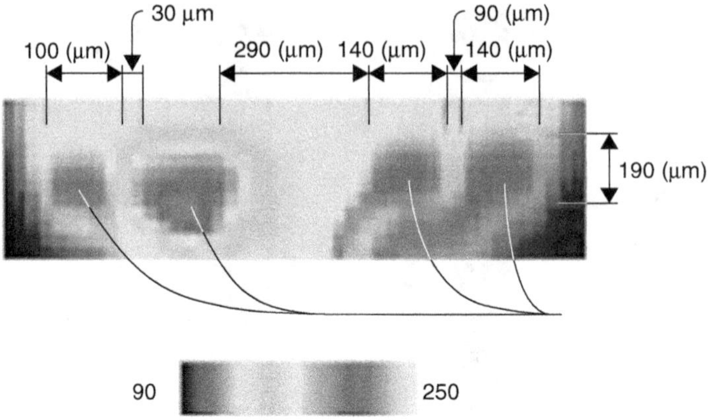

Figure 8.16. Temperature distribution on substrate (temperature is expressed in °C).

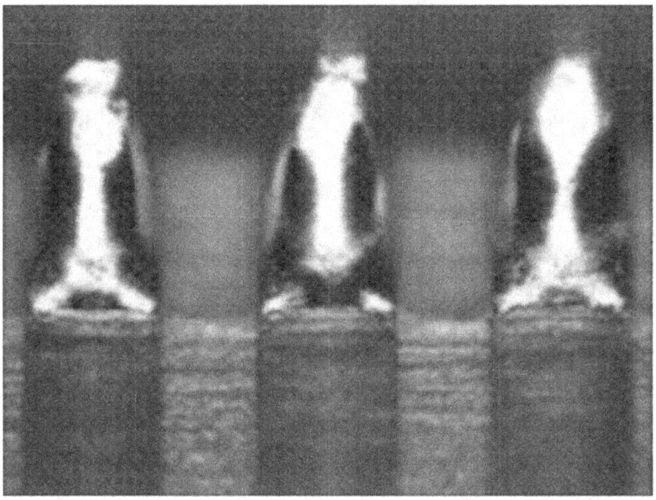

Figure 8.17. Good result of reflow soldering.

of the lens for the laser and the collet. These experiments mean that the use of laser through fibers is effective for local heating, but the selection problem of its wavelength and the projection time still remain along with the fine mechanical design. There are many issues unresolved for the reflow soldering process using microsolder balls. We expect that more analytical and experimental trials from the viewpoint of microphysics must be incorporated. In any case, we should not forget that the atmosphere around the soldering part must be free from oxygen as much as possible in order to reach a high quality of microsoldering.

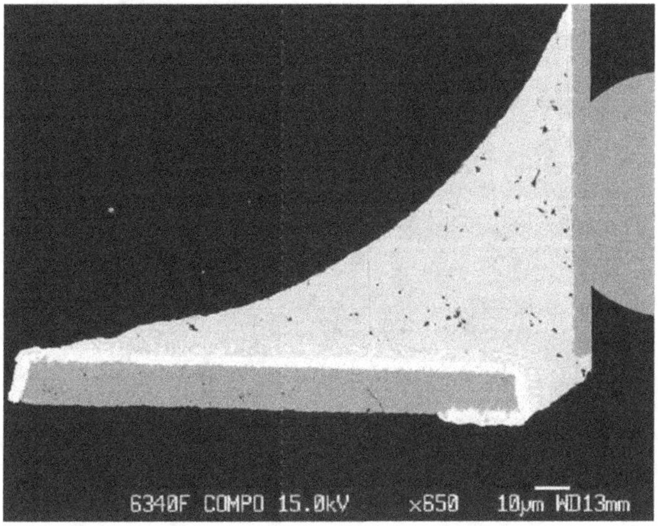

Figure 8.18. Cross section of solder.

8.4 PURSUING HIGHER ACCURACY

8.4.1 Positioning Accuracy and Placement Accuracy

In current industrial applications such as assembly (mounting) of HDD head parts, accuracy around 1 μm is enough. Nevertheless, from the manufacturers' point of view, we should continue to pursue higher accuracy. After the first installation of the machine to the factory floor, we realized that the positioning result was worse than the anticipated positioning accuracy. The differences are measured by using the vision measurement system on this machine and the result is shown in Figure 8.19. This shows that there are some phenomena that make accuracy worse while pushing parts to the submount, which must be analyzed. Currently, we assume that the difference comes from the machine itself, nonlinearlity of the force control unit, the assembly process, or other factors. For example, physical parameters of solder differ depending on the temperature. Glues between mechanical/electrical parts may work as a spring and a damper that makes positioning accuracy worse. These issues need analytical approaches from the viewpoint of microphysics, and such microtechnologies must be transferred to industry. Thus, we will call this accuracy "placement accuracy" rather than "positioning accuracy." Note that placement accuracy is the result of the assembly (placement).

8.4.2 Verification of Vision Measurement

Figure 8.20 shows the experimental results of the vision measurement of the stable object after 100 trials. The value should be zero in theory, but the result

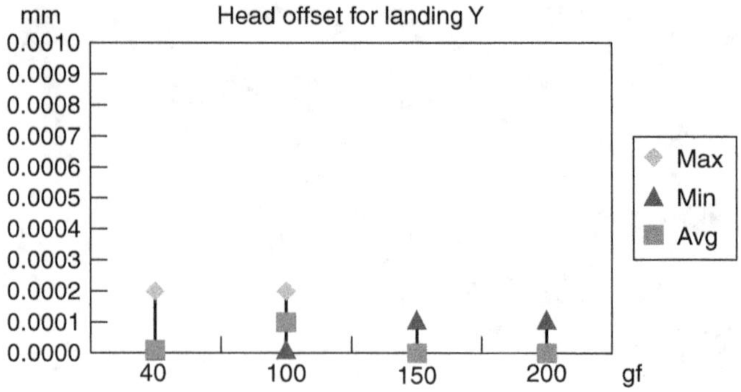

Figure 8.19. Head offset in landing.

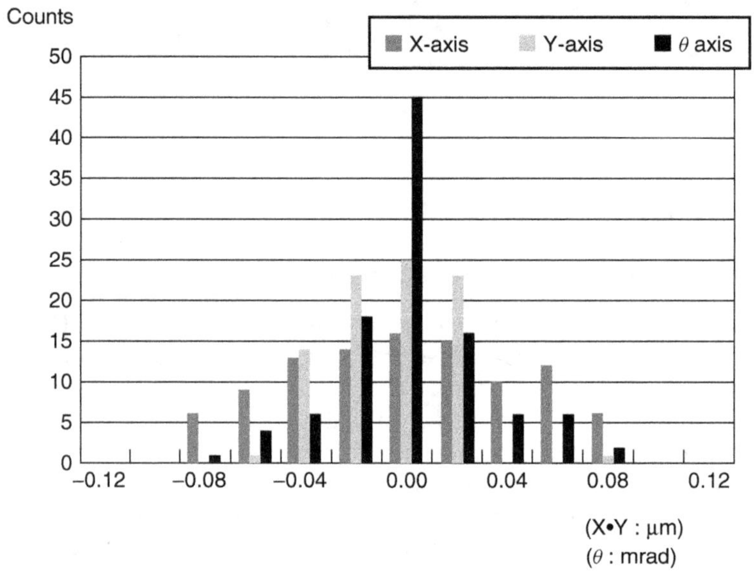

Figure 8.20. Vision measurement of the stable object [the deviation must be zero, but σ is 0.03 μm (30 nm)].

data are stochastic. In this case, the standard deviation is 0.04 μm for the X axis, 0.03 μm for Y the axis, and 0.03 mrad for θ the axis.

We considered that this result shows that the effect of airflow fluctuation between the camera and the target affects the vision measurement. If we use a kind of barrier/shield there, we can expect that the vision measurement accuracy improves. Then, we conducted experiments on whether the airflow fluctuation can affect the result of the vision measurement. Figure 8.21 shows the experimental result when the workspace is covered with a fence. This result shows that vision measurement and positioning accuracy are much higher than before, and

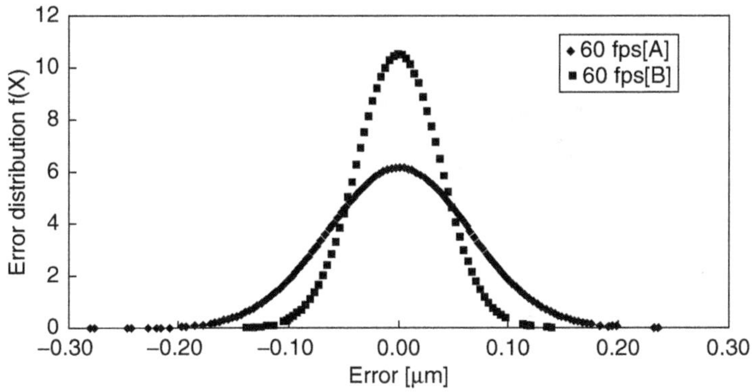

Figure 8.21. Effect of a fence around the workspace (airflow affects the vision measurement accuracy).

the airflow turbulence can no longer be neglected. Note that the theoretical measurement accuracy of vision measurement is less than 0.1 μm, which is smaller than the wavelength of visible light. It seems that we have reached the extreme of vision measurement and the results are affected by the inflection of light.

Next, instead of a fence surrounding the machine, we placed different sizes of transparent shield (5 and 20 mm), and/or paper tube (barrier) in order to avoid the effect of air turbulence (even if it is small) on the vision measurement, that is,

1. Normal FCU (force control unit) only
2. FCU +5-mm shield
3. FCU +20-mm shield
4. FCU +20-mm shield + paper tube

Transparent shield is a kind of fence and is used so that the target part is seen from the camera. Then a heater is placed below the working space and set to 400°C. The outline of the experiment is shown in Figure 8.22. The effect of such a barrier against air turbulence between the camera and working space is tested, and the result is shown in Figure 8.23. As the "barrier" for the airflow becomes bigger, the standard deviation of vision measurement becomes smaller. This effect is more remarkable when the heating temperature rises. These experiments show that protecting "the way of light" between the camera and the working space from external airflow is very effective for fine vision measurement.

8.4.3 Verification of Mechanical Structures

Next we checked whether mechanical alignment of the assembly machine itself can affect the positioning accuracy. By some adjustment of the orientation of end effector, around the *X* axis and *Y* axis as shown in Figure 8.24, the result

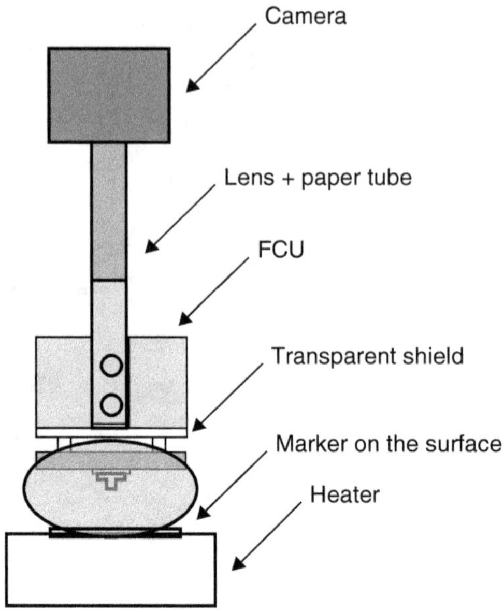

Figure 8.22. Experiment to exclude air turbulence (several types of shield/barrier are placed and tested).

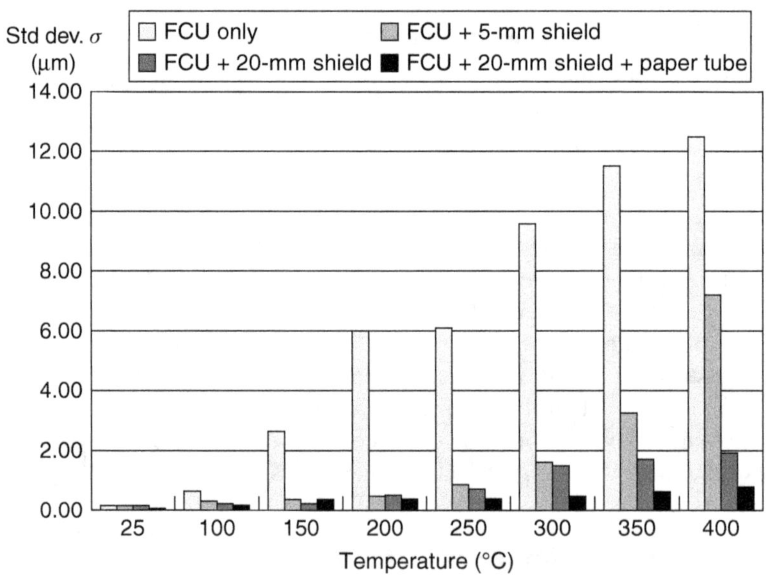

Figure 8.23. Effect of different shields/barrier to the vision measurement (the effect of shield/barrier becomes bigger when the heating temperature rises).

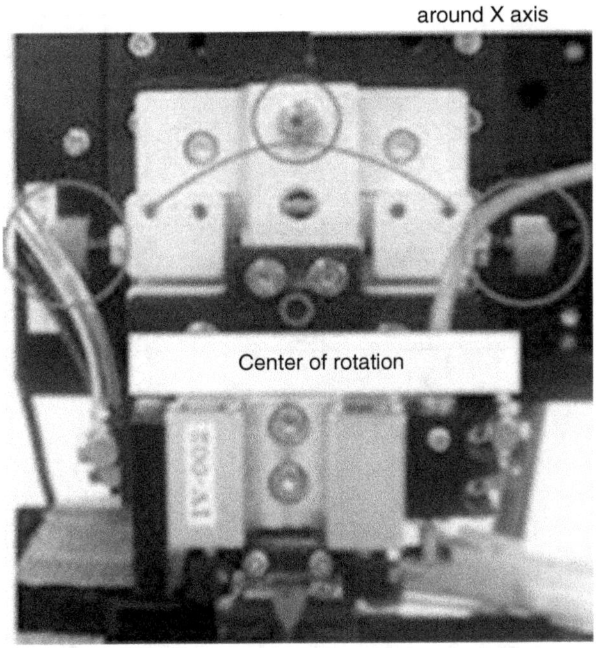

Figure 8.24. Adjustment of mechanical alignment.

of the positioning errors are measured by using different FCUs. By using the vision measurement system, the positions of the collet are measured when it is pushed onto the glass plate with 400 gf of force (the air pressure of the force control unit is 0.1 MPa). This is assuming that some target part is pushed onto the base where glue is inserted between the base and the part. This experiment is repeated by changing the value of force to 200 and 50 gf. Figure 8.25 shows the result for different FCUs. Both show nice repeatability (relative accuracy) for the same force values, but show different absolute accuracy for different force values. This is the normal result of absolute accuracy and relative accuracy. The result also shows a different performance for different force control units. This experiment shows that careful calibration is necessary and effective. However, as a whole, the results show that the mechanical alignment of the assembly machine itself does not greatly affect the final placement accuracy. Thus, although the careful adjustment of the machine alignment is necessary, it has little effect on the placement accuracy.

8.5 CONCLUSION

We showed how to achieve fine positioning, with regard to design considerations on a microassembly machine. Many ideas including repeatability, vision, and force control are applied to the microassembly machine for higher accuracy. Next

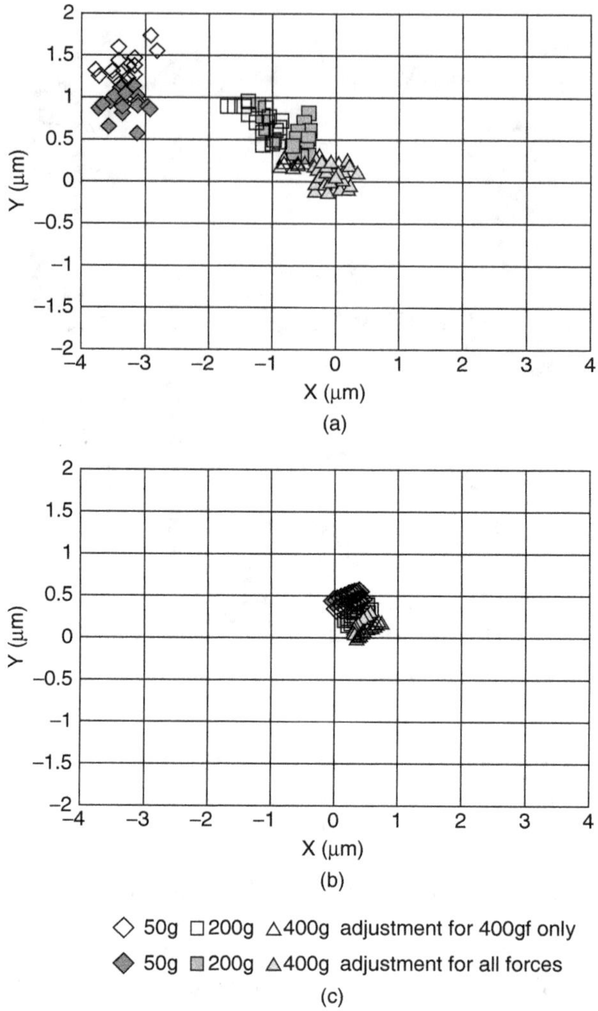

Figure 8.25. Check of placement accuracy for different FCUs showing good repeatability: (a) FCU-A, (b) FCU-B, and (c) FCU-C.

we explained an experiment on a multiple solder ball manipulation application. Last we showed our efforts toward the higher accuracy from the viewpoint of mechanical design. As a whole, we feel that the placement accuracy can be slightly better, say 0.7 μm, but the area beyond 0.5 μm is another world where knowledge of microphysics must be incorporated and experimentally evaluated.

Acknowledgment

The authors would like to thank Dr. Toshinari Akimoto at Toyo University, and Satoshi Makita and Tatsuya Kobayashi at Yokohama National University, for their efforts on making experiments. Also, they would like to thank all AJI members who joined in this activity. This work was supported in part by the city of Yokohama, the prefecture of Kanagawa, and The Electro-Mechanic Technology Advancing Foundation (EMTAF), Japan.

REFERENCES

1. J. Bert, S. Dembele, and N. Lefort-Piat, Toward the Vision Based Supervision of Microfactories through Images Mosaicing, in *Precision Assembly Technologies for Mini and Micro Products*, Springer, London, 2006.
2. T. Eriksson, H. N. Hansen, and A. Gegeckaite, Automated Assembly of Micro Mechanical Parts in a Microfactory Setup, Proc. 5th International Workshop on Microfactoris, Besancon, S2-2, Oct. 2006.
3. T. Kobayashi, Y. Maeda, S. Makita, S. Miura, I. Kunioka, and K. Yoshida, Manipulation of Micro Solder Balls for Joining Electric Components, Proc. of 2006 Int. Symposium on Flexible Automation (IFSA), Osaka, July 2006, pp. 408–411.
4. A. Matsumoto, T. Akimoto, K. Yoshida, H. Inoue, and K. Kamijo, Development of MEMS Component Assembly Machine—Application of Robotics Technology to Micromechatronics, Proc. of International Symposium on Micro-Mechanical Engineering Tsuchiura, Dec. 2003, pp. 83–88.
5. A. Matsumoto, K. Tsuiki, S. Miura, and K. Yoshida, Experimental Study of Improving the Positioning Accuracy of Micro Assembly, Proc. of the 1st CIRP-International Seminar on Assembly Systems (ISAS), Stuttgart, Nov. 2006, pp. 55–60.
6. A. Matsumoto, K. Yoshida, I. Kunioka, Y. Ozawa, S. Miura, Y. Maeda, and T. Kobayashi, Handling and Heating Problems of Micro Solder Balls for Micro Assembly, Proc. 5th International Workshop on Microfactoris, Besancon, S3-2, Oct. 2006.
7. Y. Okazaki, N. Mishima, and K. Ashida, "Microfactory - concept, history and developments", Journal of Manufacturing Science and Engineering, Vol. 126, No. 4, pp. 837–844, 2004
8. S. Perroud, A. Codourey, and Y. Mussard, PocketDelta: A Miniature Robot for Microassembly, Proc. 5th International Workshop on Microfactoris, Besancon, S2-2, P3-5, Oct. 2006.
9. K. Yoshida, H. Inoue, K. Kamijo, A. Matsumoto, and T. Akimoto, Design of Microchip Bonder to Meet Accurate MEMS-Component Assembly, Proc. of 6th Japan–France Congress on Mechatronics, Hatoyama, Sept. 2003, pp. 613–618.
10. Q. Zhou, More Confident Microhandling, Proc. 5th International Workshop on Microfactoris, Besancon, S3-1, Oct. 2006.

INDEX

www.ingramcontent.com/pod-product-compliance
Lightning Source LLC
LaVergne TN
LVHW011913200926
848484LV00002B/615
* 9 7 8 0 4 7 0 4 8 4 1 7 3 *